UNITEXT for Physics

UNITEXT for Physics series, formerly UNITEXT Collana di Fisica e Astronomia, publishes textbooks and monographs in Physics and Astronomy, mainly in English language, characterized of a didactic style and comprehensiveness. The books published in UNITEXT for Physics series are addressed to graduate and advanced graduate students, but also to scientists and researchers as important resources for their education, knowledge and teaching.

More information about this series at http://www.springer.com/series/13351

Guido Fano · S.M. Blinder

Twenty-First Century Quantum Mechanics: Hilbert Space to Quantum Computers

Mathematical Methods and Conceptual Foundations

 Springer

Guido Fano
Dipartimento di Fisica e Astronomia
Università di Bologna
Bologna
Italy

S.M. Blinder
University of Michigan
Ann Arbor, MI
USA

ISSN 2198-7882
UNITEXT for Physics
ISBN 978-3-319-86464-8
DOI 10.1007/978-3-319-58732-5

ISSN 2198-7890   (electronic)

ISBN 978-3-319-58732-5   (eBook)

Printed on acid-free paper

This Springer imprint is published by Springer Nature
The registered company is Springer International Publishing AG
The registered company address is: Gewerbestrasse 11, 6330 Cham, Switzerland

Foreword

Many years ago, my math teacher assigned me the task to give a presentation to the rest of the class. Time to prepare, 2 weeks; topic, group theory and its application to quantum mechanics. I politely observed that I had not yet had a course on quantum theory; I knew nothing about it. Teacher's reply was: "Well, you study it." His name was Guido Fano.

This is how Guido Fano had me learning quantum theory. At home, alone, with a bunch of books on my desk, Dirac's one on top of all. I have been forever grateful to him, for this and much more. If I got into theoretical physics, it's thanks to him, the trust that I found, with surprise, from him. Today, when I realize that there is something that I should know and do not know (which of course happen continuously), I hear his voice in my mind saying "Well, you study it."

But the second lesson I learned from Guido Fano, both about life and quantum theory, was even more crucial. After my presentation to the class, I thanked him and said: "I have now learned quantum mechanics." His reply was "No you haven't." How true.

And not just because a couple of weeks couldn't suffice to digest the theory, but because nearly a century from Heisenberg and Dirac hasn't sufficed the community of us all to come to terms with this theory. In the months after my presentation, I kept discovering, first from Fano himself, and then over and over for the rest of my life, how multifaceted is this theory, and how slippery our actual understanding to it. The more I have learned about the theory, the less clear it has become. When comes the moment of saying what we really have learned about the world in finding that quantum mechanics works so effectively—well, we disagree among ourselves, and there are nearly as many opinions as physicists (not counting the philosophers).

Long time later, with this new book on quantum theory, once again Guido Fano, together with his overseas colleague S.M. Blinder, opens up new sides of quantum mechanics for me. The beauty of this book, in my opinion, is that it merges very different traditions of thinking about the theory and teaching it. For a theory still so perplexing, this is what we need; in order to understand it better, to learn it better, but also to learn how to better use it.

If had to start again learning quantum theory from scratch, as I did 40 years ago, I would do as I did: at home, alone, with many texts on my desk. But I would certainly have, next to Dirac's one, this book.

Marseille, France   Carlo Rovelli
February 2017

Preface

This monograph is the result of many years of experience and contemplation by two octogenarian mathematical physicists, on opposite sides of the Atlantic, who, after all these years, are still endlessly fascinated by the marvelous intellectual edifice that has become quantum mechanics. And which, in the twenty-first century, continues its proliferation into entirely new avenues of human accomplishment, including experiment-based answers to metaphysical questions and the limitless potential of quantum computation.

The main purpose of the book is to make accessible to nonspecialists the still evolving fundamental concepts of QM and the terminology in which these are expressed. Hence our title: "Twenty-First Century Quantum Mechanics: Hilbert Space to Quantum Computers." Among the concepts which we emphasize are the following:

- The wavefunction of a particle: associated with a "cloud" of probability, such that the density of the cloud is greater in regions which have a higher probability of containing the particle.
- The Heisenberg uncertainty principle: conjugate pairs of observables, such as the position and the velocity of a particle, cannot both be precisely determined at the same time.
- Isotropic vectors: vectors with complex components, which are orthogonal to the rotation axis and thus remain invariant under rotation; these are used to construct *spinors*.
- The spin of elementary particles: the quantum counterpart of rotations of classical objects, described by spinors.
- The question of whether the fundamental laws of physics violate local realism. Locality means that the influence of one particle on another cannot exceed the speed of light. Realism means that quantum states have well-defined properties, independent of our knowledge of them.
- The possible existence of hidden variables: something analogous to the enormous number of microscopic details of molecular motions, which exhibit

themselves in the determinate macroscopic properties of matter, such as temperature, pressure, etc.

- Conceptual problems associated with measurement, superposition and decoherence in quantum systems: collapse of the wavefunction, Schrödinger's cat, and quantum entanglement.
- Quantum computers: if they can be made practicable, enormous enhancements in computing power, artificial intelligence and secure communication will follow.

Needless to say, the quantum theory raises very profound metaphysical and epistemological questions on the description about the "objective world." An important precept of Felix Klein's famous *Erlangen program* was that "A geometry is the study of invariants under a group of transformations." These ideas, subsequently elaborated in the contributions of Einstein, Dirac, and the philosopher Nozick, have led to, at least, a provisional understanding of what constitutes reality.

The need for an observer to "search for invariants" is of such generality that it must apply even in the lives of animals. Imagine a gazelle cautiously eying two lions. Suppose that she glances at the first lion, and then the second. In doing so, she must turn her head. But there must be some process in her brain that enables her to realize that, even after she turns her head (and thus registers a different image), the first lion is still there, surely an instance of "invariance." If we wanted to build a robot capable of distinguishing objects, then, when the robot's eyes move, its programming must include the capability of performing mathematical transformations among images viewed at varying angles at different times. Nature probably does not work in precisely the same way, but the fundamental conceptual features: (1) variability of images and (2) recognition of invariants, or common elements, must still be applicable.

These ideas certainly pertain to the classical view of Nature; what are their manifestations in quantum mechanics? The analog of the rotation of a classical observer is the evolution of the wavefunction. However there are two distinctly different modes of evolution in quantum mechanics. One is a continuous evolution, following the time-dependent Schrödinger equation; the second, called *collapse of the wave function*, is a random and instantaneous event brought about by a measurement or perturbation of the quantum system. This was, at least, the point of view of the founding fathers of quantum mechanics, mainly Bohr, Heisenberg, and Dirac. The most eminent critic of such ideas (a probabilistic interpretation of QM) was none other than Albert Einstein, as epitomized in his famous pronouncement: "God does not play dice with the Universe!"

A major aspect of the epistemological problem has been resolved by actual experiments (by Alain Aspect and others), motivated by the deep insights of John Stewart Bell. This has revealed a major incompatibility between the worldviews of classical and quantum physics. Bell's theorem states that it is impossible to explain the results of quantum physics using the causality of classical physics, thus negating the possible existence of local hidden variables. Quantum mechanics differs fundamentally from classical mechanics in that the underlying microscopic behavior is *not* determinate.

Measurement and decoherence: according to the traditional ("Copenhagen") interpretation of quantum mechanics, wave function collapse occurs when a measurement is performed. However there remains the problem of *when the collapse actually occurs*? In the past, some physicists thought that collapse is brought about when a conscious observer "takes note" of the new state of a system; but this point of view is now in the minority, since it is more reasonable to think that any inanimate apparatus can also make a measurement and produce a "quantum jump." A more realistic approach to this problem is to consider the microscopic system, the measuring apparatus and the environment as a single composite system. The wavefunction of the complete system must change during an exceedingly short interval of time from a "superposition of states," to just one of these states. This phenomenon is called *decoherence*; it can be proved mathematically rigorously in some models, although there is still much work to be done.

The superposition of two wavefunctions for a macroscopic object is also considered in the infamous Schrödinger's cat *Gedankenexperiment*. A cat is confined to a closed box with a Geiger counter, which detects randomly-occurring radioactive decays in a sample of radium. The Geiger counter is connected to a vial of cyanide, which is broken when a decay particle is detected, killing the unfortunate cat. Until the box is opened, its state can only be described as a "superposition," of a "live cat" and a "dead cat." According to the Copenhagen version of quantum mechanics, the cat "becomes" dead or alive only after an observer opens the box. As paradoxical as it seems, superposition of quantum states of macroscopic objects has now been achieved, for example, in a SQUID (superconducting quantum interference device).

Quantum computing proposes to apply uniquely quantum-mechanical phenomena, such as superposition and entanglement to operate on quantum units of information, called *qubits*. In contrast to classical bits, which can represent a variable with just two values, say 0 and 1, qubits can, in concept, contain an infinite continuum of information, in terms of superpositions of two basis qubits, such as $|\Psi\rangle = \alpha|0\rangle + \beta|1\rangle$. A quantum computer could, in principal, be capable of solving problems in a matter of seconds, which might take a classical computer several centuries to accomplish. Several hypothetical quantum algorithms have already been proposed for large-integer factorization and other applications. Current realizations of quantum computers are still very far from having such capabilities. But apart from potential practical applications, quantum computing remains a profound subject of fundamental interest for both computer science and quantum mechanics.

Acknowledgements

The authors wish to thank: Prof. Sergio Doplicher, for his enlightened clarifications on the concept of decoherence; Prof. Carlo Rovelli for correspondence on questions concerning the philosophy of science; Prof. Angelo Vistoli and Dr. Alessandro Malusà for their assistance in the beautiful proof presented in Appendix to Chap. 6; and, finaly, Lorenzo Felice for his skillful and imaginative sketches of most of the figures.

Bologna, Italy   Guido Fano
Ann Arbor, USA   S.M. Blinder
January 2017

Contents

About the Authors

Guido Fano graduated Magna Cum Laude in Physics at the University of Rome in 1955. He was Assistant Lecturer, later Senior Lecturer, and finally Professor of Mathematical Methods of Physics. He earned university degrees in Theoretical Physics and Mathematical Methods of Physics. He taught at several universities, including Naples, Ferrara, Bologna, and Marseille. Professor Fano is a specialist in the quantum many-body problem, with particular emphasis on the mathematical aspects; he has about 60 publications in international journals. Some particularly significant results concern the existence problem for the time-dependent Hartree–Fock equations, the use of non-orthogonal orbitals in Quantum Chemistry, and the asymptotic behavior of the Taylor expansion coefficients of some sequences of polynomials. He is the author of several textbooks, notably *Mathematical Methods of Quantum Mechanics* (McGraw-Hill, 1971).

S.M. Blinder is Professor Emeritus of Chemistry and Physics at the University of Michigan, Ann Arbor, USA. He completed his Ph.D. at Harvard University in 1958, under the supervision of Profs. W.E. Moffitt and J.H. Van Vleck (Nobel Prize in Physics, 1977). Professor Blinder has over 200 publications in both Theoretical Chemistry and Mathematical Physics. He was the first to derive a closed-form expression for the Feynman path-integral propagator for the Coulomb problem (the hydrogen atom). He is the author of several books and monographs, most notably, *Quantum Mechanics in Chemistry, Materials Science, and Biology* (Elsevier Academic Press, 2004). Professor Blinder is currently a telecommuting Senior Scientist for Wolfram Research (the developers of Mathematica and other scientific software).

Chapter 1
Twentieth-Century Quantum Mechanics

Abstract The historical development and fundamental principles of quantum mechanics are reviewed. Fundamental differences with classical statistical mechanics are emphasized. The representation of quantum phenomena in Hilbert space is introduced.

Keywords Quantum mechanics · Atomic structure · Wave-particle duality · Probability theory · Quantum states · Hilbert space

The fundamental ideas at the root of quantum mechanics are introduced in this chapter. Beginning at an elementary level, the necessary mathematical background will be developed as we go along, so that readers will be able to understand such concepts as the superposition of quantum states, time evolution of quantum systems and the Heisenberg uncertainty principle. Starting out, these will be represented in terms of a highly simplified version of Hilbert space, a "toy Hilbert space," equivalent to the two-dimensional Cartesian plane. The more advanced mathematical structure of an infinite-dimensional Hilbert space will thereby be rendered more accessible in our later work. In some cases, we will sacrifice some mathematical rigor in order to achieve greater clarity. In the words of the British author H.H. Munro (pen name Saki): "A little inaccuracy sometimes saves tons of explanation." Also, in this chapter, we will briefly allude to some of the philosophical problems created by quantum mechanics, as a prelude to more detailed discussions in later chapters.

1.1 Quantum Theory and Atomic Structure

We will first expound on a very elementary level some of the mathematical concepts which are basic in quantum mechanics (QM). These constitute, to a large extent, the theory of linear operators in a Hilbert space. This subject matter is as fundamental for QM as differential calculus was for classical mechanics. The mathematical tools of QM are not as familiar as those used in classical mechanics, at least at the

© Springer International Publishing AG 2017
G. Fano and S.M. Blinder, *Twenty-First Century Quantum Mechanics:*
Hilbert Space to Quantum Computers, UNITEXT for Physics,
DOI 10.1007/978-3-319-58732-5_1

Table 1.1 Corpuscles and waves

Corpuscular nature of light: Photoelectric effect, Compton effect, etc.	Wave nature of light: Interference and diffraction
Corpuscular nature of the electron: Millikan oil drop experiment	Wave nature of the electron: X-ray crystallography, electron diffraction

present time. But it is likely that this will not always be the case, since the process of "knowledge compression" continually enables beginning students of science to become acquainted with modern techniques earlier in their academic careers. We will briefly outline, without any attempt at historical narrative, the seminal ideas which gave rise to the quantum theory. These new concepts and experimental facts have resulted in a major revolution—a "paradigm shift," as described by Kuhn (1996)—in the transition from classical to quantum physics.

The advent of the quantum theory was made inevitable, by the necessity to rationalize experimental evidence that both light and electrons have a dual nature, behaving sometimes as particles, and at other times as waves. Let us summarize the principal defining experimental facts in a small table divided into four quadrants (Table 1.1).

1.1.1 Wave Nature of Light

If we drop a stone into a calm lake, a train of circular waves of increasing radius is produced (see Fig. 1.1). Let us suppose that we observe the waves as their radii become increasingly larger. Two fundamental parameters which describe waves are the *wavelength* λ and the *period* T. The wavelength λ is the distance between the maxima of successive wave crests (see Fig. 1.2).

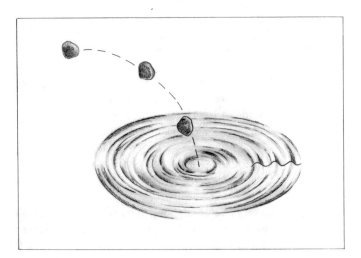

Fig. 1.1 Waves generated by a stone dropped into a body of water. Note the sinusoidal profile of the waves

Fig. 1.2 The wavelength λ
is the distance between two
successive maxima (or
minima) of a wave

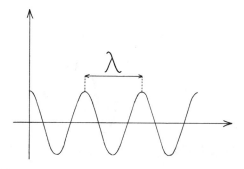

Suppose now that the wave encounters a small piece of wood floating on the water; this will cause the wood to oscillate up and down. The time between two successive maxima (highest points) is called the *period T* of the wave. The *frequency* v of the wave and its *propagation velocity c* are related by

$$v = \frac{1}{T}, \qquad c = \frac{\lambda}{T}. \tag{1.1}$$

One of the legendary giants of modern physics, James Clerk Maxwell, proposed in 1864 that light consists of propagating electromagnetic waves. This discovery is one of the most extraordinary achievements of the human intellect; a more detailed discussion will be given in Chap. 5. To begin, we consider the classic double-slit experiment, performed by Thomas Young in 1801 Young (1807), from which it could be inferred that light is a wavelike phenomenon. Consider two identical waves A, B, slightly displaced with respect to one another (see Fig. 1.3). When the two waves overlap in the same region of space, their resultant is the algebraic sum of A and B. There are two limiting possibilities (see Fig. 1.4):

(1) Constructive interference, in which the resulting wave is larger than either A or B.
(2) Destructive interference, in which the two waves annihilate one another and sum to zero.

In Young's experiment, a beam of monochromatic light illuminates a screen with two narrow vertical slits. According to Huygens's principle, every point on the slits becomes the source of a new spherical wave. An analogous effect can be observed when waves in the lake collide with a barrier with a small hole, as shown in Fig. 1.5. In this way light deviates from its original propagation direction, and is incident on a second screen, parallel to the first. This phenomenon is called *diffraction*. In Young's experiment the waves A, B produced by the two slits interfere on the second screen. Since the paths of A, B have different lengths, the two waves gives rise to bright bands when the path difference δ of A, B is an integer multiple m of the wavelength: $\delta = m\lambda$, resulting in constructive interference. However, when the path difference is an odd half integer multiple of the wavelength: $\delta = (m + \frac{1}{2})\lambda$, destructive interference produces a dark band on the screen (see Fig. 1.6). The inescapable conclusion from these experiments is that light is a wavelike phenomenon.

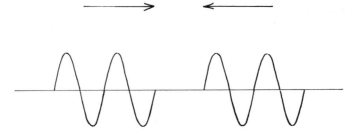

Fig. 1.3 Two identical waves approaching one another

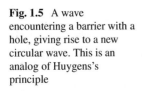

(a)   **(b)**

Fig. 1.4 **a** Constructive interference of two waves. **b** Destructive interference

Fig. 1.5 A wave
encountering a barrier with a
hole, giving rise to a new
circular wave. This is an
analog of Huygens's
principle

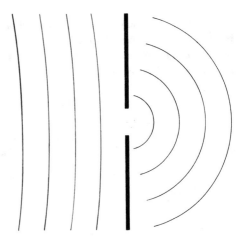

1.1.2 Corpuscular Nature of Light

Before demonstrating the remarkable fact that light also exhibits a corpuscular nature,
we briefly summarize some aspects of the interaction between matter and radiation.
It is well known that solid bodies emit radiation with maximum wavelength depend-
ing on the temperature T. If the temperature is very high, the body can become
incandescent, emitting visible light (thus the term "red hot"). Even a moderately
hot iron emits infrared radiation, which can be experienced by holding your hand
near its surface. To isolate the phenomenon of emission from irrelevant factors such

Fig. 1.6 Young's experiment showing interference of light waves

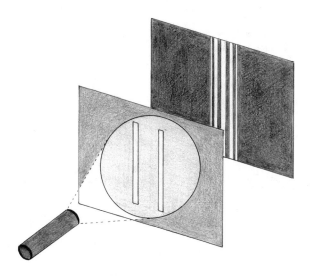

as the shape, composition or color of the emitting body, physicists have idealized the concept of a *blackbody*, which is assumed to absorb and emit radiation of all wavelengths. An approximation to a blackbody might consist of a box with perfectly absorbing walls, accessible through a small hole (called a *Hohlraum* in German). A narrow beam of radiation entering the box interacts with the dark walls through absorption and emission processes. The radiation inside consists of a superposition of many different wavelengths, while the walls are maintained at a specified temperature T. The intensity (or spectral radiance) I of the radiation emitted through the hole is proportional to the energy density ρ of the radiation inside the cavity, which is assumed to be a universal function of wavelength λ (or frequency ν) and temperature T, which can be written $\rho = \rho(\nu, T)$. As early as 1859, Kirchhoff recognized the central significance of the form of the function ρ, which must contain some important clue to the fundamental nature of light. In his words: "*Es ist eine Aufgabe von hoher Wichtigkeit, diese Funktion zu finden*" (It is very important to find this function).

Many attempts to derive such a function were based on Rayleigh's picture of radiation interacting with electric oscillators in the walls. Based on classical mechanics and electromagnetism, the Rayleigh–Jeans formula at given temperature T predicts a quadratic increase of intensity with frequency (see Fig. 1.7a). But this is manifestly contrary to everyday experience, for it would imply that the total energy emitted by the blackbody, at any temperature T, increases without limit as the frequency increases (known as the "ultraviolet catastrophe"). We would not be able to sit by a fireplace without being roasted like marshmallows by high intensity ultraviolet (and gamma) rays.

Planck (1900) conjectured a form of the function $\rho(\nu, T)$ which agreed with experiment, based on the ansatz that the energy of each electromagnetic oscillator within the cavity did not vary continuously, but rather was an integer multiple of an elementary quantum of energy ε, proportional to the frequency ν of the oscillator. He proposed that

Fig. 1.7 Blackbody
radiation: **a** Rayleigh–Jeans
quadratic behavior. **b** Planck
distribution law

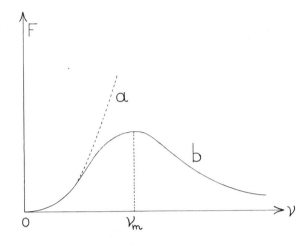

Fig. 1.7 Blackbody radiation: **a** Rayleigh–Jeans quadratic behavior. **b** Planck distribution law

$$\varepsilon = h\nu, \tag{1.2}$$

where h is a fundamental constant, now called *Planck's constant*. Setting $h = 6.55 \times 10^{-34}$ J s gave quantitative agreement with the experimental spectral energy density ρ (see Fig. 1.7b). By taking account of the frequency distribution of oscillators in three dimensions and the Boltzmann distribution at temperature T, Planck's distribution function can be derived:

$$\rho(\nu, T) = \frac{8\pi h \nu^3}{c^3} \frac{1}{e^{h\nu/kT} - 1}, \tag{1.3}$$

where c is the speed of light. We now have a more accurate value of Planck's constant: $h = 6.62606876 \times 10^{-34}$ J s.

Another phenomenon that furthered the acceptance of the quantum theory was the *photoelectric effect*: the emission of electrons from certain metals when illuminated by electromagnetic radiation, such as light, ultraviolet, X-rays, etc. It was expected, from classical theory, that increasing the intensity of the incident radiation would increase the kinetic energy of the emitted electrons; but this is *not* the case. In fact, increasing the radiation intensity *increases only the number of emitted electrons*. Furthermore no electrons are emitted if the radiation frequency is below a certain threshold. Einstein (1906) explained such behavior by assuming that the incoming radiation is not continuous, but is made up of individual *quanta*, each of energy $h\nu$, where h is again Planck's constant. Suppose that an electron receives one such quantum of radiation, and leaves the crystal to which it is bound with a binding energy ε_0. Then by conservation of energy, the kinetic energy T must be equal to the energy $h\nu$ of the quantum minus the binding energy ε_0. Einstein proposed the law:

$$T = h\nu - \varepsilon_0. \tag{1.4}$$

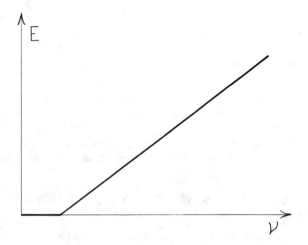

Fig. 1.8 Kinetic energy of the ejected electron in the photoelectric effect

This relation is shown in Fig. 1.8 and has been extensively tested experimentally. Since the kinetic energy must be positive, no electrons are released if the quantum energy is less than the threshold ε_0. Thus we must add to Eq. (1.4) the condition: $T = 0$ for $h\nu < \varepsilon_0$. The paradox about the number of emitted electrons is thereby resolved, since, with light of low intensity, there are fewer atoms absorbing a quantum. Increasing the intensity just produces more quanta, but all with the same energy $h\nu$. The introduction of light quanta, later called *photons*, was the most significant advance in our modern understanding of the nature of light.

Arthur Compton (1923) bombarded atomic electrons, which are relatively loosely bound, with high-energy photons. The result was that the photons were deflected from their original directions, while the electrons were ejected from their atoms with a measurable kinetic energy T. In classical mechanics, an elastic collision between two particles must obey the two fundamental principles: conservation of energy and conservation of momentum. If ν denotes the frequency of the incoming photon, and ν' that of the scattered photon, the first of the two laws implies that $h\nu = h\nu' + T$. The law of conservation of momentum, a vector relation, is a little more complicated to apply. We skip the details but note only that the momentum $\mathbf{p}_\nu$ of the incoming photon is the resultant of the momentum $\mathbf{p}_{\nu'}$ of the scattered photon and the momentum $m\mathbf{v}$ acquired by the electron (see Fig. 1.9). As a result, the frequency of the incoming radiation is shifted by $\nu - \nu'$, which can easily be computed. The *Compton effect*, which we have just described, lent further support of the Planck–Einstein theory of radiation.

The various phenomena described above presents a major metaphysical question that continues to defy a simple answer: "Is light a particle or a wave? Or somehow, both?" Possibly, the most sensible answer is that photons (and electrons) behave neither as classical particles nor waves, but in an idiosyncratic way in accordance with the laws of quantum mechanics.

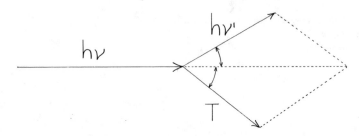

Fig. 1.9 The Compton effect: the energy of the scattered photon is equal to the energy of the incoming photon minus the energy gained by the atomic electron

1.1.3 Corpuscular Nature of the Electron

More than 2400 years ago, Democritus and Leucippus claimed that everything in the world is composed of atoms. As Feynman emphasized "all things are made of atoms" contains "an enormous amounts of information about the world." Certainly the atomic hypothesis was a critical breakthrough (or paradigm shift) for our understanding of the nature of matter.[1] In the mid-nineteenth century, Richard Laming proposed that atoms are composed by a core of matter with electrically charged balls surrounding it. In 1870, William Crookes built the first cathode ray tube, in which a flux of charged particles from a positively charged cathode impinged on a negatively charged anode. He was able to deflect the rays with a magnetic field, and concluded that they were negatively charged. In 1881, Hermann von Helmholtz postulated the existence of elementary negative and positive charges, each behaving like atoms of electricity. Now we call them *electrons* and *positive ions*. In 1890, Arthur Schuster showed that the cathode rays of the Crookes tube are deflected by an electric field, and was able to calculate the *charge to mass ratio* e/m of the electrons. J.J. Thomson gave estimates of both m and e and observed that *these values were independent of the metallic composition of the cathode.*

In 1909, Millikan and Fletcher (1913) carried out the famous "oil drop experiment" which confirmed the discrete negative charge of electrons. Very small oil droplets are observed through a microscope while they are subjected to a uniform electric field of known intensity E (see Fig. 1.10). The oil is sprayed into a chamber above a capacitor. The droplets fall by gravity into the region of the electric field. They have become electrically charged by friction with the nozzle of the sprayer; alternatively they can be subjected to ionizing radiation. Since the size and density of the droplets can be estimated, their mass M can be computed. In this way, the three forces acting on a given droplet are known: electrostatic force, gravitational force Mg, and frictional drag of the air. It was observed that the charges absorbed by the droplets were all multiples of a fundamental value, which we now recognize as the charge $-e$ of a single electron. Millikan and Fletcher found the value $e = 1.592 \times 10^{-19}$ C, within one percent of the presently accepted value, $e = 1.6021764 \times 10^{-19}$ C.

---

[1] There was no accepted coherent theory of atoms during the middle of the eighteenth century.

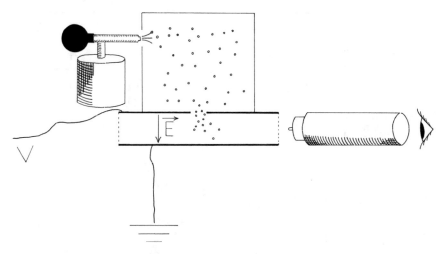

Fig. 1.10 Simplified pictorial of Millikan's oil drop experiment

1.1.4 Wave Nature of the Electron

A young French physicist, De Broglie (1924), conjectured that not only light, but also particles of matter can exhibit a wave–particle duality. Denote by $p = mv$ the *momentum* of a particle (such as an electron) with mass m traveling with velocity v. He proposed that the motion of the electron is associated with a "matter wave" with wavelength given by:

$$\lambda = \frac{h}{mv} = \frac{h}{p}. \tag{1.5}$$

De Broglie's formula can be arrived at by the following heuristic argument. In relativity theory, space and time are connected, as the components of a four-dimensional vector in *spacetime*, $(x_0, x_1, x_2, x_3) = (ct, x, y, z)$. Analogously, momentum p and energy ε can be combined into a 4-vector $(p_0, p_1, p_2, p_3) = (\varepsilon/c, p_x, p_y, p_z)$. Since time is conceptually the zeroth component of spacetime, it can be surmised that energy is equivalent to the zeroth component of 4-momentum. Recall now the wavelength–frequency relation $\lambda = c/\nu$. By the Planck–Einstein relation, $\varepsilon = h\nu$, we can write $\lambda = hc/\varepsilon = h/(\varepsilon/c)$. But ε/c is the zeroth component of 4-momentum and has the dimensions of p. Thus we can propose that $\lambda = h/p$ might be the spacelike analog of the relation $\varepsilon = h\nu$, which contains the timelike variable ε. It turns out that, in the Compton effect, the momentum of a photon is indeed given by $p = h/\lambda$.

Both the de Broglie and Planck–Einstein relations, $p = h/\lambda$ and $\varepsilon = h\nu$, are quantitative representations of the wave–particle duality: wavelength λ and frequency ν being wavelike attributes, while momentum p and energy ε are particle-like.

Von Laue (1913) and the Braggs (William Henry and William Lawrence, father and son) Bragg and Bragg (1913) showed that X-rays impinging on a crystal lattice

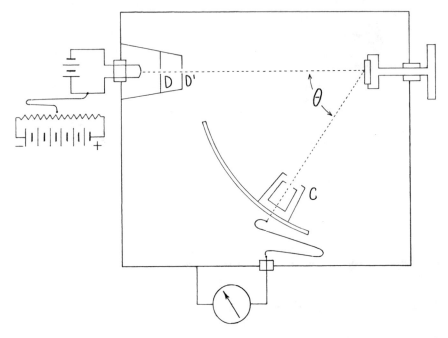

Fig. 1.11 Schematic of the Davisson and Germer experiment: a beam of electrons passes through two apertures D and D′ and is diffracted by a nickel crystal

gave waves reflected by every plane of atoms in the lattice (including oblique planes), interfering constructively or destructively according to their angle of incidence. Davisson and Germer (1927) carried out an analog of this experiment using a beam of electrons in place of X-rays. Let us describe in some detail their beautiful experiment. A beam of electrons is emitted by a cathode in a vacuum apparatus and accelerated by an electric field (see Fig. 1.11). The beam is columnated (narrowed) by passing it through the apertures D, D′ in two screens and is directed perpendicularly onto a nickel crystal. The angle θ between the scattered electrons and the normal to the crystal is measured by the electron detector C (shielded in a Faraday cage). Davisson and Germer found that for certain values of θ, the intensity of the reflected beam was enhanced, precisely analogous to the diffraction of X-rays. Indeed one can use the Bragg formula that connects the wavelength λ of the radiation with the spacing d between the planes of the atomic lattice:

$$n\lambda = 2d \sin \theta, \tag{1.6}$$

where n is an integer. Formula (1.6) is readily apparent in Fig. 1.12. The difference between the lengths of the two "rays" a, b reflected at the points P, Q, respectively, is equal to the length of the path APC. Since $\overline{AP} = \overline{PC} = d \sin \theta$, the Bragg condition (1.6) gives constructive interference between the two rays. Davisson and Germer

Fig. 1.12 The difference of the pathlengths of a and b is the sum of the distances AP and PC

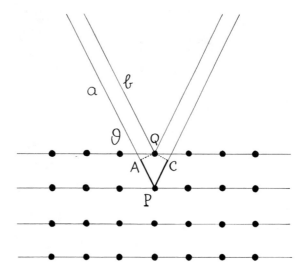

were able to compute the wavelength λ of the mysterious "matter wave" associated with the motion of the electrons. The result was in complete agreement with de Broglie's hypothesis, Eq. (1.5).

1.1.5 Early Models of Atomic Structure

In 1902, J.J. Thomson hypothesized that atoms are made up of electrons embedded in a homogeneous sphere of positive electricity (known as the "plum-pudding" model). However this model was shown to be flawed by Rutherford's famous experiments Rutherford (1911). A beam of alpha particles (doubly ionized helium atoms), emitted by a radioactive source, was directed onto a gold foil. A very small fraction of the alpha particles were reflected almost straight back, as if they had struck something very hard inside the atoms of the foil. Since alpha particles are positively charged, Rutherford proposed that the repelling force was due to the presence of a small positively charged "nucleus" in each atom. Niels Bohr subsequently proposed, in 1913, a "planetary model" of the atom, in which electrons orbit around a positive nucleus, similar to the motions of planets around the Sun. It was evident that this model egregiously violated the laws of classical mechanics and electromagnetism, since all the energy of the electrons ought to be radiated away in 10^{-10} s, as the electrons spiral into the nucleus. This has been called the "atomic Hindenburg disaster" (the Hindenberg, a hydrogen-filled dirigible, crashed and burned in a famous disaster in 1937).

The richest source of information on atomic structure is provided by the spectral lines of light emitted and absorbed by atoms. The continuous spectrum is not relevant for our present considerations (although it produces the beautiful colors of

Fig. 1.13 Emission spectrum of atomic hydrogen in the visible region

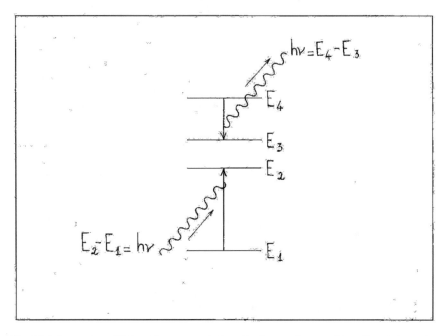

Fig. 1.14 Interaction of light with atoms: emission of a photon of energy $E_4 - E_3$ and absorption of a photon of energy $E_2 - E_1$

the rainbow). In the years 1814–1823, Fraunhofer (1817) begin a systematic study of the dark lines in sunlight, caused by absorption of light by the solar atmosphere against the continuous background of blackbody radiation. The absorption spectrum of atomic hydrogen in the visible region, shown in Fig. 1.13, consists of five lines at wavelengths of 656.3 nm (red), 486.1 nm (blue-green), 434.0 nm (blue-violet), 410.2 nm (violet) and 397.0 nm (violet). Discrete absorption or emission frequencies cannot be accounted for by classical electromagnetic theory, since the oscillation of charges generally produces a continuum of frequencies. Bohr (1913), postulated that the energy levels of the atoms form a discrete series of values $E_1, E_2, E_3, \ldots$ (see Fig. 1.14). It then becomes possible to explain the discrete spectral lines. When an electron jumps from a higher energy level E_4 to a lower energy level E_3, it emits a photon of energy

$$h\nu = E_4 - E_3. \tag{1.7}$$

These photons give rise to an emission line of frequency ν. Conversely, an electron in a lower energy level E_1 that absorbs a photon of energy $E_2 - E_1$ jumps to the E_2 energy level, producing an absorption line. Rydberg (1890) proposed the following empirical formula for the spectral lines of the hydrogen atom:

$$\frac{1}{\lambda} = R\left(\frac{1}{n^2} - \frac{1}{m^2}\right) \qquad n, m = 1, 2, 3, \ldots \quad m > n, \tag{1.8}$$

where R is a constant whose value is about 109737 cm^{-1}. Bohr proposed a planetary model of the atom in which the electronic energy levels are quantized in a new way. We will see in a moment that the Bohr model is not only in complete agreement with Eq. (1.8), but also results in an explicit formula for the Rydberg constant R, in terms of the fundamental constants h, e, m, c. And this, in our opinion, is really magical.

Let us now consider the analytical mechanics leading to Bohr's quantization postulate. In the simplest case, an electron moves around the nucleus in a *circular* orbit. Let v denote the speed of the electron, m its mass, r the radius of the orbit, and $+Ze$ the positive charge of the nucleus. Two forces, in balance, maintain the electron in its circular orbit, as shown in Fig. 1.15, namely:

(1) Coulomb attraction by the nucleus Ze^2/r^2 (note the similarity to the Newtonian gravitational force).

(2) This attraction acts as a *centripetal force* mv^2/r to restrain the electron to a circular orbit. (From another point of view, this can be regarded as a *centrifugal force* in the rotating frame of reference.)

$$\frac{Ze^2}{r^2} = m\frac{v^2}{r}. \tag{1.9}$$

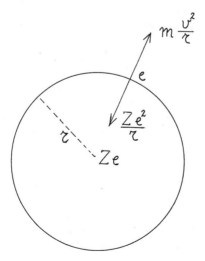

Fig. 1.15 Uniform circular motion of an electron; the centrifugal force is equal and opposite to the attraction towards the nucleus

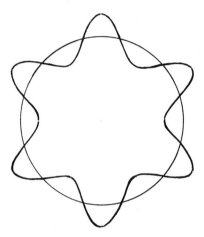

Fig. 1.16 A standing de Broglie's wave around an electronic orbit

Let us now consider the electron's de Broglie wave that "travels" around the orbit. For stationary or standing waves, the wave must overlap with itself after a complete revolution. A standing wave can be constructed only if the locations of the wave nodes do not change with time. Thus the length of the orbit must be an integer multiple $n = 1, 2, 3, \ldots$ of the wavelength λ as shown in Fig. 1.16:

$$2\pi r = n\lambda. \tag{1.10}$$

Substituting the value Eq. (1.5) of the de Broglie wavelength, we obtain $2\pi r = nh/mv$, or $rp = n\hbar$. Orbital angular momentum is defined by the vector relation $\mathbf{L} = \mathbf{r} \times \mathbf{p}$. For a circular orbit this reduces simply to $L = rp$. We obtain thereby the condition for quantization of angular momentum, which turns out to be a result of fundamental significance in quantum mechanics.

$$L = rp = n\hbar, \tag{1.11}$$

This relation leads, as we will next show, to Bohr's quantization of energy. From Eq. (1.9), we obtain $r = Ze^2/mv^2$. Substituting this expression for r into Eq. (1.11), $(Ze^2/mv^2)mv = Ze^2/v = n\hbar$; thus $v = Ze^2/n\hbar$, and finally:

$$r_n = \frac{Ze^2}{m(Z^2e^4/n^2\hbar^2)} = \frac{n^2\hbar^2}{Zme^2}. \tag{1.12}$$

It follows that the orbital radii are proportional to n^2, where n is called the *principal quantum number*. For the case of the hydrogen atom, $Z = 1$. The smallest value of the principal quantum number, $n = 1$, gives the *Bohr radius*:

$$a_0 = r_1 = \frac{\hbar^2}{me^2} = 0.529 \times 10^{-10} \text{ m.} \tag{1.13}$$

The energy E_n of the electron is the sum of its kinetic and potential energies:

$$E_n = \frac{1}{2}mv^2 - \frac{Ze^2}{r_n} = \frac{1}{2}\frac{Ze^2}{r_n} - \frac{Ze^2}{r_n} = -\frac{1}{2}\frac{Ze^2}{r_n}. \tag{1.14}$$

Substituting in the last equation the expression (1.12) for the radius r_n, we obtain the final form for the energy levels:

$$E_n = -\frac{Z^2e^4m}{2\hbar^2}\frac{1}{n^2}. \tag{1.15}$$

With $Z = 1$ and $n = 1$, we get the ground-state energy of the hydrogen atom, expressed in electron volts:

$$E_1 = -\frac{e^4m}{2\hbar^2} = -13.6 \text{ eV} \tag{1.16}$$

From Eq. (1.15) and the equation $h\nu = E_n - E_{n'}$, $(n > n')$, we obtain an explicit expression for the frequency of the photon associated with an electronic jump from the n to the n' energy level:

$$h\nu = -\frac{Z^2e^4m}{2\hbar^2}\left(\frac{1}{n^2} - \frac{1}{n'^2}\right), \quad (n > n'), \tag{1.17}$$

In terms of the wavelength λ of the photon, use $\nu = \frac{c}{\lambda}$, so that $\frac{1}{\lambda} = \frac{h\nu}{hc}$. Thus, Eq. (1.17) can be written:

$$\frac{1}{\lambda} = -Z^2 R\left(\frac{1}{n^2} - \frac{1}{n'^2}\right), \quad (n > n'). \tag{1.18}$$

with the Rydberg constant given by

$$R = \frac{2\pi^2e^4m}{ch^3}, \tag{1.19}$$

which agrees closely with the empirical value in Rydberg's formula. The similarities between the spectra of hydrogen atoms on Earth and those from stars more than 10 billion light years away from us indicate that electrons in both places have exactly the same mass m and charge e. Furthermore the speed c of light must be the same throughout the whole Universe. Again, from our naive perspective, we find this all very magical.

1.2 Probability and Statistical Mechanics

Before introducing the principles of quantum mechanics, a few concepts of probability and statistical mechanics provide a useful preliminary. It is well known that classical mechanics presents us with a precise deterministic worldview which enables, in principle, exact prediction of the evolution of a macroscopic physical system at all future times. Consider, for example, the system consisting of an artificial satellite, the Earth, the Sun, and the Moon: once the positions and the velocities at a given instant of time are known, Newton's laws allow us to determine with precision positions and velocities of the bodies at any later time. It was recognized by Poincaré that sometimes the motions depend very sensitively on the initial conditions and this can lead to chaotic behavior (the details of which are not relevant to our treatment of QM).

Let us reflect on the actual situation in the dynamics of complex systems. Newton's laws do indeed have the "absolute predictive ability" which their deterministic formulation would imply. This is true if we apply them in the appropriate domain of size and speed, possibly with the help of modern computers. The motion of a small number of macroscopic bodies (as, in the example above, an artificial satellite, the Earth, etc., neglecting interactions with far-off planets) can be predicted with fairly high accuracy. But what happens if we try to apply Newton's laws to a system with a much larger number of degrees of freedom? For example, the motion of air violently escaping from a punctured tire: it is impossible, from a practical point of view, to know with precision the positions and velocities of all the gas molecules at a given instant. Their number is too large, and the precision required in order to make reasonable predictions by integrating the equation of motion is beyond our capability. We are not dealing only with a practical difficulty, which perhaps will be solved some time in the future, but rather a conceptual difficulty. Very appropriate is a quote by Feynman (1965): "…given an arbitrary accuracy, no matter how precise, one can find a time long enough that we cannot make predictions valid for that long a time. Now the point is that this length of time is not very large …The time goes, in fact, logarithmically with the error, and it turns out that in only a very, very tiny time we lose all our information." But let us return to the above example, air leaking from a tire. In such cases, physicists generally make assumptions of a statistical nature: it suffices to assume that the huge number of air molecules, both inside and outside the tire, have uniform initial position probabilities, and a distribution of velocities in a range of magnitudes, moving in arbitrary directions. It is assumed, of course, that the pressure inside is greater than that outside. At this point, we can apply the laws of classical mechanics on a statistical level for a reasonable description of the phenomenon, what has developed into the subject of *statistical mechanics*. This approach can, for example, demonstrate how the laws of thermodynamics, which involve *macroscopic* quantities such as heat, temperature, pressure, etc., have *microscopic* analogs on a molecular level. Thus, gas temperature is a measure of the *average* kinetic energy of its molecules, pressure is an *average* measure of the frequency and intensity of the collisions of the molecules with the walls of the container, etc. The conceptual foundation of statistical mechanics is based on classical mechanics, applied to *probability distributions* in *phase space*, representing all possible microscopic configurations of a system. Since the position of a single mass point is described by 3 coordinates x, y, z, and its velocity by the 3 components v_x, v_y, v_z, a system of N mass points is represented in a phase space by a point in $6N$ dimensions. For technical reasons, it is better to specify the *momentum* $p_x = mv_x$ rather than the velocity v_x, where m is the mass of the particle, although this does not change things very much. A point in phase space completely describes, in concept, the detailed microscopic state of the system. However one should keep in mind that N is an enormous number (for one mole of gas, $N \approx 6.02 \times 10^{23}$). Of course, phase space is not be confused with the real physical space in which the particles move.

1.2.1 Introduction to Probability

As an elementary example of probability defined on a set (in the above example, it was phase space): let us assume that an American tourist has decided to go on vacation in one, and only one, European country. We ask him where he will go, and he answers: "I do not know yet, but I am 10% certain about going to England, 30% to France, 20% to Germany and 40% to Italy." You can see that he has assigned some positive numbers, between 0 and 1, to some regions of Europe. If E stands for England, F stands for France, G stands for Germany and I for Italy, the probabilities of the four *events* (vacation in England, vacation in France, etc.) are: $P(E) = \frac{1}{10}$, $P(F) = \frac{3}{10}$, $P(G) = \frac{2}{10}$, and $P(I) = \frac{4}{10}$. Of course we have $P(E) + P(F) + P(G) + P(I) = 1$, since the tourist is sure to spend his vacation in one of these countries. Some simple and very general logical properties must be satisfied by the probability function defined on a set S (in our case S is the set of the four countries): if A and B denote two disjoint subsets[2] of S, we must have $P(A \bigcup B) = P(A) + P(B)$. In our example, if $G \bigcup F$ is the set constituted by Germany *and* France, clearly $P(G \bigcup F) = \frac{2}{10} + \frac{3}{10} = \frac{5}{10}$ is the probability of vacation in Germany *or* France. Furthermore $P(S) = 1$ and $P(\phi) = 0$, where ϕ denotes the empty set.

Another elementary example: consider an urn containing N balls, of which M are red. If we withdraw a ball from the urn, and we want to calculate the probability that the ball is red, the set S now represents all N balls, and a possible subset A is formed

---

[2]We recall the meaning of *intersection* and *union* of two sets. Consider the following elementary example: let P be the set of the positive integers less or equal to 12: $P = \{1, 2, 3, \ldots, 12\}$, E the subset of the even numbers, $E = \{2, 4, 6, 8, 10, 12\}$, O the subset of the odd numbers, $O = \{1, 3, 5, 7, 9, 11\}$, and T the subset of the multiples of 3: $T = \{3, 6, 9, 12\}$. The *intersection* of E and T is composed of the numbers that are even *and* multiples of 3: $E \bigcap T = \{6, 12\}$; The *union* of E and T by the number that are even *or* multiples of 3: $E \bigcup T = \{2, 3, 4, 6, 8, 9, 10, 12\}$; since a number cannot be even and odd at the same time, the intersection $E \bigcap O = \phi$, where ϕ denotes the empty set. We say, in this case, that the sets E and O are *disjoint*. See also the geometrical representation of union and intersection in Fig. 1.17.

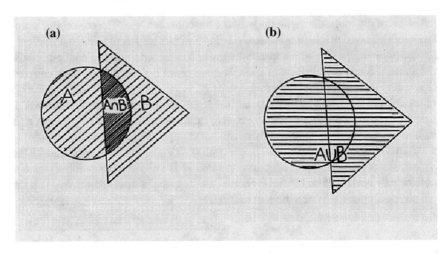

Fig. 1.17 Given two sets: a circle A and a triangle B. The shaded area shows **a** the intersection $A \bigcap B$ and **b** the union $A \bigcup B$

by the M red balls. Let us associate to any ball a probability $\frac{1}{N}$ of being extracted; then $P(A) = \frac{M}{N}$ and of course $P(S) = \frac{N}{N} = 1$. From this example we understand why a very common definition of probability is the following: In a situation where there are equally likely outcomes, the *probability of an event is the ratio of the number of favorable outcomes to the number of possible outcomes*.

In probability theory, two events A, B are said to be *independent* if $p(A \bigcap B) = p(A)p(B)$, where $A \bigcap B$ denotes the occurrence of both A and B; to clarify this concept let us consider again the example of the two urns, containing, respectively, N_1 and N_2 balls. In the first urn, there are M_1 red balls, and in the second there are M_2 red balls. Clearly, the probability p_1 of extracting a red ball from the first urn is $p_1 = \frac{M_1}{N_1}$ and the probability p_2 of extracting a red ball from the second urn is $p_2 = \frac{M_2}{N_2}$. Now if we extract two balls at random, one from the first and one from the second urn, the possible pairs are $N_1 N_2$; of these pairs $M_1 M_2$ are constituted by two red balls. Then using the rule defined above: probability p = number of favorable cases divided by the number of possible cases, we have (see Fig. 1.18) $p = M_1 M_2 / N_1 N_2 = p_1 p_2$; thus, denoting by A the event "removal of a red ball from the first urn" and by B the event "removal of a red ball from the second urn", we have $p(A \bigcap B) = p(A)p(B)$; indeed, the two events are independent.

Suppose now that we have a function f, representing a *random variable*, defined on the set S. Returning to the example of the tourist in Europe, suppose he considers the daily cost of living in his choice. Assuming that this cost is fixed for each country: for example, 40 euros for England, 60 for France, 80 for Germany and 50 for Italy. Suppose also that the tourist repeats the same travel itinerary for many years, always with the same probabilities. We can ask: "What is the *average price* the tourist will pay during his travels?" The answer is a suitable weighted sum extended over

Fig. 1.18 Points of the large rectangle represents possible choices of two balls, one from the first and one from the second urn; points of the crosshatched rectangle represent the choice of two *red balls*

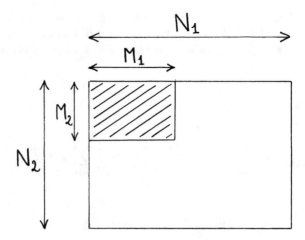

the four countries E, F, G, I. We will see in this elementary example that a probability distribution allows us to compute *averages or means of random variables*. If the set S is divided into N disjoint subsets $S_1, S_2, \ldots, S_N$, the probabilities are $P(S_1), P(S_2), \ldots, P(S_N)$, and the function f assumes the value f_i on the set S_i, the average value of f is given by:

$$\overline{f} = \sum_{i=1}^{N} f_i P(S_i),\qquad(1.20)$$

so the average daily cost is

$$\overline{f} = \frac{1}{10}40 + \frac{3}{10}60 + \frac{2}{10}80 + \frac{4}{10}50 = 58.\qquad(1.21)$$

To quantify the spread in the distribution of costs, we define a simple mathematical quantity: the *variance* σ^2 of a probability distribution, or the *standard deviation* σ, the square root of the variance. The variance is defined as the average of the squares of the deviations $f_i - \overline{f}$ (the reason for using the squares is that the average of the deviations always equals zero, with the sum of the positive and negative entries canceling out). Accordingly, we define:

$$\sigma^2 = \sum_{i=1}^{N} (f_i - \overline{f})^2 P(S_i) = \overline{(f - \overline{f})^2}.\qquad(1.22)$$

In the case of the tourist we have:

$$\sigma^2 = \frac{1}{10}(40 - 58)^2 + \frac{3}{10}(60 - 58)^2 + \frac{2}{10}(80 - 58)^2 + \frac{4}{10}(50 - 58)^2 = 156.$$
$$(1.23)$$

Therefore $\sigma = \sqrt{156} = 12.49$. σ is a measure of the *spread*; if the prices had been 59, 60, 61, 62, we would have $\sigma = 1.044$. If the N probabilities $P(S_i)$ are equal, their value should be $\frac{1}{N}$, and formulas (1.20), (1.22) become:

$$\overline{f} = \frac{1}{N} \sum_{i=1}^{N} f_i, \tag{1.24}$$

$$\sigma^2 = \frac{1}{N} \sum_{i=1}^{N} (f_i - \overline{f})^2. \tag{1.25}$$

1.2.2 Statistical Mechanics

We will see that probability distributions in phase space, of coordinates and momenta, enable us to compute average values of physical quantities of systems with many degrees of freedom, avoiding the difficulties inherent in an (impossible) detailed microscopic description. Let us begin with a very simple and elementary example: a gas composed of a very large number, N, of molecules, each of mass m, contained in a box. David Bernoulli (1738), and later Newton, proposed that the pressure of the gas on the walls of the box is due to the impact of the particles colliding with the walls. In order to make the calculation as simple as possible, we assume that the box is a cube of side a, and that the total volume occupied by the molecules themselves is negligible compared to the volume of the container, $V = a^3$. Assume, in addition, that the collisions with the walls are perfectly elastic, with every particle bouncing back and forth between the two faces orthogonal to the x axis. If v_x is the component of a particle's velocity before the collision, it will be changed to $-v_x$ after the collision, as shown in Fig. 1.19. The momentum communicated to one face equals $2mv_x$, and the time interval between successive impacts is $2a/v_x$. From classical mechanics, the force F acting on the wall equals the change in momentum $2mv_x$ divided by the time interval, so that

$$F = \frac{2mv_x}{2a/v_x} = \frac{mv_x^2}{a}. \tag{1.26}$$

Then the increment of pressure (force per unit area) p exerted by the particle is

$$p = \frac{F}{a^2} = \frac{mv_x^2}{a^3}, \tag{1.27}$$

since the area of one face is a^2.

In order to find the total pressure exerted by *all* the molecules, we must take the sum of (1.26) over all the contributions of the individual molecules. Since the values of v_x differ from one molecule to another, it is hopeless to perform the sum with all

Fig. 1.19 Elastic collision of a particle with a wall: the components of the incoming velocity **v** are (v_x, v_y), while the components of the outgoing velocity **v**′ are $(-v_x, v_y)$

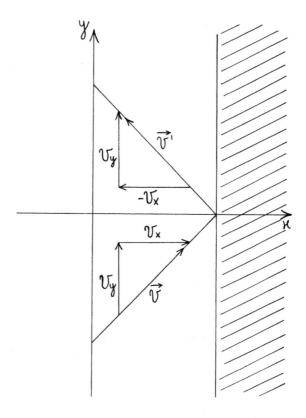

the exact details. However, we can use the definition (1.20) of *average*, where now the subset S_i is the ith molecule, and the function f is v_x^2. We have:

$$\overline{v_x^2} = \frac{1}{N} \sum_{i=1}^{N} v_x^2(i), \qquad \text{thus} \qquad \sum_{i=1}^{N} v_x^2(i) = N\overline{v_x^2}. \tag{1.28}$$

Suppose now that there is exactly one mole of gas in the container, so that the total number of molecules equals Avogadro's number $N_A \approx 6.02 \times 10^{23}$; for example, if the gas is hydrogen, this corresponds to 1 g of gas in the container. The overall pressure P exerted by the gas will then be:

$$P = N_A m \frac{\overline{v_x^2}}{a^3} = N_A m \frac{\overline{v_x^2}}{V}. \tag{1.29}$$

By extension of the Pythagorean theorem to three dimensions (see Fig. 1.20), we have $v^2 = v_x^2 + v_y^2 + v_z^2$. Ignoring the negligible effect due to gravity, the average values of v_x^2, v_y^2, v_z^2 are equal. Therefore:

Fig. 1.20 The square of the diagonal of a rectangular parallelepiped is the sum of the squares of the three sides. Thus the total velocity and its components are related by $v^2 = v_x^2 + v_y^2 + v_z^2$

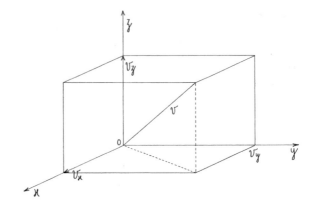

$$\overline{v^2} = 3\,\overline{v_x^2}, \tag{1.30}$$

and (1.29) becomes:

$$PV = \frac{1}{3}N_A m\overline{v^2}. \tag{1.31}$$

Experimentally, gases under normal conditions are known to obey the ideal gas law, to a good approximation. Then, for one mole of gas ($V \approx 22.4 \text{ dm}^3$) we have:

$$PV = RT \tag{1.32}$$

where R is the gas constant and T is the absolute temperature. Boltzmann's constant is given by $k_B = R/N_A = 1.38065 \times 10^{-23}\text{J K}^{-1}$. From Eqs. (1.31), (1.32) we obtain:

$$\frac{1}{2}m\overline{v^2} = \frac{3}{2}k_B T. \tag{1.33}$$

Thus *the average kinetic energy of a gas molecule is proportional to the absolute temperature*, a simple yet profound relation between a microscopic quantity—the average kinetic energy of a system of particles—and a macroscopic thermodynamic variable—the absolute temperature.

The gas law (1.32) determines the average value of the square of the velocity of the molecules, but contains no information about the detailed distribution of velocities. How many molecules have more than twice the average speed? or more than three times the average speed? Before trying to answer these questions, there is another one, concerning not the velocities, but the positions of the molecules. This can be approached by applying some results from combinatorial analysis.

1.2.3 Combinatorial Analysis

If we conceptually divide the air of a closed room into many "compartments," each of dimension 1 cubic centimeter, why do all the compartments contain approximately the same number of molecules? Or, in other words, why is the density of the air practically constant? We all breathe calmly, without worrying that suddenly a vacuum pocket will form around our heads! Why will this never happen, despite the disordered motion of molecules moving in all directions? Returning to our considerations of balls in urns, suppose we have two identical compartments, which we call L and R, respectively. Suppose we have a small number of balls, say, 4 balls, and we begin putting the balls into the two compartments. For the first ball we have just two possibilities: L or R. For simplicity we assume that L and R are *equally likely*; the relative probabilities of the two choices are equal. For the second ball we have again two possibilities, therefore we can represent the possible choices in the upper part of a graph (see Fig. 1.21), containing two bifurcations, and four possible oblique segments: LL, LR, RL, RR. For example, LL denotes the first and second ball in the L compartment, LR denotes the first ball in the left compartment, second ball in the right compartment, etc. Since the third ball has again two possibilities, the first three rows of the graph contain the following eight ways each constituted by two segments: LLL, LLR, LRL, LRR, RLL, RLR, RRL, RRR. For example, LRR denotes the first ball in the left compartment, the second and the third ball in the right compartment, etc. Finally adding the fourth ball gives 16 possibilities, which we order in the following way (see Fig. 1.21):

(A) all the balls are in L: LLLL.
(B) 3 balls are in L and 1 in R: RLLL, LRLL, LLRL, LLLR.
(C) 2 balls are in L and 2 in R: LLRR, LRLR, LRRL, RLLR, RLRL, RRLL.
(D) 1 ball is in L and 3 in R: LRRR, RLRR, RRLR, RRRL.
(E) All the balls are in R: RRRR.

In combinatorial analysis, we often encounter the symbol $n!$, called n *factorial*, equal to the product of all positive integers less than or equal to n. For example,

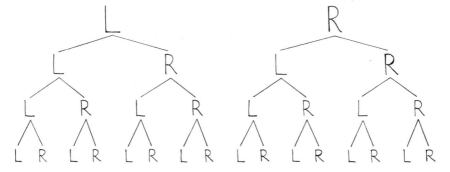

Fig. 1.21 Sixteen possible paths that start from an initial choice (L or R) and fall along the graph, giving rise to sixteen possible composite possibilities: LLLL, LLLR, etc

$5! = 5 \times 4 \times 3 \times 2 \times 1 = 120$. By definition, $0! = 1! = 1$. The number of ways of choosing k objects from a set of n objects equals $\frac{n!}{k!(n-k)!}$, the binomial coefficient denoted by $\binom{n}{k}$. Thus, in the above example, the number of ways of distributing 4 balls into 2 compartments are given by: (A) $\binom{4}{4} = 1$, (B) $\binom{4}{3} = 4$, (C) $\binom{4}{2} = 6$, (D) $\binom{4}{1} = 4$, (E) $\binom{4}{0} = 1$. The total number of possibilities equals $1 + 4 + 6 + 4 + 1 = 16 = 2^4$. Now to return to the problem of the distribution of molecules in the air of a tightly closed room. If we mentally divide the room into two equal parts, L and R, we can repeat the same line of reasoning as before, the only difference being that in place of balls now we have molecules, and the number of "compartments," possible partitions of space, is huge. The factorial $n!$ increases very steeply, much more rapidly than exponentially. For example $20! = 2\,432\,902\,008\,176\,640\,000$. Since the number of molecules in a room is usually much larger than Avogadro's number, the factorials involved become exceedingly huge numbers, and a new domain of behavior comes into play. From a practical point of view, it can be decreed that *microscopic configurations that are very unlikely become impossible, and the most probable configuration become certain*. In philosophical terminology, this is sometimes referred to as *moral certainty*.

In the example of the balls in four compartments, the probability of the partition with equal numbers, such as 2 in compartment L and 2 in compartment R, is $\frac{6}{16}$, slightly larger than the others. But in the case of molecules, with the room divided into two equal parts, the configurations with nearly equal densities of air overwhelm all the other possibilities. Here a subtle question arises: denoting by N_L, N_R, the number of molecules in L and R respectively, and by N the total number of molecules $N = N_L + N_R$, what is the probability that $\frac{N_L - N_R}{N} > 0.1$ or > 0.01? In order to answer this type of question, we must relate the global behavior of the air to some property of its *individual* molecules. That can be done since our model of the container of air is based on an *independent particle model*. It is a generally true in physics that problems in which particles behave independently are relatively easy to solve, but become very difficult when interactions between the particles have to be taken into account. Let us again return to the colored balls; the combinatorial calculus is obviously the same as with molecules. We introduce a "small" random variable x_i, with the values $x_i = 1$ if the ith ball is in the L box, and $x_i = 0$ if it is in the R box. Clearly the "*global*" random variable

$$X = \sum_{k=1}^{N} x_k,$$
(1.34)

is equal to the final number of balls in the L compartment. Since we have assumed that the choices L and R are equally likely, the average $\overline{x}_k = \frac{1+0}{2} = \frac{1}{2}$ and from (1.34) we have:

$$\overline{X} = \sum_{k=1}^{N} \frac{1}{2} = \frac{N}{2}.$$
(1.35)

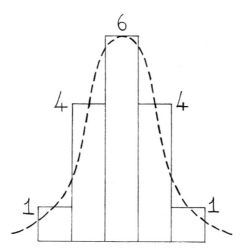

Fig. 1.22 Graph obtained by plotting five rectangles of height $\binom{4}{4} = 1$, $\binom{4}{3} = 4$, $\binom{4}{2} = 6$, $\binom{4}{1} = 4$, $\binom{4}{0} = 1$, respectively. The graph approaches a Gaussian or normal distribution (bell curve) as the number of rectangles is increased

As a check, let us compute $\overline{X}$ for $N = 4$, using the partitions A, B, C, D, E. We have:

$$\overline{X} = 4P(A) + 3P(B) + 2P(C) + 1P(D) + 0\,P(E) =$$
$$4 \times \frac{1}{16} + 3 \times \frac{4}{16} + 2 \times \frac{6}{16} + 1 \times \frac{4}{16} + 0 \times \frac{1}{16} = 2. \tag{1.36}$$

A graph of the five probabilities $P(A), P(B), \ldots, P(E)$ is shown in Fig. 1.22. It closely resembles the famous "bell curve" or *normal distribution*, which is supposed to model the distribution of IQs and other characteristics of a human population. The standard deviation σ measures the half-width of the curve. Using the definition Eq. (1.22), we have:

$$\sigma^2 = 2^2 P(A) + 1^2 P(B) + 0^2 P(C) + 1^2 P(D) + 2^2 P(E) =$$
$$4 \times \frac{1}{16} + 1 \times \frac{4}{16} + 1 \times \frac{4}{16} + 4 \times \frac{1}{16} = 1. \tag{1.37}$$

However 4 balls (or 4 molecules) and $2^4 = 16$ possible cases are too few to draw any general conclusion. Let us consider the general case of N balls and 2^N possibilities. Equation (1.35) tells us that in the L half of a room the average number of molecules is exactly half of the total, which is quite reasonable. How about the standard deviation? From (1.22) we have:

$$\sigma^2 = \overline{\left(X - \sum_{k=1}^{N} \overline{x_k}\right)^2} = \overline{\left(\sum_{k=1}^{N} x_k - \sum_{k=1}^{N} \frac{1}{2}\right)^2} =$$

$$\overline{\left(\sum_{k=1}^{N} \left(x_k - \frac{1}{2}\right)\right)^2} = \overline{\sum_{k=1}^{N} \left(x_k - \frac{1}{2}\right) \sum_{j=1}^{N} \left(x_j - \frac{1}{2}\right)}. \tag{1.38}$$

The double sum $\sum_{k=1}^{N} \sum_{j=1}^{N}$ can be split into two terms: the first with $k \neq j$ and the second with $k = j$:

$$\sigma^2 = \overline{\sum_{k \neq j}^{N} \left(x_k - \frac{1}{2}\right)\left(x_j - \frac{1}{2}\right)} + \overline{\sum_{k=1}^{N} \left(x_k - \frac{1}{2}\right)^2}. \tag{1.39}$$

It is easy to check that the first term vanishes. Write the 4 contributions to the sum of a given pair k, j:

$x_k = 1, x_j = 1$: $(x_k - \frac{1}{2})(x_j - \frac{1}{2}) = \frac{1}{4}$,
$x_k = 1, x_j = 0$: $(x_k - \frac{1}{2})(x_j - \frac{1}{2}) = -\frac{1}{4}$,
$x_k = 0, x_j = 1$: $(x_k - \frac{1}{2})(x_j - \frac{1}{2}) = -\frac{1}{4}$,
$x_k = 0, x_j = 0$: $(x_k - \frac{1}{2})(x_j - \frac{1}{2}) = \frac{1}{4}$.

Since $\frac{1}{4} - \frac{1}{4} - \frac{1}{4} + \frac{1}{4} = 0$, we are left with:

$$\sigma^2 = \overline{\sum_{k=1}^{N} \left(x_k - \frac{1}{2}\right)^2} = \overline{\sum_{k=1}^{N} \left(x_k^2 - x_k + \frac{1}{4}\right)} = \sum_{k=1}^{N} \frac{1}{4} = \frac{N}{4}. \tag{1.40}$$

(We have used the equality $x_k^2 - x_k = 0$, which holds for $x_k = 1$ or $x_k = 0$). If we set $N = 4$, we find $\sigma^2 = 1$, in agreement with (1.37). Furthermore

$$\sigma = \frac{1}{2}\sqrt{N} \tag{1.41}$$

and

$$\frac{\sigma}{\overline{X}} = \frac{\frac{1}{2}\sqrt{N}}{N/2} = \frac{1}{\sqrt{N}}. \tag{1.42}$$

The expression $\sigma/\overline{X}$ is sometimes called the *relative fluctuation*. Formula (1.42) is of extraordinary importance for our understanding of the Universe, and even influences our daily lives.

Look again at Fig. 1.22. If we draw analogous graphs for increasingly larger N, the probability peak around the average value $N/2$ becomes ever more narrow, since the limit of $\frac{1}{\sqrt{N}}$ for $N \to \infty$ vanishes. Coming back to the problem of the fluctuations of the air in the L half of a room, we can take, for example, $N = 10^{26}$, so the ratio (1.42) becomes 10^{-13}. This means that the density fluctuations of the air, due to statistical factors alone, are vanishingly small. Thus the reader can breathe a sigh of relief and continue to breathe easily, without fearing that vacuum pockets (or even a significant variation in the air density) are suddenly going to form in front of his mouth!

1.2.4 Gaussian Integrals

We digress briefly to consider the evaluation of integrals containing Gaussian functions. We need these now in order to treat the distribution of molecular velocities; later they will occur in consideration of the quantum mechanical harmonic oscillator. The indefinite integral $\int e^{-x^2} dx$ (or antiderivative of the function e^{-x^2}) cannot be expressed in terms of elementary functions. (One can, however, define the related *error function* by $\mathrm{erf}(x) = \frac{2}{\sqrt{\pi}} \int_0^x e^{-t^2} dt$.) It turns out, however, that certain *definite* integrals with infinite limits can be evaluated without knowing the corresponding indefinite integrals. Consider first the definite integral

$$I = \int_{-\infty}^{+\infty} e^{-x^2} dx. \tag{1.43}$$

The square of this integral is given by

$$I^2 = \left(\int_{-\infty}^{+\infty} e^{-x^2} dx \right)^2 = \left(\int_{-\infty}^{+\infty} e^{-x^2} dx \right) \left(\int_{-\infty}^{+\infty} e^{-y^2} dy \right) =$$
$$\int_{-\infty}^{+\infty} \int_{-\infty}^{+\infty} e^{-(x^2+y^2)} dx \, dy. \tag{1.44}$$

Note that, in writing the square, that the second factor must be written with a *different* dummy variable, y. The last double integral can be interpreted as an integration over the entire xy-plane ($-\infty \leq x, y \leq +\infty$) of the function $e^{-(x^2+y^2)}$, expressed in Cartesian coordinates. The trick now is to transform to polar coordinates, with $x = r \cos \phi$, $y = r \sin \phi$ and $r^2 = x^2 + y^2$. The element of area transforms from $dx \, dy$ to $r \, dr \, d\phi$ with the variable ranges $0 \leq r \leq \infty, 0 \leq \phi \leq 2\pi$. Since the integrand is now independent of ϕ, we can integrate immediately to get a factor 2π, so that Eq. (1.46) reduces to

$$I^2 = 2\pi \int_0^\infty e^{-r^2} r \, dr. \tag{1.45}$$

We note now that the indefinite integral of the above, with the added factor r, *can* be evaluated. With a change of variable $u = r^2$, $du = 2r \, dr$, we find

$$2 \int e^{-r^2} r \, dr = \int e^{-u} du = -e^{-u} \tag{1.46}$$

and the corresponding definite integral with, $0 \leq u \leq \infty$, equals $e^{-\infty} - (-e^0) = 1$. Thus $I^2 = \pi$, and we obtain finally

$$I = \int_{-\infty}^{+\infty} e^{-x^2} dx = \sqrt{\pi}. \qquad (1.47)$$

This remarkable (and totally amazing) result was evidently first derived by Euler in 1729.

In applications, we often need the following slightly generalized Gaussian integral:

$$I_\alpha = \int_{-\infty}^{+\infty} e^{-\alpha x^2} dx \qquad (1.48)$$

with $\alpha > 0$. This integral can be reduced to the case with $\alpha = 1$ making a change of variables to $t = \sqrt{\alpha} x$, $dx = \frac{1}{\sqrt{\alpha}} dt$. Then we have:

$$I_\alpha = \int_{-\infty}^{+\infty} e^{-\alpha x^2} dx = \sqrt{\frac{\pi}{a}}. \qquad (1.49)$$

Another generalization of Gaussian integrals, which we will encounter, are the forms

$$\int_{-\infty}^{+\infty} x^n e^{-\alpha x^2}, \quad n = 2, 4, \ldots \qquad (1.50)$$

Note that when n is odd, the integral equals zero, since the integrand is an odd function. The trick involves differentiation of the integral with respect to the parameter α. For example,

$$\frac{\partial}{\partial \alpha} \int_{-\infty}^{+\infty} e^{-\alpha x^2} dx = \int_{-\infty}^{+\infty} (-x^2) e^{-\alpha x^2} dx = \frac{\partial}{\partial \alpha} \left(\sqrt{\frac{\pi}{\alpha}} \right). \qquad (1.51)$$

This works out to

$$\int_{-\infty}^{+\infty} x^2 e^{-\alpha x^2} dx = \frac{\pi^{1/2}}{2\alpha^{3/2}}. \qquad (1.52)$$

Differentiating once more with respect to α gives

$$\int_{-\infty}^{+\infty} x^4 e^{-\alpha x^2} dx = \frac{3\pi^{1/2}}{4\alpha^{5/2}}. \qquad (1.53)$$

This trick was famously exploited by Feynman, in deriving a number of useful results in quantum electrodynamics. However, the maneuver can actually traced back to Leibniz. It is rigorously valid, provided that the integrand and its x- and α-derivatives are continuous in the domain of integration.

1.2.5 Maxwell–Boltzmann Distribution

Having settled the questions about the average positions of the molecules, we turn to the distribution of velocities. This problem is more difficult, since, while it is true that the positions of a molecule are equally probable, the velocities are not. As before, we denote by m the mass of a molecule, and by v its velocity. Since the kinetic energy is $E = \frac{1}{2}mv^2$, it will be sufficient to find the distribution of energies $\rho(E)$. This is actually a *reduced* distribution in phase space, having implicitly integrated over all dynamical variables other than molecular translational kinetic energy. Let us consider one molecule of an ideal gas, and denote by $\rho_1(E)dE$ the probability that the energy of the molecule is in the interval E to $E + dE$. Since the total probability is 1, we must have $\int_0^\infty \rho_1(E)dE = 1$. Consider now two molecules, with the energy of the first molecule lying in the infinitesimal interval E_1 to $E_1 + dE_1$, and the energy of the second in the interval E_2 to $E_2 + dE_2$. The joint probability is then given by:

$$p(A \bigcap B) = p(A)p(B) = \rho_1(E_1)\rho_1(E_2)dE_1dE_2. \tag{1.54}$$

This result can be obviously generalized for the probability distribution of N molecules:

$$\rho(E_1, E_2, ...E_N) = \rho_1(E_1)\rho_1(E_2)\ldots\rho_1(E_N). \tag{1.55}$$

Let us assume that the total energy

$$E_1 + E_2 + \ldots + E_N = E, \tag{1.56}$$

is constant, meaning that the air in the room is effectively a closed system.

Taking the logarithm of (1.55) we find:

$$\log \rho = \log \rho_1(E_1) + \log \rho_1(E_2) + \log \rho_1(E_3) + \ldots + \log \rho_1(E_N). \tag{1.57}$$

The logarithm of the probability distribution is an additive function, as are the mass, the number of molecules, and the energy. This suggests that $\log \rho$ is a linear function of the energy, such that:

$$\log \rho_1(E_i) = \alpha - \beta E_i, \quad \text{for } i = 1, 2, \ldots, N, \tag{1.58}$$

with a minus sign in front of βE_i, since for large E_i the probability should approach zero. From (1.58) it follows that:

$$\rho_1(E_i) = A\, e^{-\beta E_i}, \tag{1.59}$$

where A and β are appropriate constants. Taking the sum over i of (1.58), we get:

$$\log \rho = \sum_{i=1}^{N} \log \rho_1(E_i) = \alpha N - \beta E, \tag{1.60}$$

where E is the total energy. It follows that:

$$\rho(E) = Z^{-1}e^{-\beta E}, \tag{1.61}$$

where Z^{-1} is the conventional notation for another constant. This is called the *Boltzmann distribution* Boltzman (2005), which is of central significance in the physics of thermal phenomena and in astrophysics. Since the total probability must be equal to 1, from Eq. (1.59) we must have:

$$\int \rho(E)d^3v = A\int e^{-\beta m(v_x^2+v_y^2+v_z^2)/2}d^3v = 1. \tag{1.62}$$

The symbol d^3v denotes that the integration is extended to the whole three-dimensional space of velocities: $d^3v = dv_x\, dv_y\, dv_z$. Thus Eq. (1.62) reduces to A times the product of three identical Gaussian integrals:

$$A\int_{-\infty}^{+\infty} e^{-\beta m v_x^2/2}\, dv_x \int_{-\infty}^{+\infty} e^{-\beta m v_y^2/2}\, dv_y \int_{-\infty}^{+\infty} e^{-\beta m v_z^2/2}\, dv_z =$$
$$A\left(\int_{-\infty}^{+\infty} e^{-\beta m v^2/2}\, dv\right)^3 = 1. \tag{1.63}$$

The integrals in Eq. (1.63) are of the form of Eq. (1.48). The three integrals in Eq. (1.63) are equal, so that we can write:

$$A\left(\int_{-\infty}^{+\infty} e^{-\beta m v^2/2}dv\right)^3 = A\left(\frac{2\pi}{m\beta}\right)^{3/2} = 1. \tag{1.64}$$

Thus, the constant A, normalizing the distribution, is determined:

$$A = \left(\frac{m\beta}{2\pi}\right)^{3/2}. \tag{1.65}$$

We are now able to compute the average value $\overline{E}$ of the kinetic energy $E = \frac{1}{2}m(v_x^2 + v_y^2 + v_z^2)$ of a single molecule. We find:

$$\overline{E} = A\int \frac{1}{2}m(v_x^2 + v_y^2 + v_z^2)e^{-\beta m(v_x^2+v_y^2+v_z^2)/2}d^3v \tag{1.66}$$

The triple integral can be evaluated by transforming from Cartesian to spherical polar coordinates in velocity space (v, θ, ϕ), such that

$$d^3v = dv_x\, dv_y\, dv_z \rightarrow v^2 \sin\theta dv\, d\theta\, d\phi, \tag{1.67}$$

where v represents the magnitude of the velocity vector (the *speed*), with $v^2 = v_x^2 + v_y^2 + v_z^2$, while θ, ϕ specify the direction in three-dimensional space. Since the integrand is independent of the angles, integration over θ and ϕ gives a factor 4π, so that Eq. (1.66) reduces to

$$\overline{E} = A \int_0^\infty \frac{1}{2} m v^2 e^{-\beta m v^2/2} 4\pi v^2 dv = \frac{3}{2\beta}. \tag{1.68}$$

Comparing this result to $\overline{E} = \frac{3}{2} k_B T$ (Eq. 1.33), we find $\beta = \frac{1}{k_B T}$. Substituting from (1.65) for A, we arrive at the famous *Maxwell–Boltzmann distribution* of molecular velocities:

$$f(v) = 4\pi \left(\frac{m}{2\pi k_B T} \right)^{3/2} v^2 e^{-mv^2/2k_B T} \tag{1.69}$$

A plot of the distribution $f(v)$ is shown in Fig. 1.23. It has the form of a skewed Gaussian function, with $f(0) = 0$. The half-width of the curve is $w = \sqrt{\frac{2k_B T}{m}}$, not very different from $u = \sqrt{\overline{v^2}}$. In fact, $u = 1.225\,w$. Using (1.69) we can compute how many molecules have velocities belonging to a given interval. For example, for a sample of one billion molecules, there are:

427 million with velocity $0 < v < w$,
526 million with velocity $w < v < 2w$,
45 million with velocity $2w < v < 3w$,
0.4 million with velocity $3w < v < 4w$, etc.

The purpose of this section was to review some basic concepts of statistical mechanics and probability theory because, as we will see, the results of quantum mechanics are of a fundamentally probabilistic character. Statistical mechanics allows us to understand the microscopic causation underlying the behavior of macroscopic systems. These causes are so hidden from our senses that it was necessary for the genius of Maxwell, Boltzmann, Gibbs, and others to discover and describe the microscopic world of molecular motion. We understand now how some common

Fig. 1.23 Plot of Maxwell–Boltzmann distribution of velocities $f(v)$ at two temperatures, with $T_2 > T_1$

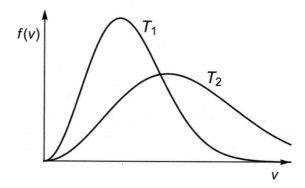

experimental facts, for example, that heat always flows spontaneously from hotter to colder bodies, or that air escapes from a puncture in a high-pressure tire, are simply probabilistic events that effectively become certainties because of the statistical behavior of enormously large numbers of molecules. It is definitely true that macroscopic systems exhibit a sort of classical "hidden variable" behavior. We will see that, in quantum mechanics, the probabilistic character is inherent in the fundamental structure of the theory, as confirmed by a very large number of experimental results (Bell's theorem, etc.). Up to now nobody has been able to definitively observe any "hidden variables" which might make QM a deterministic theory, reducible merely to statistical behavior analogous to classical systems. It is the present consensus that quantum mechanical hidden variables do not exist, although the foundations of quantum mechanics still involve some very profound and controversial questions of a metaphysical nature.

1.3 The Birth of Quantum Mechanics

The celebrated experiments by Young and Fresnel and the marvelous theoretical synthesis of Maxwell demonstrated the fundamental nature of light as an electromagnetic wave propagating through space. Schrödinger (1926), inspired by de Broglie's ideas, wrote an equation for the "matter waves" accompanying the motion of electrons. This was, in concept, analogous to Maxwell's equations for light waves (although Maxwell's equations are relativistically invariant, while Schrödinger's original equation was nonrelativistic, but this is not relevant for our purposes). Successful applications of the Schrödinger equation came soon afterward. It became possible to describe both the motion of free electrons (such as those considered by de Broglie) and electrons bound in matter. Schrödinger demonstrated that in analogy with the way that a violin string or a drumhead vibrates, with a set of discrete frequencies, so too must the energy levels of electrons in atoms exhibit discrete values. This was already known from the old Bohr theory, but Schrödinger arrived at these results by a much more rational approach, rather than by ad hoc assumptions. Schrödinger's theory can be applied, in principle, to a much larger class of problems, including atomic and molecular structure, solid-state physics and the molecular biology of life processes. The only limitations are difficulties in carrying out very lengthy and laborious computations. Quoting Dirac[3]:

> The fundamental laws necessary for the mathematical treatment of a large part of physics and the whole of chemistry are thus completely known, and the difficulty lies only in the fact that application of these laws leads to equations that are too complex to be solved.

From the very beginnings of QM (originally called wave mechanics), the physical interpretation of the waves associated with the motion of electrons was controversial, even though the spatial conformation and temporal evolution of these waves

---

[3]Dirac PAM (1929) Proc Roy Soc A (London) 123, 714–733.

are unambiguously determined by the Schrödinger equation. In the case of light, the physical quantities that vibrate and propagate in space are clearly the electric and magnetic fields, but it was not as easy to answer the question "what is vibrating?" in the case of waves that accompany the motion of an electron. At first, Schrödinger thought it was a charge density distributed in space, but soon it became clear that this interpretation was flawed. Since this is a crucial point in our development, let us briefly describe a beautiful experiment of electron interference that has been carried out with a modern electronic microscope. In this apparatus, electric and magnetic fields deflect the trajectories of beams of electrons, analogous to the way that optical lenses control the trajectories of light rays. It is then possible to observe interference on a photographic plate between two electron beams emitted from the same source, but which have followed different trajectories; we can observe typical interference patterns (see Fig. 1.24), taken from Merli et al. (1976)[4] there are fringes which are more intense, where a larger number of electrons has hit the plate, and darker where fewer electrons struck. This behavior contradicts Schrödinger's primitive interpretation, that the electron instead of being localized at one point, is "spread out", giving a cloud of charge distributed in a region of space, similar to an electromagnetic wave. In fact, with an electron beam of very low intensity, we can observe the arrival of *individual electrons* as pointlike flashes of light. With a low-intensity exposure over a long period of time, the electrons randomly fall here and there, but in the end, they accumulate in interference fringe patterns, analogous to those in Young's experiment with light waves. The sequence of pictures in Fig. 1.24 shows how the image is built up from flashes of individual electrons. In some places on the photographic plate, the probability of an electron striking is larger, in other places it is smaller. Like the bullet marks in target practice, in which, by increasing the number of shots, hits become more numerous near the bull's eye, so too do the electrons create fringe patterns. Now in the case of target practice, the difference of the various bullet marks can be explained by small variations in the initial conditions, such as minute differences in the marksman's steady hand, while nothing of this sort can be identified in the case of electrons. In fact the "electronic probability clouds" around the nucleus are identical for all hydrogen atoms (even those in the stars!).

Richard Feynman described two-slit-diffraction experiments, for both electrons and photons, as "the experiment with the two holes." He considered this to be the "central mystery of quantum mechanics."[5] See Fig. 1.25 for a cartoon capturing the conceptual problem. In the words of Yogi Berra (an American baseball player known for witty quips), "When you come to a fork in the road, take it!"

At this point, we state the celebrated *Born interpretation* of the wavefunction, which can be considered one of the fundamental postulates of QM: the waves representing solutions of the Schrödinger equation do not describe electronic charge density, but are rather *probability waves* of electron distribution. More precisely, the wave function is a pre-probability or probability amplitude since we must square its

---

[4]In 2002, *Physics World* voted this double-slit experiment with single electrons as "the most beautiful experiment in physics" of all time.

[5]Feynman RP (1965) The Character of Physical Law, BBC Publications, London.

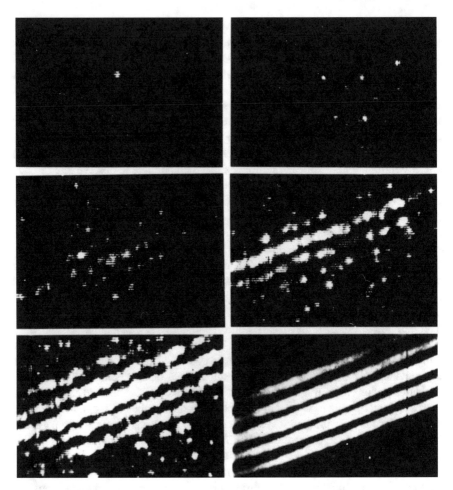

Fig. 1.24 The experiment of Merli, et al. Electron interference fringe patterns displayed on a TV monitor with increasing current densities. Reproduced from American Journal of Physics **44**, 306-307 (1976), Fig. 1.1, with permission of the American Association of Physics Teachers. Enhanced photograph courtesy of Prof. Giulio Pozzi

absolute value to obtain an observable probability density. In the microscopic world, we cannot imagine particles in the same way as we usually picture macroscopic objects in everyday life. Instead, we must picture *probability clouds* that move and fluctuate in space; where the cloud is more "dense" it is more likely to observe the particle. Outside of the cloud, no particle will be found. A small, dense cloud corresponds to a well-localized particle, while a more diffuse cloud distributed over a larger region of space corresponds to a delocalized particle. As a very important consequence: before actually carrying out a measurement of a particle's position, the question "where is the particle located?" is meaningless.

Fig. 1.25 Cartoon parodying the counterintuitive nature of quantum mechanics. Drawn by Lorenzo Felice. Inspired by the famous cartoon of Charles Addams on the cover of Aharonov Y, Rohrlich D (2005) *Quantum Paradoxes: Quantum Theory for the Perplexed*, Wiley, New York

The term "wave mechanics" was suggested by the oscillation and propagation of "probability clouds," closely resembling the behavior of classical wave propagation. However, we are faced with a completely different situation, which becomes paradoxical if we insist on maintaining the classical concepts of position and velocity. In fact, from de Broglie's relation $\lambda = \frac{h}{mv}$, which was, in fact, the inspiration for the Schrödinger equation, it follows that if the velocity v is small, the wavelength of the "cloud" becomes large, and therefore the electron cannot be localized in a small region of space. Conversely, if the particle is well localized, its "average wavelength" must also be small, and therefore the possible values of its velocity (or momentum) are broadly spread. This is one aspect of the famous *Heisenberg uncertainty principle*: it is not possible to simultaneously measure the exact position and exact momentum of a particle. In some books on QM the uncertainty principle is explained by emphasizing the "perturbation" of the observed object by the measuring instrument. This statement contains some grain of truth but it is somewhat misleading. Often we read that "if we measure the position of a particle with great accuracy, we perturb it in such a way that the momentum becomes highly uncertain, and, conversely, if we try to accurately measure its momentum, the position becomes uncertain." But, in our view, we must imagine the particle, not as a mass point, but as a nebulous cloud. Then, we have clouds (which resemble the waves created on the surface of a lake by a falling stone) in which the wavelength (related to momentum) is well defined, but the position is diffuse. Where is the perturbation of a measurement apparatus now? A more logical analysis is the impossibility for the cloud to assume the two contrasting shapes at the same time. It must either be delocalized like a wave or localized like a particle. A "perturbation" is nothing more than an attempt to prepare the cloud in one such form and show that it can no longer be described in terms of the other. These concepts should become clearer in the following pages, where we will show how *Hilbert space* provides a coherent description of the fundamental state of a quantum system.

1.3.1 Hilbert Space

During the years of the "Cambrian explosion" of QM, physicists (notably Dirac and Wigner) and mathematicians (von Neumann, Weyl) recognized that there was a particular mathematical structure that could provide a logical basis of the new theory. It will turn out that this new formalism is as fundamental for the new mechanics as the concepts of differential calculus were for Newtonian mechanics. As first emphasized by von Neumann (1932), this mathematical structure features vectors in a Hilbert space H, and linear operators acting on these vectors. In our development of QM in Hilbert space, we will proceed by small, intuitively accessible steps. The strictly axiomatic method, in which theorems (and their applications) are developed from a set of axioms, is more difficult for nonspecialists to follow.

To each "probability wave," physicists associate a detailed "condition" or *state* for the system in question. This holds true both for a single particle and for a system of many particles. This is analogous to the classical case, which we described above, in which the state of the system is represented by a point in the appropriate phase space. As it turns out, the quantum mechanical description of a physical system is considerably more complicated than the classical description, essentially because it is necessary to describe with accuracy the shape of a cloud rather than the position and velocity of a pointlike particle. But, apart from this complication, Hilbert space is for the quantum description quite analogous to phase space for the classical description. To begin:

> **Postulate**: Any state of a quantum system corresponds to a point in an abstract multidimensional (usually infinite-dimensional) Hilbert space.

Therefore, any possible *probability cloud* is represented by a corresponding *point* in Hilbert space. When the cloud undergoes some change, the point will correspondingly move along some trajectory in Hilbert space. The following paragraphs will introduce a more pictorial description of a Hilbert space. We might, of course, give the mathematical definition right away: "a Hilbert space is a real or complex abstract vector space endowed with an inner product." But this sort of definition would be readily understood only by those who are already familiar with higher algebra. Instead, we will develop, in a concrete physical application, a highly simplified version, in which the Hilbert space is reduced to a simple Cartesian plane with two coordinates x, y. We

Fig. 1.26 Correspondence between points P of the plane and vectors OP from the origin

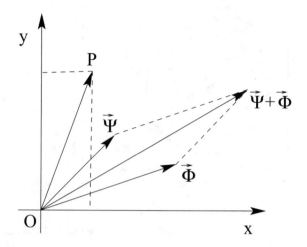

will call this a "toy Hilbert space." This can be classified as a *vector space*, since we can associate every point P in this plane with a vector OP, from the origin to a point in P, as shown in Fig. 1.26. The two coordinates of P are called the *components* of OP. Figure 1.26 shows two vectors $\vec{\Psi}$, $\vec{\Phi}$ in the plane. Vectors in Hilbert space can be manipulated in much of the same way as the more familiar vectors in Euclidean space. For example, a vector sum could be constructed using the parallelogram rule, as shown in Fig. 1.26.

We describe, as a highly simplified example, the "probability waves" or "clouds" associated with an electron in the vicinity of two protons, localized at points A and B. These are actually the ingredients of the hydrogen molecular ion, H_2^+. We consider two hypothetical states: Ψ_A, in which the electron is localized around proton A and Ψ_B, in which it is localized around proton B. These two states are sketched in Fig. 1.27. Now let us propose a correspondence between the wave Ψ_A and a vector with coordinates $(1, 0)$ in our toy Hilbert space (the Cartesian plane). Similarly, let the wave Ψ_B correspond to the vector $(0, 1)$, as shown in Fig. 1.28. Since there is a unique correspondence between probability waves and points (or vectors) in our Hilbert space, we will denote the vector $(1, 0)$ by $\vec{\Psi_A}$, and the vector $(0, 1)$ by $\vec{\Psi_B}$. Thus we have now created a unique representation of probability clouds Ψ as vectors in Hilbert space.[6]

Now let us introduce a new set of points S in the Cartesian plane that lie on the circumference of a circle of radius 1, about the origin (see Fig. 1.29). These points have coordinates $(\cos\theta, \sin\theta)$, where θ is the angle between the vector OS and the x-axis. When we vary θ the point S revolves around the circumference, since $\cos^2\theta + \sin^2\theta = 1$. The point S represents a "mixed" wave, obtained by interference (or better, superposition) of wave Ψ_A with wave Ψ_B. Correspondingly,

---

[6]To be technically correct, the negative of a vector, say $-\vec{\Psi_A}$, represents the same state as $\vec{\Psi_A}$, differing only by a *phase factor* (-1) of magnitude 1. Only when there is interference between two waves does their phase difference becomes significant, as in the superpositions Ψ_+ and Ψ_-.

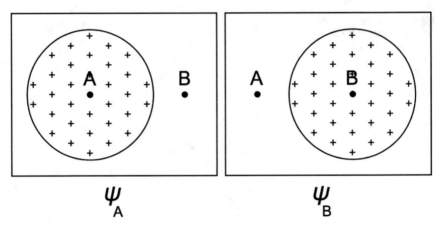

Fig. 1.27 "Probability waves" Ψ_A and Ψ_B associated with an electron localized around protons A and B, respectively

Fig. 1.28 The vectors $(1, 0)$ and $(0, 1)$ in Hilbert space, representing the quantum states Ψ_A and Ψ_B, respectively

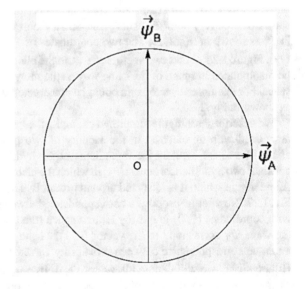

$\overrightarrow{OS} = \cos\theta\,\overrightarrow{\Psi_A} + \sin\theta\,\overrightarrow{\Psi_B}$. The physical situation corresponds to a wave which is delocalized: not concentrated, either around center A or around center B, but rather diffusely distributed around both centers. Two particular examples of superpositions are the normalized vectors $\Psi_+ = \frac{1}{\sqrt{2}}(\Psi_A + \Psi_B)$ and $\Psi_- = \frac{1}{\sqrt{2}}(\Psi_A - \Psi_B)$. The probability waves for these states are sketched in Figs. 1.30 and 1.31, respectively. Note that Ψ_- has regions where the wave amplitude is positive, indicated by plus signs (+), and regions of negative amplitude, indicated by minus signs (−), so that phase factors are evidently important here. The superposition Ψ_+, called the *bonding orbital*, is a simple approximation to the ground electronic state of H_2^+, while Ψ_-, the *anti-*

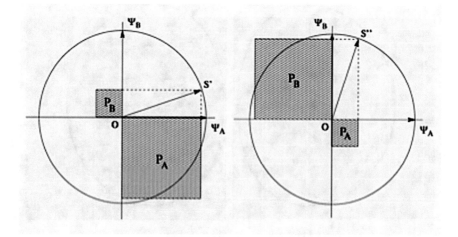

Fig. 1.29 The wave represented by the vector OS' or OS'' is the superposition of the waves Ψ_A and Ψ_B, with $P_A = P(A)$, $P_B = P(B)$. *Left* the vector OS' for $P_A > P_B$. *Right* OS'' for $P_B > P_A$

Fig. 1.30 Probability wave obtained by "constructive interference" in superposition of Ψ_A and Ψ_B, The factor $1/\sqrt{2}$ gives a normalized function Ψ_+

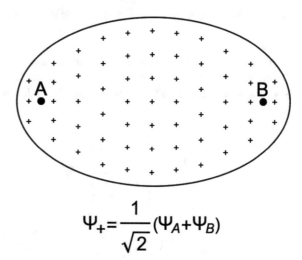

$$\Psi_+ = \frac{1}{\sqrt{2}}(\Psi_A + \Psi_B)$$

bonding orbital, is a *repulsive state*, in which no stable molecule is formed. Looking again at Fig. 1.29, we note that OS' and OS'' are the orthogonal projections of point S onto the x, y axes, respectively. Therefore the lengths of OS' and OS'', denoted $|OS'|$ and $|OS''|$, respectively, are the coordinates of S. This leads to a fundamental interpretative postulate of QM, in terms of our toy Hilbert space:

> **Postulate**: Any point S on the circumference of radius 1 corresponds to a possible state of the electron. The square of the length, $|OS'|^2$, is equal to the probability $P(A)$ that the electron is localized around A, while $|OS''|^2$ is equal to the probability $P(B)$ that the electron is localized around B.

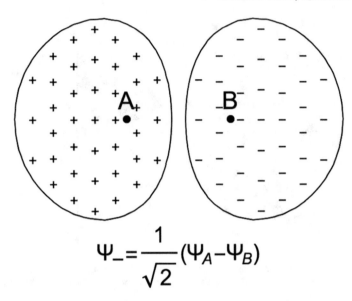

$$\Psi_- = \frac{1}{\sqrt{2}}(\Psi_A - \Psi_B)$$

Fig. 1.31 Probability wave obtained by "destructive interference" of Ψ_A and Ψ_B

At this point some readers will perhaps experience a feeling of puzzlement or disorientation and might ask: "But what is this circle, what are these vectors, that have suddenly shown up?" The following discussion might make the geometrical

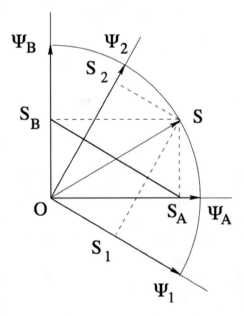

Fig. 1.32 Geometric construction to show conservation of probability

construction in Fig. 1.32 appear more plausible, in particular, the representation of "clouds" by vectors. This will be discussed further in Chap. 5. However, such explanations cannot go beyond a certain limit of plausibility; any mathematical postulate, any recipe, even if it conflicts with everyday intuition, can be accepted by physicists if it is part of a consistent formalism and if it explains a large number of experimental facts. Let us show that our postulate is consistent with the property that the sum of all the probabilities adds up to 1. Indeed, by applying the Pythagorean theorem to the right triangle $OS_A S_B$ of Fig. 1.32, we find:

$$|OS_A|^2 + |OS_B|^2 = |S_A S_B|^2 = |OS|^2 = 1, \tag{1.70}$$

so that, $P(A) + P(B) = 1$. In Fig. 1.29, a vector to the point S' on the circumference indicates that it is likely that the electron would be found closer to atom A, while a vector to point S'' indicates that it is likely that the electron is closer to B.

To summarize, the fundamental interpretative postulate of quantum mechanics implies that the probability in an individual measurement of a physical observable, such as position, velocity, etc., does not give a predictable result, since QM is inherently of an irreducibly statistical and probabilistic character. *The maximal knowledge, the maximal information, that we can possibly have of a physical system, is represented by its state vector OS.* (Physicists often use the term *state* to designate vectors of unit length in Hilbert space.) And this information does not allow us to predict the actual results of individual microscopic events, but only their probabilities: specifically, the average number of times these results show up in a statistical analysis of a large number of repeated measurements on identical systems.

How does the probabilistic nature of QM compare with the situation in classical statistical mechanics? Classically, molecules have a certain probability of being found in given regions of space, of moving with a certain range of velocities, etc.; but, at least in principle, there is correlation between this apparently probabilistic behavior and the detailed motions of individual molecules. Thus, we have seen that the pressure of a gas can be calculated from a suitable average over phase space. We actually know, in principle, "what happens behind the scenes." For example, pressure is the result of a large number of collisions of individual molecules with the walls of the container. By contrast, a detailed understanding of this nature does not exist for quantum mechanics; there do not appear to be any *hidden variables* which, once known, might explain why the electron is found on atom A, rather than on atom B (as in the above example); or, in the more realistic case of Fig. 1.24, why an electron ends up on a particular interference fringe rather than on another. We will return to this question later, in much greater detail.

In the following two chapters, we will introduce some advanced mathematical tools, which will enable a deeper understanding of the structure of QM. We will continue to exploit our toy Hilbert space and gently transition to the "grown-up" version of Hilbert space.

Chapter 2
Mathematical Methods in Quantum Mechanics

Abstract The mathematical methods used in quantum mechanics are developed, with emphasis on linear algebra and complex variables. Dirac notation for vectors in Hilbert space is introduced. The representation of coordinates and momenta in quantum mechanics is analyzed and applied to the Heisenberg uncertainty principle.

Keywords Vectors · Matrices · Hilbert space · Heisenberg uncertainty principle

2.1 Vector Analysis

From a geometric point of view, any point P in the Cartesian x-y plane can be associated with a *vector* $\overrightarrow{OP}$, from the origin O to the point P. The corresponding algebraic interpretation of a vector is an ordered pair of real numbers (the coordinates of P). We will write either $\mathbf{v} = (v_1, v_2)$ or $\overrightarrow{v} = (v_1, v_2)$. Both the boldface and the arrow notation are extensively used in the literature and we will use whichever one looks better in a formula. The origin O is the vector $(0, 0)$. The space of all these vectors is denoted by $\mathbb{R}^2$. The superscript 2 reminds us that two coordinates are sufficient to determine $\mathbf{v}$. Once we adopt the convention that all our vectors start from the origin, the terms vectors and points are equivalent. The generalization from two to three dimension is straightforward: a vector $\overrightarrow{OP}$ in three-dimensional space is specified by three real numbers. The origin O is now $(0, 0, 0)$. This space, containing all sets of ordered triples of real numbers, is denoted by $\mathbb{R}^3$. A vector in 3-space is shown in Fig. 2.1.

Vectors in classical physics are used to represent forces, velocities, etc. What is the mathematical counterpart of the physical concept of doubling or tripling a force? It is easy to see that this is equivalent to doubling or tripling the coordinates of the endpoint P. Therefore $2\mathbf{v}$ has the same direction of $\mathbf{v}$, but is twice as long. Its coordinates are $(2v_1, 2v_2)$. More generally, for any real number a, we can define:

$$a\mathbf{v} = (av_1, av_2). \tag{2.1}$$

G. Fano and S.M. Blinder, *Twenty-First Century Quantum Mechanics:
Hilbert Space to Quantum Computers*, UNITEXT for Physics,
DOI 10.1007/978-3-319-58732-5_2

43

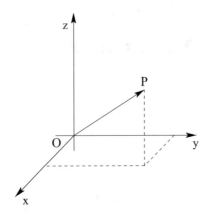

Fig. 2.1 A three-dimensional vector OP

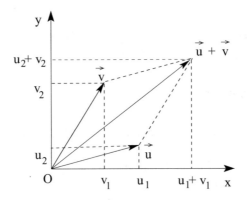

Fig. 2.2 Vectors $\mathbf{u} = (u_1, u_2)$, $\mathbf{v} = (v_1, v_2)$ and their sum, using the parallelogram rule

From elementary physics, we know also that forces can be added by means of the parallelogram rule (see Fig. 2.2). Given two vectors $\mathbf{u} = (u_1, u_2)$, $\mathbf{v} = (v_1, v_2)$, what are the coordinates of the sum $\mathbf{u} + \mathbf{v}$? It is easy to see that the following definition:

$$\mathbf{u} + \mathbf{v} = (u_1 + u_2, v_1 + v_2), \tag{2.2}$$

is in agreement with the parallelogram rule.

We have shown, both from geometric and algebraic points of view, the two fundamental operations of the vector space $\mathbb{R}^2$: multiplication by a real number, and summation of two vectors. Following are some properties implied by the fundamental operations of a *real vector space* (a, b denote real numbers, while $\mathbf{0} = (0, 0)$ is the null vector):

$$\mathbf{u} + \mathbf{v} = \mathbf{v} + \mathbf{u}, \tag{2.3}$$

$$\mathbf{u} + \mathbf{0} = \mathbf{u}, \tag{2.4}$$

$$a(\mathbf{u} + \mathbf{v}) = a\mathbf{u} + a\mathbf{v}, \tag{2.5}$$

$$(a + b)\mathbf{u} = a\mathbf{u} + b\mathbf{u}. \tag{2.6}$$

Let us now introduce the very important concept of *linear combination* of vectors. Given two vectors **u** and **v** and two real numbers a and b the vector

$$\mathbf{w} = a\mathbf{u} + b\mathbf{v} \tag{2.7}$$

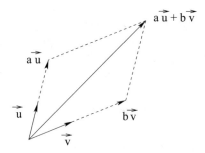

Fig. 2.3 Linear combinations of the fixed two-dimensional vectors **u**, **v** generate the whole plane $\mathbb{R}^2$ by running over all the coefficients a, b

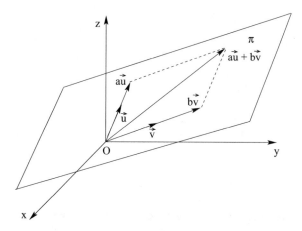

Fig. 2.4 Linear combinations of the three-dimensional vectors **u**, **v** generate the plane π, by varying the coefficients a, b

is called a linear combination of **u** and **v**, with coefficients a and b. It is evident that, except for the particular case in which **u** and **v** are parallel, two vectors with varying coefficients a, b can cover the entire plane $\mathbb{R}^2$ (see Fig. 2.3). Since **u** and **v** generate, with their linear combinations, the entire plane $\mathbb{R}^2$, we say that **u** and **v** form a *basis* in $\mathbb{R}^2$. Vectors of the form $\overrightarrow{OS} = \cos\theta\,\mathbf{u} + \sin\theta\,\mathbf{v}$ are a particular case of (2.7), representing vectors of unit length (*unit vectors*). If the two vectors **u**, **v** live in a space of larger dimension, such as $\mathbb{R}^3$ (ordinary 3D physical space), the linear combination (2.7) still belongs to the plane containing **u** and **v**, and the set generated by **u** and **v** is the plane π, shown in Fig. 2.4. The construction of Fig. 2.3 is still valid, and the plane π still contains the two basis vectors **u** and **v**. Furthermore when $a = b = 0$ we get $\mathbf{w} = (0, 0, 0)$, the null vector sitting at the origin. Therefore π must contain the origin O. The plane π is an example of *linear subspace* of $\mathbb{R}^3$ since it is both a subset of $\mathbb{R}^3$ and it is itself a linear space (indeed, if two vectors belong to π, their sum also belongs to π, etc.). Linear subspaces will play an important role in QM. The only nontrivial linear subspaces of $\mathbb{R}^2$ are straight lines through the origin O.

We have seen that in the fundamental postulate of QM the *distance* of a point S from the origin O (the *length* of the vector $\overrightarrow{OS}$) plays an important role. For an arbitrary vector $\mathbf{v} = (v_1, v_2)$ we know by the Pythagorean theorem that its length (which we denote by $|\mathbf{v}|$ or v) is given by:

$$v = |\mathbf{v}| = \sqrt{v_1^2 + v_2^2} \tag{2.8}$$

If $\mathbf{v} = (v_1, v_2, v_3)$ is a vector belonging to the space $\mathbb{R}^3$, its length being given by a simple generalization of Eq. (2.8):

$$v = |\mathbf{v}| = \sqrt{v_1^2 + v_2^2 + v_3^2} \tag{2.9}$$

The *angle* between two vectors is also important. If ϕ denotes the angle between the vectors $\mathbf{u} = (u_1, u_2)$ and $\mathbf{v} = (v_1, v_2)$ of the plane, the following relation holds:

$$u\,v\,\cos\phi = u_1 v_1 + u_2 v_2 \tag{2.10}$$

In three-dimensional space, a similar formula holds: denoting by $\mathbf{u} = (u_1, u_2, u_3)$, $\mathbf{v} = (v_1, v_2, v_3)$ two vectors of $\mathbb{R}^3$, and with the same meaning of the angle ϕ, it is possible to prove that:

$$u\,v\,\cos\phi = u_1 v_1 + u_2 v_2 + u_3 v_3 \tag{2.11}$$

The reader will note that the expressions (2.10), (2.11) are quite similar. Indeed these expressions are more fundamental than the concept of "angle between two vectors," which cannot be visualized in dimension higher then three. These expressions define the *scalar product* $\mathbf{u} \cdot \mathbf{v}$ of two vectors. Thus in $\mathbb{R}^2$ we have $\mathbf{u} \cdot \mathbf{v} = u_1 v_1 + u_2 v_2$, in

$\mathbb{R}^3$ we have $\mathbf{u} \cdot \mathbf{v} = u_1 v_1 + u_2 v_2 + u_3 v_3$, etc. When writing the scalar product, the two vectors $\mathbf{u}$, $\mathbf{v}$ can be represented by the symbols $\|u_1, u_2\|$, $\left\|\begin{matrix} v_1 \\ v_2 \end{matrix}\right\|$, called a *row vector* and a *column vector*, respectively. When a row vector is placed in front of a column vector, you can perform vector multiplication using a "row times column" sum, as follows:

$$\|u_1, u_2\| \left\|\begin{matrix} v_1 \\ v_2 \end{matrix}\right\| = \mathbf{u} \cdot \mathbf{v} = u_1 v_1 + u_2 v_2 \tag{2.12}$$

The scalar product has the following properties:

1. The scalar product of a vector $\mathbf{v}$ with itself is equal to the square of its length:

$$\mathbf{v} \cdot \mathbf{v} = v_1^2 + v_2^2 = |\mathbf{v}|^2 = v^2. \tag{2.13}$$

2. The commutative property:

$$\mathbf{u} \cdot \mathbf{v} = \mathbf{v} \cdot \mathbf{u}. \tag{2.14}$$

3. The distributive property:

$$\mathbf{u} \cdot (\mathbf{v} + \mathbf{w}) = \mathbf{u} \cdot \mathbf{v} + \mathbf{u} \cdot \mathbf{w}. \tag{2.15}$$

4. Multiplying $\mathbf{u}$ or $\mathbf{v}$ by a real number a, gives the same multiple of the scalar product $\mathbf{u} \cdot \mathbf{v}$:

$$(a\mathbf{u}) \cdot \mathbf{v} = \mathbf{u} \cdot (a\mathbf{v}) = a\, \mathbf{u} \cdot \mathbf{v} \tag{2.16}$$

The scalar product is related to the *projection* of a vector onto a straight line. Consider a vector $\mathbf{v}$ in the plane $\mathbb{R}^2$ and a straight line r through the origin O. Denote by $\mathbf{u}$ a unit vector (whose length is equal to 1) directed along r. The scalar product $\mathbf{u} \cdot \mathbf{v}$ is equal to $u\, v \cos \phi$, where ϕ is the angle between $\mathbf{v}$, $\mathbf{u}$ (for simplicity we assume $\cos \phi \geq 0$). Therefore $\mathbf{u} \cdot \mathbf{v}$ is equal to the length of the vector $\mathbf{v}'$ obtained by projecting $\mathbf{v}$ onto r (see Fig. 2.5). Clearly then:

$$\mathbf{v}' = (\mathbf{u} \cdot \mathbf{v})\mathbf{u}. \tag{2.17}$$

Equations (2.13)–(2.17) can readily be generalized to a higher dimensional space. For example, in the four-dimensional space $\mathbb{R}^4$, which is the set of ordered *quadruples* of real numbers, the scalar product of two vectors $\mathbf{u} = (u_1, u_2, u_3, u_4)$, $\mathbf{v} = (v_1, v_2, v_3, v_4)$ is given by:

$$\mathbf{u} \cdot \mathbf{v} = u_1 v_1 + u_2 v_2 + u_3 v_3 + u_4 v_4 \tag{2.18}$$

Setting $\mathbf{u} = \mathbf{v}$ we obtain the square of the length of $\mathbf{u}$:

$$|\mathbf{u}|^2 = u_1^2 + u_2^2 + u_3^2 + u_4^2. \tag{2.19}$$

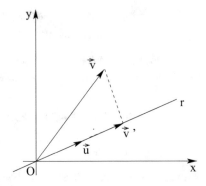

Fig. 2.5 Projection $\mathbf{v}'$ of vector $\mathbf{v}$ on the straight line r

Let us return to the simple case of the plane $\mathbb{R}^2$, and consider two orthogonal non-null vectors $\mathbf{u}$, $\mathbf{v}$; from Eq. (2.10) we have:

$$\mathbf{u} \cdot \mathbf{v} = u\, v\, \cos\phi = 0 \qquad (2.20)$$

since the cosine of a right angle vanishes. The same happens in dimension 3. When the dimension of the vector space is greater than 3, Eq. (2.20) can be regarded as the *definition* of orthogonality of two vectors. Finally, we consider the important concept for the spaces $\mathbb{R}^2$, $\mathbb{R}^3$, ... of an *orthonormal basis*. Recall that the vectors $\overrightarrow{\Psi_A} = (1, 0)$ and $\overrightarrow{\Psi_B} = (0, 1)$ in our toy Hilbert space also were both of unit length and mutually orthogonal (see Fig. 1.28), therefore:

$$\overrightarrow{\Psi_A} \cdot \overrightarrow{\Psi_A} = |\overrightarrow{\Psi_A}|^2 = 1; \quad \overrightarrow{\Psi_B} \cdot \overrightarrow{\Psi_B} = |\overrightarrow{\Psi_B}|^2 = 1; \quad \overrightarrow{\Psi_A} \cdot \overrightarrow{\Psi_B} = 0. \qquad (2.21)$$

The same property holds for the vectors $\overrightarrow{\Psi_1}$ and $\overrightarrow{\Psi_2}$, which were also of unit length and orthogonal. Since any vector $\mathbf{v} = (v_1, v_2)$ of $\mathbb{R}^2$ can be written as a linear combination of $\overrightarrow{\Psi_A}$ and $\overrightarrow{\Psi_B}$, for example, $\mathbf{v} = v_1 \overrightarrow{\Psi_A} + v_2 \overrightarrow{\Psi_B}$, we can say that they form a *basis*. In $\mathbb{R}^3$, the vectors $\mathbf{i} = (1, 0, 0)$, $\mathbf{j} = (0, 1, 0)$, $\mathbf{k} = (0, 0, 1)$ form an orthonormal basis, since they are of unit length, pairwise orthogonal, and any vector $\mathbf{v} = (v_1, v_2, v_3)$ can be written as the linear combination of $\mathbf{i}, \mathbf{j}, \mathbf{k}$:

$$\mathbf{v} = v_1 \mathbf{i} + v_2 \mathbf{j} + v_3 \mathbf{k} \qquad (2.22)$$

In general, in an n-dimensional space $\mathbb{R}^n$ (the set of all n-tuples of real numbers), n orthonormal vectors are required to form a *basis*; it is then possible to write *any* vector of the space as a linear combination of basis vectors.

2.2 Matrices in Quantum Mechanics

The earliest formulation of QM, developed around 1925 by Heisenberg, Born and Jordan, was called *matrix mechanics*. Classical observables, such as the position q or momentum p of a particle, were represented, in this theory, not by simple numbers, but rather by arrays Q and P containing an infinite number of rows and columns. The numbers appearing in an array might, for example, be related to the frequencies of radiation observed in a transition between two energy levels of an atom. Indeed the dimension of the array is equal to the number of these levels, and for a complete theory of even a simple system, such as the hydrogen atom, this number is infinite. The theory of infinite matrices is not at all simple, and this is a part of the reason that matrix mechanics is the less popular formulation of QM. Let us consider a simple situation, in which a quantum system has just two levels, which corresponds perfectly to our toy Hilbert space; it will be instructive to see how physical observables are represented in this model. All the formulas we have already encountered will turn out to have counterparts in the general case, almost without modification.

Consider the following 2×2 array of real numbers:

$$\left\| \begin{array}{cc} A_{11} & A_{12} \\ A_{21} & A_{22} \end{array} \right\|, \tag{2.23}$$

which is called a 2×2 *matrix*. The numbers $A_{11}, A_{12}, \ldots$ are called *matrix elements*. In the following, unless otherwise specified, we refer to objects such as the array (2.23) as a matrix of dimension 2×2.

We can also think of this matrix as an "operator," since it determines a transformation among the vectors of $\mathbb{R}^2$. Let us see how this happens. Given a vector $\mathbf{v} = (v_1, v_2)$, we can produce a new vector $\mathbf{w} = (w_1, w_2)$ using the formulas:

$$w_1 = A_{11}v_1 + A_{12}v_2, \qquad w_2 = A_{21}v_1 + A_{22}v_2. \tag{2.24}$$

A useful mnemonic for Eq. (2.24) is to consider the first row of the matrix as the row vector $\mathbf{A}_1 = \|A_{11}, A_{12}\|$, and the second row as the row vector $\mathbf{A}_2 = \|A_{21}, A_{22}\|$; then Eq. (2.24) can be written using the row times column products:

$$w_1 = \|A_{11}, A_{12}\| \left\| \begin{array}{c} v_1 \\ v_2 \end{array} \right\| = \mathbf{A}_1 \cdot \mathbf{v}, \qquad w_2 = \|A_{21}, A_{22}\| \left\| \begin{array}{c} v_1 \\ v_2 \end{array} \right\| = \mathbf{A}_2 \cdot \mathbf{v}. \tag{2.25}$$

Equation (2.24) can be written symbolically as:

$$\mathbf{w} = A\mathbf{v} \tag{2.26}$$

We say that the vector $\mathbf{w}$ is the *image* of $\mathbf{v}$ under the *mapping* A. Physicists use the term *operator* to denote the mapping A and, for them, the terms "matrix" and "operator" are used interchangeably (of course physicists are less meticulous than mathematicians). As an elementary example, let $A = \begin{Vmatrix} 3 & 5 \\ 7 & 2 \end{Vmatrix}$ and $\mathbf{v} = (6, 4)$. Then $\mathbf{w} = (3 \times 6 + 5 \times 4, 7 \times 6 + 2 \times 4) = (38, 50)$.

A simple, but essential, matrix is the following:

$$I = \begin{Vmatrix} 1 & 0 \\ 0 & 1 \end{Vmatrix} \tag{2.27}$$

It has the property that it maps any vector into itself: $I\mathbf{v} = \mathbf{v}$. It is called the *identity matrix* or simply the *identity*. The matrix with all elements vanishing is called the *null* matrix. We will use the same symbol (in capital letters) for a matrix and the corresponding mapping. An important class of matrices represent rotations. For example, let R denote the matrix:

$$R = \begin{Vmatrix} \cos\theta & -\sin\theta \\ \sin\theta & \cos\theta \end{Vmatrix} \tag{2.28}$$

Equation (2.24) becomes:

$$\begin{aligned} w_1 &= \cos\theta \, v_1 - \sin\theta \, v_2 \\ w_2 &= \sin\theta \, v_1 + \cos\theta \, v_2 \end{aligned} \tag{2.29}$$

It is easy to verify that for any vector $\mathbf{v}$ the vector $\mathbf{w} = R\mathbf{v}$ is obtained by a counterclockwise rotation through an angle θ. Indeed, if α denotes the angle between $\mathbf{v}$ and the x axis of the Cartesian plane, we have: $v_1 = |\mathbf{v}| \cos\alpha$, $v_2 = |\mathbf{v}| \sin\alpha$, and substituting in Eq. (2.29) we have: $w_1 = |\mathbf{v}| \cos(\alpha + \theta)$, $w_2 = |\mathbf{v}| \sin(\alpha + \theta)$ (see Fig. 2.6). It is simple to verify that multiplying the matrix (2.28), which corresponds to a rotation through an angle θ, by an analogous matrix with θ replaced by α, gives another rotation matrix with the angle $\theta + \alpha$, in agreement with the interpretation of successive applications of the two rotations.

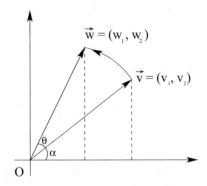

Fig. 2.6 The vector $\mathbf{w}$ is obtained from the vector $\mathbf{v}$ by rotation by an angle θ

A characteristic property of the mapping defined by Eq. (2.24) is *linearity*. This means that sums of vectors are sent into sums of images (geometrically, parallelograms are sent into parallelograms), linear combinations are sent into linear combinations, and so forth. In formulas, for all vectors $\mathbf{u}$, $\mathbf{v}$ and all real numbers c:

$$A(\mathbf{u} + \mathbf{v}) = A\mathbf{u} + A\mathbf{v}, \qquad A(c\mathbf{v}) = c(A\mathbf{v}). \tag{2.30}$$

These algebraic properties are essential in QM, consistent with the notion that matrices constitute a generalization of real numbers. Mathematicians tell us that, given a suitable definition of addition and multiplication, matrices form a *ring*, as do the real and complex numbers[1]; thus it is not entirely surprising that physical quantities can be represented by matrices.

We define the *sum* C of two matrices A, B, written $C = A + B$. if the matrix elements of C are sums of the corresponding matrix elements of A, B:

$$C_{ik} = A_{ik} + B_{ik}, \quad (i, k = 1, 2). \tag{2.31}$$

We can multiply a matrix A by a real number c:

$$(cA)_{ik} = cA_{ik}, \quad (i, k = 1, 2). \tag{2.32}$$

The simple rules of algebra also apply to matrices, for example:

$$(A + B)\mathbf{v} = A\mathbf{v} + B\mathbf{v}, \qquad (cA)\mathbf{v} = c(A\mathbf{v}). \tag{2.33}$$

Let us now define the *product* of two matrices. The idea is that the product of the two successive linear mappings A, B, on a vector, can be done by first applying B, then A:

$$(AB)\mathbf{v} = A(B\mathbf{v}) \qquad \text{for any vector } \mathbf{v}. \tag{2.34}$$

In Fig. 2.7, we see that if the mapping B sends $\mathbf{u}$ in $\mathbf{v}$, and the mapping A sends $\mathbf{v}$ in $\mathbf{w}$, then $C = AB$ sends $\mathbf{u}$ directly into $\mathbf{w}$; these are pictorial representations of the operations: $\mathbf{v} = B\mathbf{u}$, $\mathbf{w} = A\mathbf{v}$, and $\mathbf{w} = C\mathbf{u}$. In terms of matrix elements, the corresponding matrix products are given by

$$C_{11} = A_{11}B_{11} + A_{12}B_{21}, \ C_{12} = A_{11}B_{12} + A_{12}B_{22},$$
$$C_{21} = A_{21}B_{11} + A_{22}B_{21}, \ C_{22} = A_{21}B_{12} + A_{22}B_{22}. \tag{2.35}$$

Again the multiplication rule "rows times columns" applies. Writing $\mathbf{A}_1 = \|A_{11}, A_{12}\|$, $\mathbf{A}_2 = \|A_{21}, A_{22}\|$, two row vectors, and $\mathbf{B}_1 = \left\| \begin{matrix} B_{11} \\ B_{21} \end{matrix} \right\|$, $\mathbf{B}_2 = \left\| \begin{matrix} B_{12} \\ B_{22} \end{matrix} \right\|$, two column vectors, Eq. (2.35) can be written as:

---

[1] A field is a ring in which multiplication is commutative and every nonzero element has a multiplicative inverse. Thus real and complex numbers are also fields, while matrices are just rings.

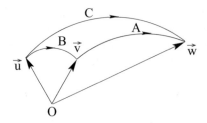

Fig. 2.7 The mapping $C = AB$ is obtained by successive applications of the mappings B, then A

$$C_{11} = \mathbf{A}_1 \cdot \mathbf{B}_1, \quad C_{12} = \mathbf{A}_1 \cdot \mathbf{B}_2,$$
$$C_{21} = \mathbf{A}_2 \cdot \mathbf{B}_1, \quad C_{22} = \mathbf{A}_2 \cdot \mathbf{B}_2. \tag{2.36}$$

In general, an $m \times n$ matrix is a rectangular array of numbers with m rows and n columns. For example, if $m = 2$ and $n = 3$ we have the matrix A:

$$A = \left\| \begin{matrix} A_{11} & A_{12} & A_{13} \\ A_{21} & A_{22} & A_{23} \end{matrix} \right\|. \tag{2.37}$$

We denote by A_{ij} the matrix element in the ith row and the jth column. Given a second matrix B, the matrix product AB requires that the number n of columns of A must match the number of rows of B; thus B must be a $n \times l$ matrix, l being arbitrary. In the general case, the matrix elements $(AB)_{ik}$ are given by:

$$(AB)_{ik} = \sum_{j=1}^{n} A_{ij} \, B_{jk}, \quad i = 1, 2, \ldots, m, \quad k = 1, 2, \ldots, l \tag{2.38}$$

For example, if A is the matrix (2.37) and B is the 3×2 matrix:

$$B = \left\| \begin{matrix} B_{11} & B_{12} \\ B_{21} & B_{22} \\ B_{31} & B_{32} \end{matrix} \right\|, \tag{2.39}$$

the matrix multiplication row times columns thus gives:

$$AB = \left\| \begin{matrix} A_{11}B_{11} + A_{12}B_{21} + A_{13}B_{31} & A_{11}B_{12} + A_{12}B_{22} + A_{13}B_{32} \\ A_{21}B_{11} + A_{22}B_{21} + A_{23}B_{31} & A_{21}B_{12} + A_{22}B_{22} + A_{23}B_{32} \end{matrix} \right\|, \tag{2.40}$$

so that AB is a 2×2 square matrix. The geometrical meaning of the "operators" A, B and AB is the following: B maps vectors belonging to $\mathbb{R}^2$ into $\mathbb{R}^3$, while A maps vectors of $\mathbb{R}^3$ into $\mathbb{R}^2$; therefore AB maps vectors of $\mathbb{R}^2$ into vectors of $\mathbb{R}^2$.

The application of a square matrix $m \times m$ to a vector in $\mathbb{R}^m$ is a particular case of Eq. (2.38); for example, setting $n = m = 2$, $B_{11} = v_1$, $B_{21} = v_2$, we obtain:

$$\begin{aligned}(AB)_{11} &= A_{11}v_1 + A_{12}v_2, \\ (AB)_{21} &= A_{21}v_1 + A_{22}v_2.\end{aligned} \tag{2.41}$$

Thus:

$$AB = \left\| \begin{matrix} A_{11} & A_{12} \\ A_{21} & A_{22} \end{matrix} \right\| \left\| \begin{matrix} v_1 \\ v_2 \end{matrix} \right\| = \left\| \begin{matrix} A_{11}v_1 + A_{12}v_2 \\ A_{21}v_1 + A_{22}v_2 \end{matrix} \right\|. \tag{2.42}$$

Many of the familiar formulas of elementary algebra still apply; for example, the associative property $C(AB) = (CA)B$; the distributive property $(A + B)C = AC + BC$, etc., but a new feature appears: the commutative property does not hold! It is *not* true, in general, that $AB = BA$, as in elementary arithmetic. This fact has profound consequences in QM. (It is, in fact, the root of the uncertainty principle.) Let us give an example of two non-commuting matrices:

$$A = \left\| \begin{matrix} 0 & 1 \\ 0 & 0 \end{matrix} \right\|, \qquad B = \left\| \begin{matrix} 0 & 0 \\ 1 & 0 \end{matrix} \right\|. \tag{2.43}$$

Their products are then given by:

$$AB = \left\| \begin{matrix} 1 & 0 \\ 0 & 0 \end{matrix} \right\|, \qquad BA = \left\| \begin{matrix} 0 & 0 \\ 0 & 1 \end{matrix} \right\|. \tag{2.44}$$

Since matrices represent *operations*, it is not unexpected that they sometimes do not commute. In everyday life, we can experience situations in which the order of operations is important: such as writing a letter and sealing it in an envelope. The result in elementary mathematics, that multiplying first by a and then by b, gives the same result as multiplying first by b and then by a, turns out to be rather exceptional in higher mathematics. The *commutator* of two matrices is defined by

$$[A, B] = AB - BA. \tag{2.45}$$

If the commutator equals 0, then the matrices A and B commute: $AB = BA$.

The *inverse* A^{-1} of a matrix A is defined by the following property:

$$AA^{-1} = A^{-1}A = I. \tag{2.46}$$

where I denotes the identity matrix; A^{-1} corresponds to the *inverse* transformation. For example, the inverse of a rotation matrix through an angle θ in a counterclockwise sense, is a rotation matrix through the same angle in a clockwise sense; in order to obtain R^{-1} it suffices to substitute $-\theta$ in the place of θ into Eq. (2.28). For the case of real numbers, the inverse (here, meaning the reciprocal) always exists except for the number zero. For 2×2 matrices, the inverse exists unless the following expression

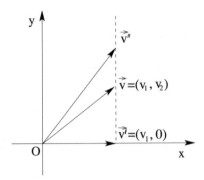

Fig. 2.8 There are an infinite number of vectors **v** whose projection on the x axis is equal to **v**$'$

vanishes: $D = A_{11}A_{22} - A_{12}A_{21}$. The geometric meaning of D (which is a 2×2 determinant) is the ratio of the area of the parallelogram of two images $A\mathbf{v}$, $A\mathbf{u}$ to the area of the parallelogram of **v**, **u**. It is a scale dilatation of the space $\mathbb{R}^2$ under the action of the operator A. It is necessary that $D \neq 0$ for the inverse of A to exist.

An important class of operators (or matrices) that usually do *not* admit an inverse[2] are *projection operators*. The simplest projection operator can be represented by the matrix:

$$P = \begin{Vmatrix} 1 & 0 \\ 0 & 0 \end{Vmatrix}. \qquad (2.47)$$

If $\mathbf{v} = (v_1, v_2)$ is an arbitrary vector, the image $P\mathbf{v} = (v_1, 0)$ is the vector **v**$'$ obtained by projecting **v** onto the x axis, as shown in Fig. 2.8. The reason P does not admit an inverse is that there exist an infinite number of vectors **v**$''$ whose images $P\mathbf{v}''$ coincide with v'. These vectors **v**$''$ have their free end on a straight line parallel to the y axis. Mathematicians say that the mapping P is not *injective*. The inverse P^{-1} does not exist since it is ill-defined: which vector do we choose? **v**$'$, **v**$''$, ...? Even in the simple space $\mathbb{R}^2$ there are many projection operators. Given any straight line **r** through the origin, let us denote by P_r the projection operator onto the line **r**, as shown in Fig. 2.9. The matrix corresponding to P_r is easily found: Let $\mathbf{u} = (c, s)$ be a unit vector directed along the line **r**; since $|\mathbf{u}| = 1$, $c^2 + s^2 = 1$. Therefore:

$$P_r = \begin{Vmatrix} c^2 & cs \\ cs & s^2 \end{Vmatrix}. \qquad (2.48)$$

An example of a projection operator in $\mathbb{R}^3$ is P_L, defined as follows: given a vector $\mathbf{v} = (v_1, v_2, v_3)$ belonging to $\mathbb{R}^3$, and a plane L through the origin O, $P_L\mathbf{v}$ is the vector obtained taking the orthogonal projection of **v** on the plane L, as shown in Fig. 2.10.

---

[2]Only the identity I is a projection operator that admits an inverse.

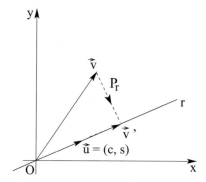

Fig. 2.9 The action of the projection operator P_r in two dimensions

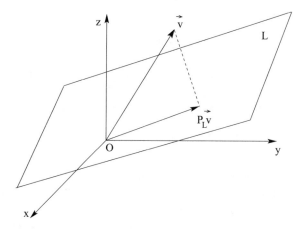

Fig. 2.10 The action of the projection operator P_L in three dimensions

Some matrices can be associated with physical observables. Apart from some subtle points that we will discuss later, only matrices A such that $A_{12} = A_{21}$ are possible candidates. (We are still limiting ourselves to *real* matrices.) These matrices are *symmetric matrices*, for example,

$$\left\| \begin{matrix} a & b \\ b & c \end{matrix} \right\|, \tag{2.49}$$

which is the most general 2×2 symmetric matrix. Of course, the matrix R of Eq. (2.28) is not symmetric, since $R_{12} = -\sin\theta$, while $R_{21} = \sin\theta$.

We have noted above that matrices can be regarded as generalizations of numbers (which can be considered 1×1 matrices). But in some cases, a matrix A can behave like a number in another way. This happens when it operates on particular vectors, called *eigenvectors*. For these vectors, the application of A is equivalent to the multiplication by a real number λ. More precisely, we shall say that a vector $\mathbf{v} = (v_1, v_2)$ (excluding $(0, 0)$) is an *eigenvector* of the matrix A corresponding to the eigenvalue λ, if the following relation holds:

$$A\mathbf{v} = \lambda \mathbf{v}. \tag{2.50}$$

For example, if $\lambda = 3$, the image vector $A\mathbf{v}$ is three times longer than $\mathbf{v}$, if $\lambda = \frac{1}{2}$, $A\mathbf{v}$ is half of $\mathbf{v}$, and so forth. The important point is that the *direction* of the eigenvector remains unchanged. Equation (2.50) for a symmetric matrix A is equivalent to the two scalar equations:

$$\begin{aligned} A_{11}v_1 + A_{12}v_2 &= \lambda v_1, \\ A_{12}v_1 + A_{22}v_2 &= \lambda v_2. \end{aligned} \tag{2.51}$$

Symmetric matrices have the remarkable property (which is at the root of their utility in QM) of admitting an *orthonormal basis* of eigenvectors. In our toy space, for any symmetric matrix A there exist two vectors $\overrightarrow{\Psi_1}, \overrightarrow{\Psi_2}$ such that:

$$A\overrightarrow{\Psi_1} = \lambda_1 \overrightarrow{\Psi_1}, \quad A\overrightarrow{\Psi_2} = \lambda_2 \overrightarrow{\Psi_2}. \tag{2.52}$$

Since these equations do not imply any restriction on their lengths, $\overrightarrow{\Psi_1}$ and $\overrightarrow{\Psi_2}$ can be chosen with unit lengths. Furthermore it can be shown that if λ_1 is different from λ_2, $\overrightarrow{\Psi_1}$ and $\overrightarrow{\Psi_2}$ are *orthogonal* to each other.

Matrices, such as the 2×2 with $A_{12} = A_{21} = 0$, are called *diagonal*. They have the nice property that the eigenvectors are directed along the coordinate axis. We will denote them by $\overrightarrow{\Psi_A}, \overrightarrow{\Psi_B}$, in agreement with our notation in toy Hilbert space. As an example, let Q be the matrix:

$$Q = \begin{Vmatrix} 3 & 0 \\ 0 & 2 \end{Vmatrix}. \tag{2.53}$$

We can easily verify that

$$Q\overrightarrow{\Psi_A} = 3\overrightarrow{\Psi_A}, \quad Q\overrightarrow{\Psi_B} = 2\overrightarrow{\Psi_B}, \tag{2.54}$$

where $\overrightarrow{\Psi_A} = (1, 0)$ and $\overrightarrow{\Psi_B} = (0, 1)$. Two diagonal matrices always commute. As an example, consider the matrix:

$$R = \begin{Vmatrix} 5 & 0 \\ 0 & 7 \end{Vmatrix} \tag{2.55}$$

and compute the *commutator* $QR - RQ$; one finds a matrix with all elements equal to zero, namely, the null matrix, which we denote by 0. Therefore $QR - RQ = 0$. Of course, R admits as eigenvectors the same eigenvectors as Q, so that the vectors $\overrightarrow{\Psi_A}$ and $\overrightarrow{\Psi_B}$ are the same; only the eigenvalues of Q and R are different. This is a general rule: if two symmetric matrices commute, *they possess a common set of orthonormal eigenvectors*.

Returning to Eq. (2.51) for a symmetric matrix, we can now find the eigenvalues λ. Assuming, for simplicity, that v_2 is not equal to zero, we can divide both equations by v_2. Denoting by r the quotient v_1/v_2, we get:

$$A_{11}r + A_{12} = \lambda r,$$
$$A_{12}r + A_{22} = \lambda. \tag{2.56}$$

Solving for r in the second equation and substituting in the first, we obtain an equation determining the eigenvalues:

$$(A_{11} - \lambda)(A_{22} - \lambda) - A_{12}^2 = 0, \tag{2.57}$$

which is a simple quadratic equation in the variable λ. In the most general case, it has two solutions λ_1, λ_2, and for each solution, the relation $A_{12}r + A_{22} = \lambda$ determines a possible value of r, which gives the *direction* of the corresponding eigenvector. (For simplicity, we neglect here the possibility of degeneracy, when more than one eigenvector corresponds to the *same* eigenvalue.) As an example, if all the matrix elements A_{ik} are equal to 1, the matrix A is simply:

$$A = \begin{Vmatrix} 1 & 1 \\ 1 & 1 \end{Vmatrix}. \tag{2.58}$$

and the eigenvalue Eq. (2.57) reduces to $(1 - \lambda)^2 = 1$, so that $(1 - \lambda) = \pm 1$, and the eigenvalues have the values 0 and 2. For $\lambda = 0$ we get $r = -1$, thus the eigenvector is $\overrightarrow{\Psi_1} = (1, -1)$, and for $\lambda = 2$ we get $r = 1$ and the eigenvector $\overrightarrow{\Psi_2} = (1, 1)$. Note that $\overrightarrow{\Psi_1}$ and $\overrightarrow{\Psi_2}$ are orthogonal, as they should be. However, their lengths are not equal to 1 (actually to $\sqrt{2}$). *Normalizing* the eigenvectors by dividing $\overrightarrow{\Psi_1}$ and $\overrightarrow{\Psi_2}$ by $\sqrt{2}$, we obtain two *orthonormal* eigenvectors (which we still denote by $\overrightarrow{\Psi_1}$, $\overrightarrow{\Psi_2}$, since they also obey Eq. (2.52) and no confusion need arise):

$$\overrightarrow{\Psi_1} = \left(\frac{1}{\sqrt{2}}, -\frac{1}{\sqrt{2}} \right), \quad \overrightarrow{\Psi_2} = \left(\frac{1}{\sqrt{2}}, \frac{1}{\sqrt{2}} \right). \tag{2.59}$$

Let us compute the commutator $QA - AQ$ of the matrices (2.53), (2.58). We find, using Eq. (2.35):

$$QA = \begin{Vmatrix} 3 & 0 \\ 0 & 2 \end{Vmatrix} \begin{Vmatrix} 1 & 1 \\ 1 & 1 \end{Vmatrix} = \begin{Vmatrix} 3 & 3 \\ 2 & 2 \end{Vmatrix}, \tag{2.60}$$

$$AQ = \begin{Vmatrix} 1 & 1 \\ 1 & 1 \end{Vmatrix} \begin{Vmatrix} 3 & 0 \\ 0 & 2 \end{Vmatrix} = \begin{Vmatrix} 3 & 2 \\ 3 & 2 \end{Vmatrix}. \tag{2.61}$$

Evidently, QA and AQ have rows and columns interchanged, so the commutator $QA - AQ$ does not vanish. The noncommutativity of Q, A and the differing eigenvectors of Q and A are, in fact, related. Indeed, we can state the following theorem:

Theorem 2.1   *Two symmetric matrices admit a common basis of orthonormal eigenvectors if and only if they commute.*

2.3   Quantum Mechanics in Toy Hilbert Space

In Sect. 1.3, we introduced a "toy Hilbert space," an extremely simplified representation for a two-state quantum system, whereby quantum states can be represented by unit vectors in the 2D Cartesian plane, with coordinates x, y. Physical observables are correspondingly represented by real symmetric 2×2 matrices. The theorem at the end of the last section is relevant to a fundamental interpretative postulate of QM:

> **Postulate 1**: To any possible state of a physical system there corresponds a vector $\overrightarrow{OS}$ of length 1. To any physical quantity F there corresponds a symmetric matrix (also denoted by F). The possible results of a measurement of F on any state are the eigenvalues of the matrix F. If $\overrightarrow{\Psi}$ is a normalized eigenvector of F corresponding to the eigenvalue λ, so that
>
> $$F\overrightarrow{\Psi} = \lambda\overrightarrow{\Psi}, \tag{2.62}$$
>
> then $|\overrightarrow{\Psi} \cdot \overrightarrow{OS}|^2$ is the probability that the result of a measurement of F is λ.

To this we add:

> **Postulate 2**: After the measurement of F, if the result is λ, the state vector $\overrightarrow{OS}$ coincides with the eigenvector $\overrightarrow{\Psi}$, thus verifying Eq. (2.62).

In order to explain the motivation for Postulate 2, we quote from Dirac (Dirac 1958):

> From physical continuity, if we make a second measurement immediately after the first, the result of the second measurement must be the same as that of the first. Thus after the first measurement has been made, there is no indeterminacy in the result of the second …This conclusion must still hold if the second measurement is not actually made.

Physical quantities like F are called *observables* by physicists. They correspond to symmetric matrices (more precisely, Hermitian matrices, see Sect. 2.6). Likewise, the corresponding physical states are represented by *state vectors*.

Suppose now that the matrices Q and A correspond to position q and velocity v, respectively. The matrix Q has the eigenvectors $\overrightarrow{\Psi_A}$, $\overrightarrow{\Psi_B}$, while A has the eigenvectors $\overrightarrow{\Psi_1}$ and $\overrightarrow{\Psi_2}$. In our toy Hilbert space, the possible results of a measurement of q are 3 and 2 (the eigenvalues of Q), while the possible results of a measurement of v are 0 and 2 (the eigenvalues of A). Furthermore, the expression $|\overrightarrow{\Psi_1} \cdot \overrightarrow{OS}|^2$ is the square of the projection of $\overrightarrow{OS}$ on the straight line determined by $\overrightarrow{\Psi_1}$, etc. The pair of "axis" $\overrightarrow{\Psi_1}$, $\overrightarrow{\Psi_2}$ is "rotated" with respect to the "axes" $\overrightarrow{\Psi_A}$, $\overrightarrow{\Psi_B}$ by virtue of the fact that the commutator $QA - AQ$ does not vanish. By Postulate 2, if we first measure Q, the state vector $\overrightarrow{OS}$ will "jump" to either $\overrightarrow{\Psi_A}$ or $\overrightarrow{\Psi_B}$. In either case, a subsequent measurement of A will be uncertain. Physicists say that the observables Q and A are *not compatible*. When physicists realized that the matrices corresponding to very simple observables such as q and p (position and momentum of a particle) do not commute, it is not surprising that this possibility was initially regarded with skepticism. Actually, the matrices representing q and p are of infinite dimension, but the geometry of our toy Hilbert space is still a valid analogy. As a consequence of the mathematical structure of the theory, q and p do not admit common eigenvectors, similar to the situation we found for the matrices Q, A above. No state vector exists such that we can obtain with certainty (probability 1) a value of q and a value of p. The conclusion follows that the observables q and p cannot be simultaneously measured. An analogous result applies for any pair of non-commuting observables; and, since symmetric matrices do not, in general, commute, indeterminacy relations are quite commonplace, rather than an exception. Other than position and velocity, some well-known cases of non-commuting observables include two different components of angular momentum, as well as operators representing time and energy.

An important quantity in QM is the *average* or *expectation value* of an observable in a given state. Suppose the observable F is represented by the simple diagonal matrix:

$$F = \left\| \begin{matrix} F_A & 0 \\ 0 & F_B \end{matrix} \right\|. \tag{2.63}$$

We consider a completely general state vector $\overrightarrow{OS} = (x, y)$, requiring only the normalization condition $x^2 + y^2 = 1$, such that S lies on a circle of radius 1 centered at the origin. As always, we suppose, that $\overrightarrow{OS}$ represents the state of the system. Let us perform a measurement of F. We already know that the eigenvalues of F are the numbers F_A, F_B, and therefore the probability of finding the value F_A is x^2, and the probability of finding the value F_B is y^2. Then the *average* $\overline{F}$ of the results of a measurement of F can be computed by the standard formula of statistics, $\overline{F} = \sum_i F_i P_i$, and we can write:

$$\overline{F} = F_A x^2 + F_B y^2. \tag{2.64}$$

Alternatively, by taking the scalar product of the vectors $\overrightarrow{OS}$ and $F\overrightarrow{OS}$, we get the same result. In fact, the vector obtained applying the operator F to $\overrightarrow{OS}$ is simply $(F_A x, F_B y)$; taking the scalar product of this vector with $\overrightarrow{OS} = (x, y)$ we obtain the right-hand side of Eq. (2.64), whereby

$$\overline{F} = \overrightarrow{OS} \cdot F\overrightarrow{OS}. \tag{2.65}$$

The last formula, which has been obtained in a very particular case, is actually a completely general and very elegant result.

Postulate 1 does not say anything about the time evolution of the state vector. Actually, the motion in the Hilbert space of vector $\overrightarrow{OS}$ is determined by the *time-dependent Schrödinger equation*. In our toy space, the path of point S is simply the circumference of a circle of radius 1. Clearly, in spaces of higher dimension, this path is more complicated. We must imagine a point S moving *continuously* (without sudden jumps), maintaining its unit distance from the origin, just like a mass point constrained to the circumference of a circle. In the original formulation of QM, sudden jumps might occur when a measurement is made (see Postulate 2). We will come back to this subtle (and controversial) point later, exemplified by the question: "Are there quantum jumps?".

In the work we have done thus far, observables F have been independent of time (as have both the eigenvectors and the eigenvalues), while the vector state $\overrightarrow{OS}$ carries all the dependence on time. This is known as the *Schrödinger picture*. It is not difficult to formulate an alternative interpretation in Hilbert space, which corresponds to the same physical situation, but uses a *fixed* state vector but time-dependent operators. The idea is to rotate the eigenvectors of F back in such a way that their relative position with $\overrightarrow{OS}$ (which now is fixed) is the same as in the Schrödinger picture. We need first the following result:

Lemma 2.1 *If R is a rotation matrix, and* **A**, **B** *are arbitrary vectors, the scalar product of* **A** *with* R**B** *is equal to the scalar product of* **B** *with* R^{-1}**A**.

An algebraic proof is elementary. The matrix of R^{-1} is obtained from the matrix R simply by changing the sign of θ in Eq. (2.28). The following intuitive argument is perhaps more direct: consider the angle ϕ between the vectors **A**, **B** (see Fig. 2.11). Rotating **B** in a counterclockwise sense through an angle θ, we obtain the vector R**B**, while the angle between **A** and R**B** becomes $\theta + \phi$. And if we keep **B** fixed and rotate the vector **A** back in a clockwise sense (by applying R^{-1}) through an angle θ, the angle between **B** and R^{-1}**A** remains equal to $\phi + \theta$. From Eq. (2.10), we see that the scalar product of two vectors depends on the lengths of the vectors and the angle between them. But rotations do not change lengths, and since the angle is $\theta + \phi$ in both cases, the Lemma is proved.

Let us denote by $\overrightarrow{\Psi}(t)$ the state vector $\overrightarrow{OS}$ as a function of time t. Suppose, for simplicity, that at $t = 0$ the state vector coincides with the x axis, $\overrightarrow{\Psi}(0) = (1, 0)$, and at time $t = T$, the state vector $\overrightarrow{\Psi}(T)$ makes an angle $\theta(T)$ with the x axis; in other words, during the time from $t = 0$ and $t = T$ the state vector is *rotated* through an angle θ. Therefore the mapping from $\overrightarrow{\Psi}(0)$ to $\overrightarrow{\Psi}(T)$ can be obtained by means of the rotation matrix (2.28) and we have:

$$\overrightarrow{\Psi}(T) = R(T)\overrightarrow{\Psi}(0). \tag{2.66}$$

In the last equation, we have written $R(T)$ to emphasize the dependence of the rotation R on time T. In the more general situation, the analog of Eq. (2.66) provides the

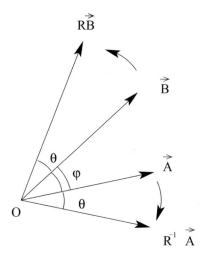

Fig. 2.11 The scalar products $\langle R\mathbf{B}|\mathbf{A}\rangle$ and $\langle \mathbf{B}|R^{-1}\mathbf{A}\rangle$ are equal

solution of the time-dependent Schrödinger equation, once the initial wave function $\vec{\Psi}(0)$ is specified. Substituting (2.66) in place of $\overrightarrow{OS}$ into Eq. (2.65) we get:

$$\overline{F} = R(T)\vec{\Psi}(0) \cdot F R(T)\vec{\Psi}(0). \tag{2.67}$$

Let us apply Lemma 1 with $\mathbf{A} = F R(T)\vec{\Psi}(0)$ and $\mathbf{B} = \vec{\Psi}(0)$. This allows us to move the rotation operator $R(T)$ to the other side of the scalar product, then replacing R by R^{-1}. We obtain:

$$\overline{F} = \vec{\Psi}(0) \cdot R(T)^{-1} F R(T)\vec{\Psi}(0). \tag{2.68}$$

We call the time-dependent operator $F(T) = R(T)^{-1} F R(T)$ *operator F in the Heisenberg picture*, and we write:

$$\overline{F} = \vec{\Psi}(0) \cdot F(T)\vec{\Psi}(0). \tag{2.69}$$

Of course, the state vector in the Heisenberg picture is $\vec{\Psi}(0)$. We see from formulas (2.69), (2.65) that the expression for $\overline{F}$ is the same in the two pictures. However, in the Heisenberg picture, the time dependence has been entirely transferred to the operator representing the observable. This is quite analogous to the picture in classical mechanics, in which we seek the changes in observables with time, as described by equations of motion.

A final topic we want to introduce for our toy Hilbert space is the *density matrix*. For any matrix A_{ij}, the sum of the elements of the main diagonal is called the *trace* of A, denoted by Tr A. Therefore:

$$\mathrm{Tr}\, A = A_{11} + A_{22}. \tag{2.70}$$

Given a unit vector $\overrightarrow{OS} = (x, y)$, the projection operator P on the line determined by $\overrightarrow{OS}$ is given, using Eq. (2.48) and setting $c = x$, $s = y$:

$$P = \left\| \begin{matrix} x^2 & xy \\ xy & y^2 \end{matrix} \right\|. \tag{2.71}$$

Given the operator F, the trace of the product $P\, F$ is given by:

$$\mathrm{Tr}(P\, F) = \mathrm{Tr} \left\| \begin{matrix} x^2 F_{11} + xy F_{12} & x^2 F_{12} + xy F_{22} \\ xy F_{11} + y^2 F_{12} & xy F_{12} + y^2 F_{22} \end{matrix} \right\| = x^2 F_{11} + 2xy F_{12} + y^2 F_{22}. \tag{2.72}$$

Let us prove that the last expression constitutes a generalization of Eq. (2.64) when F is not diagonal. First compute (2.65): the coordinates of the vector $F\,\overrightarrow{OS}$ are $(F_{11}x + F_{12}y, F_{12}x + F_{22}y)$; then take the scalar product of this vector with $\overrightarrow{OS} = (x, y)$, giving precisely the expression (2.72). Therefore:

$$\mathrm{Tr}(P\, F) = \langle \overrightarrow{OS} | F\, \overrightarrow{OS} \rangle = \overline{F}. \tag{2.73}$$

Note that if F equals the identity I, Eq. (2.27) reduces to

$$\mathrm{Tr}(P\, I) = \mathrm{Tr}\, P = x^2 + y^2 = 1. \tag{2.74}$$

Consider now two orthogonal states $\overrightarrow{\Psi}_1$, $\overrightarrow{\Psi}_2$ and suppose that there is a probability p_1 that the state vector $\overrightarrow{OS}$ of a physical system coincides with $\overrightarrow{\Psi}_1$, and a probability p_2 that it coincides with $\overrightarrow{\Psi}_2$. Note that in the actual case the probabilities p_1 and p_2 are not the fundamental probabilities of QM (which Heaven only knows!). Here, p_1 and p_2 might represent classical probabilities of distinct physical situations, as we encounter, in classical statistical mechanics. In any event, we must have $p_1 + p_2 = 1$. In order to obtain the average value of an observable F, we compute a double average: first we find the two quantum averages $\langle \overrightarrow{\Psi}_1 | F\, \overrightarrow{\Psi}_1 \rangle$, $\langle \overrightarrow{\Psi}_2 | F\, \overrightarrow{\Psi}_2 \rangle$, and then we average these results, making use of the probabilities p_1, p_2; at the end of this procedure we get:

$$\overline{F} = p_1 \langle \overrightarrow{\Psi}_1 | F\, \overrightarrow{\Psi}_1 \rangle + p_2 \langle \overrightarrow{\Psi}_2 | F\, \overrightarrow{\Psi}_2 \rangle. \tag{2.75}$$

Now let P_1, P_2 be the projection operators onto the straight lines determined by $\overrightarrow{\Psi}_1$, $\overrightarrow{\Psi}_2$. Using Eq. (2.73) we have $\langle \overrightarrow{\Psi}_1 | F \overrightarrow{\Psi}_1 \rangle = \mathrm{Tr}(P_1 F)$, and $\langle \overrightarrow{\Psi}_2 | F \overrightarrow{\Psi}_2 \rangle = \mathrm{Tr}(P_2 F)$. Therefore, introducing the *density matrix* $\rho = p_1 P_1 + p_2 P_2$ and using the fact that taking the trace is a *linear* operation, we can write:

$$\overline{F} = p_1 \mathrm{Tr}(P_1 F) + p_2 \mathrm{Tr}(P_2 F) = \mathrm{Tr}[\,(p_1 P_1 + p_2 P_2)F\,] = \mathrm{Tr}(\rho F). \tag{2.76}$$

Equation (2.76) constitutes a generalization of Eq. (2.73); knowledge of the density matrix ρ allows us to compute averages of any observable F; therefore ρ determines the *state* of the system in way analogous to the state vector $\overrightarrow{OS}$. When $p_1 = 1$ and $p_2 = 0$, or $p_1 = 0$ and $p_2 = 1$, this reduces to the previous case; we say that the system is in a *pure state*. The more general state defined by $\rho = p_1 P_1 + p_2 P_2$ is called a *mixed state*. Let us verify that p_1, p_2 are the eigenvalues of ρ, and $\overrightarrow{\Psi_1}$, $\overrightarrow{\Psi_2}$, its eigenvectors. Denoting by $\mathbf{0}$ the null vector, we have:

$$P_1 \overrightarrow{\Psi_1} = \overrightarrow{\Psi_1}, \qquad P_2 \overrightarrow{\Psi_1} = \overrightarrow{0}, \tag{2.77}$$

and therefore:

$$\rho \overrightarrow{\Psi_1} = (p_1 P_1 + p_2 P_2) \overrightarrow{\Psi_1} = p_1 \overrightarrow{\Psi_1} + p_2 \overrightarrow{0} = p_1 \overrightarrow{\Psi_1}. \tag{2.78}$$

In the same way we can prove that $\rho \overrightarrow{\Psi_2} = p_2 \overrightarrow{\Psi_2}$. Since there is no restriction on the pair of orthogonal vectors $\overrightarrow{\Psi_1}$, $\overrightarrow{\Psi_2}$, we see that the most general density matrix is a symmetric matrix whose eigenvalues are positive numbers p_1, p_2 less than or equal to 1, and such that $p_1 + p_2 = 1$. Using the relation $x^2 + y^2 = 1$, it can be verified that the projection operator P, given by Eq. (2.71), is *idempotent*, meaning that it obeys the relation $P^2 = P$. This condition is, in fact, a defining characteristic of a pure state.

2.4 The Hilbert Space of Real Wavefunctions

We have now acquired an understanding of the toy model, but it may still not be entirely clear why wave functions representing "clouds of probability" have anything to do with vectors of the plane $\mathbb{R}^2$. The answer of a mathematician might again run: "Both $\mathbb{R}^2$ and the set of wave functions of a physical system are vector spaces endowed with a scalar product." However, to show that wave functions do indeed belong in a Hilbert space, we will follow a more elementary, less abstract, line of development: we will exhibit an "analogy" between the vectors of $\mathbb{R}^2$, $\mathbb{R}^3$, ..., and the set $C(a, b)$ of continuous wavefunctions defined on an interval $[a, b]$ of the real axis. Actually, physical wave functions $\psi(x, y, z)$ are defined on points (x, y, z) of three-dimensional space; indeed $|\psi(x, y, z)|^2$ is actually the *probability density* in the clouds drawn in Chap. 1. To simplify the mathematics, we can imagine that our physical system to be one-dimensional, so that the wavefunctions depend on just a single variable x, on a line segment $[a, b]$.

Two of the fundamental operations of a vector space, given in Eqs. (2.1) and (2.2), have obvious analogs for our set of functions: given two continuous functions $f(x)$, $g(x)$, their sum is the function $f(x) + g(x)$, as shown in Fig. 2.12. Also, just as we can multiply the coordinates of a vector; by a real number λ, to obtain a new vector $\lambda \mathbf{f}$ in the same direction as the original, we can likewise multiply a function $f(x)$, to give the analogous scaled function $\lambda f(x)$. The operations of addition and multiplication

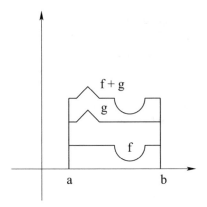

Fig. 2.12 Sum of two functions f, g

by a real number suggests the terminology *linear combination of two functions* $f(x)$, $g(x)$, with real coefficients a, b, namely, the function $af(x) + bg(x)$. The set of such linear combinations can be thought of as a *three-dimensional subspace of* $C(a, b)$, provided $f(x)$, $g(x)$ do not have the same "direction," meaning that $f(x)$ is not simply a multiple of $g(x)$. This two-dimensional subspace can be thought of as a plane through the origin. What is the origin? The object analogous to the null vector $(0, 0)$ of $\mathbb{R}^2$ is a function which equals zero everywhere: $f(x) = 0$ for all x.

A more challenging question is: what constitutes the *coordinates* of a function, which are somehow the analogs of the components of a vector? Later we will give a more rigorous answer to this question; for the moment, we tentatively settle for a more heuristic approach, which will enable us to understand the meaning of the scalar product of two wavefunctions. Let f be a continuous function defined on [a,b]; see Fig. 2.13, where the graph of $f(x)$ is shown. We choose n equally spaced points $x_1, x_2, \ldots, x_n$ in the interval (so that $x_1 = a$ and $x_n = b$), and we compute the values of the functions $f(x_1)$, $f(x_2)$, $\ldots$, $f(x_n)$. We can then imagine these numbers to be

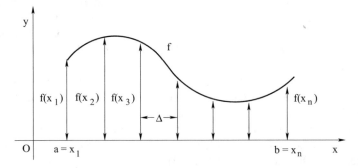

Fig. 2.13 A function $f(x)$ is approximated by a vector in $\mathbb{R}^n$, $\mathbf{f} = (f(x_1), f(x_2), \ldots, f(x_n))$

the *coordinates* of a vector belonging to $\mathbb{R}^n$. For example, let us take $a = 0$, $b = 3$, $n = 4$ and consider the simple function $f(x) = x^2$. Then $x_1 = 0$, $x_2 = 1$, $x_3 = 2$, $x_4 = 3$, and $f(x_1) = 0$, $f(x_2) = 1$, $f(x_3) = 4$, $f(x_4) = 9$. We have obtained the vector $(0, 1, 4, 9)$ of $\mathbb{R}^4$. You may ask: How do we choose the number n? Indeed this number is arbitrary, since given a function on an interval, we can compute it at as many points as we want. This is an inherent weakness in the identification of the values $f(x_1), f(x_2), \ldots, f(x_n)$ as "coordinates of f." However, let us boldly proceed nonetheless, and try to find a tentative definition of the *scalar product of two functions* which is analogous to the definitions (2.10), (2.11) of the scalar product of two vectors?

Given two functions f and g, both continuous on the interval [a,b], let us compute these functions on the equally spaced points $x_1 = a$, $x_2, \ldots, x_n = b$, as above. We will obtain, in this way, two vectors $\mathbf{f}$, $\mathbf{g}$ belonging to $\mathbb{R}^n$:

$$\begin{aligned}\mathbf{f} &= (f(x_1), f(x_2), \ldots, f(x_n)), \\ \mathbf{g} &= (g(x_1), g(x_2), \ldots, g(x_n)),\end{aligned} \tag{2.79}$$

whose scalar product is given by (see the analog in $\mathbb{R}^4$, Eq. (2.18)):

$$f(x_1)g(x_1) + f(x_2)g(x_2) + \cdots + f(x_n)g(x_n). \tag{2.80}$$

However this expression depends on the number n, which is arbitrary. Since our knowledge of a function becomes more precise when we know more and more values $f(x_i)$ (think, for example, of these values as data in an experiment), we can *take the limit* of Eq. (2.80) as n approaches infinity. Even for the elementary case of two constant functions, say $f(x) = 2$ and $g(x) = 3$, this limit is not finite. On the other hand, there exists an expression which is similar to Eq. (2.80) that admits a finite limit when $n \to \infty$, and will provide us with a more rigorous definition. Denoting by Δ_n the distance between two consecutive points, we have $\Delta_n = x_2 - x_1 = x_3 - x_2, \ldots = \frac{(b-a)}{(n-1)}$. It is easy to see that the limit:

$$\lim_{n \to \infty} \Delta_n \left[f(x_1)g(x_1) + f(x_2)g(x_2) + \cdots + f(x_n)g(x_n) \right] \tag{2.81}$$

exists and corresponds to the definition of a Riemann integral:

$$\int_a^b f(x)g(x)\,dx. \tag{2.82}$$

Thus the expression (2.82) provides us with a consistent definition of *scalar product of two real functions* $f(x)$, $g(x)$. To emphasize the analogy with vectors, we can write:

$$\mathbf{f} \cdot \mathbf{g} = \int_a^b f(x)g(x)\,dx. \tag{2.83}$$

The importance of this definition of scalar product can hardly be overestimated. It allows us to continue using a geometric language in the space of wavefunctions, and suggests an intuitive picture of physical states, even when we are referring, not to physical space, but to an abstract function space. We continue to define the *norm* or *length* of the vector **f** using the expression:

$$|\mathbf{f}| = \sqrt{\mathbf{f} \cdot \mathbf{f}} = \sqrt{\int_a^b dx \, f(x)^2}. \qquad (2.84)$$

As an example, let us evaluate the scalar product of the functions x and $1 + x^2$ on the interval $[0, 1]$. We must perform the integration:

$$\int_0^1 (1 + x^2)x = \left[\frac{x^2}{2} + \frac{x^4}{4}\right]_0^1 dx = \frac{3}{4}. \qquad (2.85)$$

Table 2.1 shows the analogies between vectors in $\mathbb{R}^n$ and the corresponding functional relations. On the left of the table, we show expressions involving the vectors **v**, **w**, ...; on the right are the analogs, in terms of the functions $f(x)$, $g(x)$ or **f**, **g**, ... Such correspondences will be particularly useful in Dirac's bra/ket formulation of QM.

Table 2.1 Analogies between vectors and functions

Vectors	Functions
Components of a vector $\mathbf{v} = (v_1, v_2, \ldots, v_n)$	Values of a function $f(x_1), f(x_2), \ldots, f(x_n)$
Sum of two vectors $\mathbf{v} + \mathbf{w}$	Sum of two functions $f(x) + g(x)$
Linear combination of two vectors $c_1\mathbf{v} + c_2\mathbf{w}$	Linear combination of two functions $c_1 f(x) + c_2 g(x)$
Scalar product of two vectors $\mathbf{v} \cdot \mathbf{w} = v_1 w_1 + v_2 w_2 + \cdots + v_n w_n$	Scalar product of two functions $\mathbf{f} \cdot \mathbf{g} = \int_a^b f(x)g(x)\,dx$
Norm of a vector $|\mathbf{v}| = \sqrt{\mathbf{v} \cdot \mathbf{v}} = \sqrt{v_1^2 + v_2^2 + \cdots + v_n^2}$	Norm of a function $|\mathbf{f}| = \sqrt{\mathbf{f} \cdot \mathbf{f}} = \sqrt{\int_a^b dx \, f(x)^2}$
Linearity of the scalar product $\langle c\,\mathbf{v}|\mathbf{w}\rangle = c\,\langle\mathbf{v}|\mathbf{w}\rangle$ $\langle\mathbf{v}|\mathbf{w}_1 + \mathbf{w}_2\rangle = \langle\mathbf{v}|\mathbf{w}_1\rangle + \langle\mathbf{v}|\mathbf{w}_2\rangle$	Linearity of the scalar product $\langle c\,\mathbf{f}|\mathbf{g}\rangle = c\,\langle\mathbf{f}|\mathbf{g}\rangle$ $\langle\mathbf{f}|\mathbf{g}_1 + \mathbf{g}_2\rangle = \langle\mathbf{f}|\mathbf{g}_1\rangle + \langle\mathbf{f}|\mathbf{g}_2\rangle$
Distance between two vectors $d = |\mathbf{v} - \mathbf{w}| = \sqrt{\langle\mathbf{v} - \mathbf{w}|\mathbf{v} - \mathbf{w}\rangle} = $ $\left[(v_1 - w_1)^2 + (v_2 - w_2)^2 + \cdots + (v_n - w_n)^2\right]^{\frac{1}{2}}$	Distance between two functions $d = |\mathbf{f} - \mathbf{g}| = \sqrt{\langle\mathbf{f} - \mathbf{g}|\mathbf{f} - \mathbf{g}\rangle} = $ $\left[\int_a^b dx \, (f(x) - g(x))^2\right]^{\frac{1}{2}}$

The proof of the relations: $\langle c\mathbf{f}|\mathbf{g}\rangle = c\langle\mathbf{f}|\mathbf{g}\rangle$, and $\langle\mathbf{f}|\mathbf{g}_1 + \mathbf{g}_2\rangle = \langle\mathbf{f}|\mathbf{g}_1\rangle + \langle\mathbf{f}|\mathbf{g}_2\rangle$ is elementary also in the case of functions, since

$$\int_a^b [cf(x)]g(x)\,dx = c\int_a^b f(x)g(x)\,dx, \tag{2.86}$$

and

$$\int_a^b f(x)[g_1(x) + g_2(x)]\,dx = \int_a^b f(x)g_1(x)\,dx + \int_a^b f(x)g_2(x)\,dx. \tag{2.87}$$

The analogy between the distance between vectors and the "distance" between functions deserves a word of comment: if distance d is very small the coordinates of the vectors $\mathbf{v}$, $\mathbf{w}$ are almost equal, since the sum of the positive numbers $(v_1 - w_1)^2$, $(v_2 - w_2)^2$, ..., cannot be small unless every one of these contributions is small. In an analogous way, for functions, a very small value of d means, by and large, that the graphs of the functions $f(x)$, $g(x)$ are very close together. (There might be exceptions, in which the difference of the functions is large in small intervals on the x-axis.)

Another case to be considered is the existence of functions that do *not* have finite norm. A simple example is the function $f(x) = \frac{1}{\sqrt{x}}$, defined on the *open* interval $(0, 1)$, that is, the interval excluding the endpoints 0, 1. In fact, $\int_0^1 f(x)^2\,dx = \int_0^1 \frac{1}{x}\,dx = \infty$ or, better, $\lim_{\epsilon \to 0} \int_\epsilon^1 \frac{1}{x}\,dx = \infty$, since the function x^{-1} becomes very large for small x. Such behavior is excluded from our formalism, since we have restricted functions to be continuous and well defined in the whole interval $[a, b]$ ($\frac{1}{x}$ is not defined for $x = 0$). However, in physics, the interval $[a, b]$ is often the entire x axis, so that our integration $\int_a^b$ becomes $\int_{-\infty}^{+\infty}$. Therefore, even some very simple functions such as x^2, x^4, etc., must be excluded since their integrals diverge to ∞. However, since a wave function $f(x)$ gives the probability amplitude of finding a particle at point x, it is reasonable to assume that this amplitude goes to zero when x becomes very large (for example, an electron bound to an atom has practically zero probability of being found on the Moon). Coming back to the purely mathematical aspects of the theory (while leaving aside certain mathematical subtleties), we will define as a *Hilbert space*, denoted by $L^2(a, b)$, the set of functions $f(x)$ such that

$$\int_a^b [f(x)]^2\,dx < \infty, \tag{2.88}$$

meaning that the integral must be *finite*. Therefore, $\frac{1}{x}$, for example, does *not* belong to $L^2(0, 1)$. In many physical applications, we will have $a = -\infty$, $b = +\infty$, with the corresponding Hilbert space denoted by $L^2(-\infty, +\infty)$.

We usually assume that functions belonging to the Hilbert space correspond to vectors of *finite* length. The scalar product must then also be finite. As an example, the wave function:

Table 2.2 Vector products and integrals of functions

Two vectors $\mathbf{v}$, $\mathbf{w}$ are are orthogonal if $\mathbf{v} \cdot \mathbf{w} = 0$, so that $v_1 w_1 + v_2 w_2 + \cdots + v_n w_n = 0$	Two functions f, g orthogonal if $\mathbf{f} \cdot \mathbf{g} = 0$, so that $\int_a^b f(x)g(x)\,dx = 0$
A vector $\mathbf{v}$ is normalized if $|\mathbf{v}|^2 = v_1^2 + v_2^2 + \cdots + v_n^2 = 1$	A function $f(x)$ is normalized if $|\mathbf{f}|^2 = \int_a^b f(x)^2\,dx = 1$
A basis of n orthonormal vectors in $\mathbb{R}^n$ is a set of n vectors $\mathbf{v}^{(1)}, \mathbf{v}^{(2)}, \ldots, \mathbf{v}^{(n)}$ such that $\mathbf{v}^{(i)} . \mathbf{v}^{(j)} = \begin{cases} 1 \text{ if } i=j \\ 0 \text{ if } i \neq j \end{cases}$	A basis of orthonormal functions in L^2 is a sequence of (∞) functions $f_1(x), f_2(x), \ldots, f_n(x) \ldots$ such that $\mathbf{f}_i . \mathbf{f}_j = \int_a^b f_i(x) f_j(x)\,dx = \begin{cases} 1 \text{ if } i=j \\ 0 \text{ if } i \neq j \end{cases}$
Expansion of a vector in an orthonormal basis: $\mathbf{v} = \sum_{i=1}^n v_i \mathbf{e}^{(i)} \quad v_i = \mathbf{e}^{(i)} \cdot \mathbf{v}$	Expansion of a function in an orthonormal basis: $f(x) = \sum_{i=1}^\infty c_i f_i(x)$ $c_i = \mathbf{f}_i \cdot \mathbf{f} = \int_a^b f_i(x) f(x)\,dx$

$$\frac{1}{\sqrt[4]{\pi}}\, e^{-(x-a)^2/2} \tag{2.89}$$

has norm equal to one and represents a "cloud" of probability localized around the point $x = a$, decreasing rapidly as $|x - a|$ becomes large. To complete our analogy between vectors and functions, there is no problem in extending the concept of orthogonality to functions; by the definition of scalar product, we can say that two functions $f(x)$, $g(x)$ are *orthogonal* in $L^2(a, b)$ if $\langle \mathbf{f}|\mathbf{g} \rangle = \int_a^b f(x)g(x)\,dx = 0$. With this definition in mind, we define an *orthonormal set of functions* $f_1(x)$, $f_2(x),\ldots, f_n(x)$, such that all functions have "length" 1 and are orthogonal to one other, as in the simple example of the vectors $(1, 0)$, $(0, 1)$ in the plane $\mathbb{R}^2$ or the vectors $\mathbf{i} = (1, 0, 0)$, $\mathbf{j} = (0, 1, 0)$ and $\mathbf{k} = (0, 0, 1)$ in the space $\mathbb{R}^3$. We know (see Eq. (2.22)) that any vector $\mathbf{v} = (v_1, v_2, v_2) \in \mathbb{R}^3$ can be written as the linear combination $\mathbf{v} = v_1\mathbf{i} + v_2\mathbf{j} + v_3\mathbf{k}$. Furthermore, the components v_1, v_2, v_3 satisfy the relations:

$$\mathbf{v} \cdot \mathbf{i} = v_1, \qquad \mathbf{v} \cdot \mathbf{j} = v_2, \qquad \mathbf{v} \cdot \mathbf{k} = v_3, \tag{2.90}$$

which can be generalized to any orthonormal basis of n vectors in $\mathbb{R}^n$. In other words, the scalar product of any vector $\mathbf{v}$ with the ith basis vector, gives the magnitude of the ith "coordinate" of $\mathbf{v}$ with respect to the ith "axis." This result suggests other analogous definitions and formulas, which we summarize in Table 2.2.

According to the last row of the table, for the case of vectors in $\mathbb{R}^n$, the equality $\mathbf{v} = \sum_{i=1}^n v_i \mathbf{e}^{(i)}$ has an obvious meaning. It implies that n orthonormal vectors do form a *basis* on which we can expand any vector; we know that in order to have such a basis we need two vectors in $\mathbb{R}^2$ or three vectors in $\mathbb{R}^3$, etc. What happens in the Hilbert space $L^2(a, b)$? There must then exist sequences of an infinite number of orthonormal functions. For example, if we take $a = 0$ and $b = 2\pi$, the following functions:

$$f_1(x) = \frac{1}{\sqrt{2\pi}}, \quad f_2(x) = \frac{\cos x}{\sqrt{\pi}}, \quad f_3(x) = \frac{\sin x}{\sqrt{\pi}},$$

$$f_4(x) = \frac{\cos 2x}{\sqrt{\pi}}, \quad f_5(x) = \frac{\sin 2x}{\sqrt{\pi}}, \quad \dots, \tag{2.91}$$

do form an orthonormal system in $L^2(0, 2\pi)$. The function $f(x)$ is now represented by an infinite sum over these basis functions (this might be recognized as a Fourier series):

$$f(x) = \sum_{i=1}^{\infty} c_i f_i(x). \tag{2.92}$$

There remain questions of convergence and such, but we will not worry about these. If, indeed, Eq. (2.92) does hold for some orthonormal system of functions, such as the sequence (2.91), this set of functions is called *complete* and thereby provides a *basis* for expanding any admissible function in the Hilbert space. Given a function $f(x)$, the coefficients c_i (for $i = 1, 2, 3 \dots$) are the best candidates to be designated "coordinates" of the vector $\mathbf{f}$; this interpretation has a more rigorous foundation than the one we have introduced earlier, when we cited the values $f(x_1), f(x_2), \dots, f(x_n)$. In Dirac's formalism, the two interpretation can be unified in an elegant (but not entirely rigorous) way, which is beyond the scope of our coverage. The limit implied by the infinite summation appearing in Eq. (2.92) must be understood in the following sense: the Hilbert space *distance* $|\mathbf{f} - \mathbf{f}_n|$ between the function $f_n(x) = \sum_{i=1}^{n} c_i f_i(x)$ and the function $f(x)$ tends to zero when $n \to \infty$.

Now that we have a better understanding of what the Hilbert space is, we can further extend our analogy between vectors and functions, and ask: what are the "matrices" or better the *linear operators* relevant to QM which act in the Hilbert space $L^2(-\infty, +\infty)$, analogous to the way 2×2 matrices act on vectors in the plane? Let us give two key examples of operators defined in this Hilbert space (leaving aside mathematical subtleties, that are treated, for example, in Fano 1971):

1. The operator that multiplies any function $\psi(x)$ by the variable x. This operator is the famous q operator of QM, which represents the position of a particle. The result of the application of q to ψ is the function $x\psi(x)$, as follows:

$$(q\psi)(x) = x\psi(x), \tag{2.93}$$

where $(q\psi)(x)$ is the image function $q\psi$ computed at the point x. For example, q maps x^n into x^{n+1}, $\sin x$ into $x \sin x$, etc.

2. The operator that takes the derivative of a function $\psi(x)$. Denoting this operator by $\frac{d}{dx}$, we write:

$$\frac{d}{dx} \psi(x) = \frac{d\psi(x)}{dx}. \tag{2.94}$$

This means that $\frac{d}{dx}$ maps x^n into nx^{n-1}, $\sin x$ into $\cos x$, etc.

Clearly, the operators q and $\frac{d}{dx}$ are *linear*. They satisfy the analog of Eq. (2.30). Thus for $\frac{d}{dx}$, since the derivative of the sum of two functions is the sum of the derivatives, we have, for example,

$$\frac{d}{dx}(f + g) = \frac{d}{dx}f + \frac{d}{dx}g. \tag{2.95}$$

The operators q and $\frac{d}{dx}$ are of primary importance in QM, since the first represents the *position* of a particle, and the second is proportional to its momentum. An important fact about these two operators, is that they *do not commute*. Let us denote by D the operator $\frac{d}{dx}$. For "any" function $\psi(x)^3$ we have:

$$q\,D\psi = q\frac{d}{dx}\psi = x\frac{d\psi}{dx},$$
$$D\,q\psi = \frac{d}{dx}q\psi = \frac{d}{dx}(x\psi) = \psi + x\frac{d\psi}{dx}. \tag{2.96}$$

Subtracting the two equations, we obtain, for "any" ψ, $Dq\psi - qD\psi = \psi$, or:

$$Dq - qD = I. \tag{2.97}$$

Powers of the operators q and D are easy to compute. For example, the functions $q^2\,\psi$ and $D^2\,\psi$ are:

$$q^2\psi = x(x\psi) = x^2\psi,$$
$$D^2\psi = \frac{d}{dx}\frac{d}{dx}\psi = \frac{d^2}{d^2x}\psi. \tag{2.98}$$

2.5 Complex Variables

To extend our repertoire of mathematical proficiency, this section will review some aspects of complex numbers and complex functions. (Our apologies to readers already well versed in this subject.) Mathematicians define an algebraic structure called a *field* as a set of (usually) numbers, along with two operations, which can be identified with addition and multiplication (subtraction and division are implicitly included), and satisfies the associative and distributive laws. The most commonly encountered fields are the real numbers, the rational numbers and, the subject of this section, the complex numbers. Complex analysis turns out to be mandatory for understanding the full mathematical structure of quantum mechanics. It is not strictly necessary for classical mechanics or electrodynamics, although complex variables can provide very useful enhancements to their mathematical formulation.

---

[3] The quotation marks refer to some mathematical conditions that the function $\psi(x)$ must fulfill: in essence, ψ must be differentiable almost everywhere and $D\psi$ must remain in the Hilbert space.

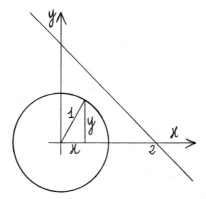

Fig. 2.14 The circle Γ and the straight line Σ do not intersect; therefore their equations admit no simultaneous solution in the field of real numbers

To begin, consider a circle Γ in the Cartesian plane and a straight line Σ lying outside Γ (see Fig. 2.14), for example, the circle with center at the origin and radius 1, represented by the equation

$$x^2 + y^2 = 1, \tag{2.99}$$

and the straight line represented by the equation:

$$x + y = 2. \tag{2.100}$$

The simultaneous equations for the circle Γ and the line Σ, Eqs. (2.99) and (2.100), therefore do *not* have any real simultaneous solutions. Let us nevertheless solve (2.100) for y, to get $y = 2 - x$, and substitute this into (2.99). We obtain $x^2 + (2 - x)^2 = 1$; therefore:

$$2x^2 - 4x + 3 = 0. \tag{2.101}$$

The solutions of the quadratic equation $ax^2 + bx + c = 0$ are:

$$x = \frac{-b \pm \sqrt{b^2 - 4ac}}{2a}. \tag{2.102}$$

In our case $a = 2$, $b = -4$, and $c = 3$. Thus

$$x = \frac{4 \pm \sqrt{16 - 24}}{4} = \frac{2 \pm \sqrt{-2}}{2}. \tag{2.103}$$

The square root of a negative number appears. This appears contradictory: since the square of a real number is always positive (for example $(+2) \times (+2) = +4$ and $(-2) \times (-2) = +4$), the argument of a square root should always be a positive number. For many centuries, it was believed that roots of negative numbers have no meaning, consistent with the nonexistence of points common to a circle and a nonintersecting straight line. But in 1572, Rafael Bombelli in his book *L'Algebra*, gave meaning to the expression $\sqrt{-1}$.

We will denote $\sqrt{-1}$ by the usual symbol i and call it the *imaginary unit*. Thus, by definition $i^2 = -1$, and (2.103) becomes $x = \frac{2 \pm \sqrt{2}\,i}{2}$ We are now dealing with a new kind of numbers, which we call *complex numbers*. If $z = a + ib$, with a and b real; a is called the *real part* and b the *imaginary part* of z. (In the above case $a = 1$ and $b = \pm \frac{\sqrt{2}}{2}$). Complex numbers have the following properties:

(1) $a + ib = c + id$ if and only if $a = c$ and $b = d$.
(2) $(a + ib) + (c + id) = (a + c) + i(b + d)$.
(3) $(a + ib)(c + id) = ac - bd + i(ad + bc)$.

The multiplication law (3) is consistent with the usual properties of real numbers with the addition of a new rule: $i^2 = -1$. Furthermore, Items (1) and (2) suggest a representation of the complex number $x + iy$ by the vector (x, y) in the Cartesian plane R^2. The x-axis now serves as the *real axis*, while the y-axis is the *imaginary axis*, since it consists of points of type $(0, y)$. The x-y plane is now called the *complex plane* or an *Argand diagram*. Figure 2.15 shows the vector corresponding to the complex number $z = a + ib$, while Fig. 2.16 shows the vector corresponding to the complex number $-z = -a - ib$. Item (2) above implies that the parallelogram rule applies to the sum of two complex numbers (see Fig. 2.17).

A suggestive property is the following: if we multiply the complex number $a + ib$ times the imaginary unit i, the corresponding vector is rotated by $\frac{\pi}{2}$ (90 degrees). Indeed the complex number $i(a + ib) = -b + ia$ corresponds to the vector $(-b, a)$ which is rotated by $\frac{\pi}{2}$ with respect to (a, b) (see Fig. 2.18). The "vectors" $i, i^2 = -1$, $i^3 = -i, i^4 = +1$ are related by successive rotations by $\frac{\pi}{2}$ (see Fig. 2.19). The *complex conjugate* z^* (alternatively written $\bar{z}$ in many texts) of the number $z = a + ib$

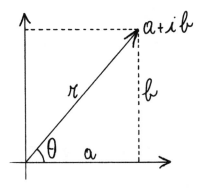

Fig. 2.15 Identification of the complex number $a + ib$ with the point (a, b) in the Cartesian plane

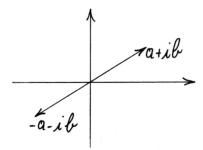

Fig. 2.16 Identification of the complex number $-z = -a - ib$ with the point $(-a, -b)$ of the Cartesian plane

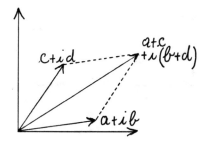

Fig. 2.17 The parallelogram's rule holds for the sum of two complex numbers $a + ib$ and $c + id$

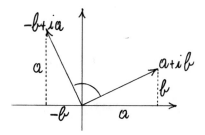

Fig. 2.18 The complex number $i(a + ib) = -b + ia$ corresponds to the vector $(-b, a)$ which is rotated by 90 degrees with respect to (a, b)

is defined as $z^* = a - ib$. Notice that $a + ib$ and $a - ib$ are symmetric with respect to reflection in the real axis (see Fig. 2.20). Clearly, $(a - ib)^* = a + ib$, so that $(z^*)^* = z$. If we multiply $a + ib$ times its complex conjugate $a - ib$, we obtain the square of the length of the vector (a, b); indeed,

$$(a + ib)(a - ib) = a^2 - iab + iab + b^2 = a^2 + b^2. \qquad (2.104)$$

The *modulus r* of the complex number $a + ib$ is defined as its length, $r = \sqrt{a^2 + b^2}$. If θ denotes the angle measured counterclockwise from the real axis to (a, b), we

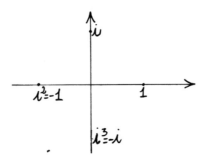

Fig. 2.19 The complex numbers i, i^2, i^3, i^4, correspond, respectively, to the points $(0, 1), (-1, 0)$, $(0, -1), (1, 0)$. This is a simple example of a cyclic group (designated $\mathbb{Z}_4$)

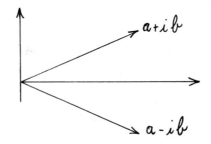

Fig. 2.20 The complex conjugate $a - ib$ of $a + ib$ can be obtained by reflection through the real axis

have (see Fig. 2.15):

$$a = r \cos \theta, \quad b = r \sin \theta. \tag{2.105}$$

Thus

$$z = a + ib = r(\cos \theta + i \sin \theta). \tag{2.106}$$

The angle θ is called *argument* of z. The real numbers r and θ uniquely determine the complex number z. For example, if $r = 1$ and $\theta = \frac{\pi}{4}$ (45 degrees), $z = \frac{1}{\sqrt{2}}(1 + i)$. In general, complex numbers of modulus 1 are represented by points on the unit circle (with center at the origin and radius 1). For $z = a + ib$, the angle θ is given by:

$$\theta = \arctan \frac{b}{a}, \tag{2.107}$$

with θ determined up to multiples of 2π (360 degrees). For example, the pair r, θ and the pair $r, \theta + 2\pi$ correspond to the *same* complex number.

The following very useful and beautiful formula can be used in place of (2.106):

$$z = a + ib = re^{i\theta}, \tag{2.108}$$

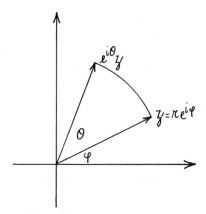

Fig. 2.21 $e^{i\theta}z$ can be obtained from z by rotation by the angle θ

where e is Euler's constant, $e = 2.7182818\ldots$ (the base of natural logarithms). The usual algebraic properties of the exponential function: $e^0 = 1$, $e^a e^b = e^{a+b}$, $e^{-a} = 1/e^a$, etc., still hold even when the exponent is imaginary or complex. However, the corresponding geometric representation is entirely different from the real case. A complex number $e^{i\theta}$ can alternatively be regarded as an *operator*. Given the complex number $z = re^{i\varphi}$, $e^{i\theta}z = ze^{i\theta}e^{i\varphi} = re^{i(\theta+\varphi)}$ is a number with the same modulus r but a new argument $\varphi + \theta$ (see Fig. 2.21). The factor $e^{i\theta}$ "rotates" z by an angle θ, similar to the way an orthogonal matrix, such as Eq. (2.28) rotates a two component vector.

2.6 Complex Vector Spaces and Dirac Notation

In order to make our development as simple as possible up to now, we have considered only *real* vector spaces, totally avoiding *complex numbers*. However, complex quantities turn out to be mandatory for complete understanding of the fundamental equations of QM, in particular, the Schrödinger equation itself. We have already noted that in QM, the operator D is proportional to the momentum $p = mv$ of a particle. In fact, the proportionality factor is $-i\hbar$, where $i = \sqrt{-1}$, the *imaginary unit*. We will also, in this section, be introducing *Dirac's bra/ket notation*, invented in 1939 by P.A.M. Dirac (1958), one of the founding fathers of quantum mechanics. This is now a standard notation for describing quantum states, using angle brackets and vertical bars to represent abstract vectors and linear operations. The notation has also become popular in other mathematical applications.

We denote by $\mathbb{C}^n$ the set of the ordered n-tuples $(z_1, z_2, \ldots, z_n)$ of *complex numbers*. An element $z = (z_1, z_2, \ldots, z_n)$ of such a set is now what we designate a *vector*. Thus, for $n = 2$, a vector $(z_1, z_2) = (x_1 + iy_1, x_2 + iy_2)$ of $\mathbb{C}^2$ is determined by 4 *real numbers*. In the following, unless explicitly stated, we will consider the complex space $\mathbb{C}^2$. In Dirac notation, a vector $z = (z_1, z_2)$ will be denoted by $|z\rangle$, in place of **z**. The *null* vector $|0\rangle$ is the vector with all components equal to zero. Therefore, in $\mathbb{C}^n$, $|0\rangle = (0, 0, \ldots, 0)$. Dirac denoted vectors, such as $|\psi\rangle$, representing quantum states, as "kets." Adjoint vectors (associated with the complex conjugate of a wavefunction), such as $\langle\phi|$, were called "bras." The product of a bra and a ket is a *bracket*, representing a scalar product $\langle\phi|\psi\rangle$. This connects with the notation we have already introduced for scalar products.

In $\mathbb{C}^2$ space, all the familiar linear properties still apply,. The linear combination $z = a|u\rangle + b|v\rangle$ is, in general, constructed with *complex a, b*. We can visualize the sum $|z\rangle = |u\rangle + |v\rangle$ as in Fig. 2.2, but to determine $|z\rangle$ now requires 4 real numbers (although our physical space remains $\mathbb{R}^3$, not $\mathbb{R}^4$). For example, if $u_1 = 1 + i$, $u_2 = 2 + 3i$, $v_1 = 1 - 2i$, $v_2 = 3 - 2i$, then $|u\rangle + |v\rangle = (2 - i, 5 + i)$. The scalar product used in quantum mechanics is a generalization of the form of Eqs. (2.11) and (2.18), used in real spaces. Instead, an *Hermitian* scalar product is defined in $\mathbb{C}^n$, as follows:

Definition 2.1 Given two vectors $|z\rangle = (z_1, z_2, \ldots, z_n)$, $|w\rangle = (w_1, w_2, \ldots, w_n)$, the *Hermitian scalar product* $\langle z|w\rangle$ (sometimes written $\langle z|w\rangle_H$) is defined by:

$$\langle z|w\rangle = z_1^* w_1 + z_2^* w_2 + \ldots \tag{2.109}$$

Clearly, if $|z\rangle$, $|w\rangle$ have only real components, $\langle z|w\rangle$ reduces to $\mathbf{z} \cdot \mathbf{w}$. The Hermitian scalar product is not necessarily symmetrical: $\langle z|w\rangle$ is not, in general, equal to $\langle w|z\rangle$. Instead:

$$\langle v|u\rangle = \langle u|v\rangle^*. \tag{2.110}$$

The *norm* or length $|u|$ of a vector $|u\rangle$ is defined by a formula analogous to Eqs. (2.8) and (2.9):

$$|u| = +\sqrt{\langle u|u\rangle}. \tag{2.111}$$

It is still, of course, a nonnegative real number. It is also easy to verify that $|u| = 0$, if and only if $|u\rangle = |0\rangle$. The Hermitian scalar product $\langle u|v\rangle$ is linear with respect to $|v\rangle$, but *antilinear* with respect to $\langle u|$, thus,

$$\langle u|cv\rangle = c\langle u|v\rangle, \quad \langle cu|v\rangle = c^*\langle u|v\rangle, \tag{2.112}$$

and

$$\langle u|v + w\rangle = \langle u|v\rangle + \langle u|w\rangle. \tag{2.113}$$

The notion of *orthonormal basis* can be extended. The vectors **u**, **v** form an orthonormal basis in $\mathbb{C}^2$ if they are of unit length and orthogonal:

$$\langle u|u\rangle = \langle v|v\rangle = 1, \quad \langle u|v\rangle = 0. \tag{2.114}$$

The definition of the scalar product, as integrals over complex-valued functions, must be generalized from the form of Eq. (2.83). Using Dirac notation and extending the integration over the whole real axis, we now write

$$\langle f|g\rangle = \int_{-\infty}^{+\infty} f(x)^* g(x)\, dx. \tag{2.115}$$

If $|f|$ and $|g|$ are finite, the scalar product $\langle f|g\rangle$ is also finite, as implied by the Cauchy–Schwarz inequality:

$$|f|\,|g| \geq |\langle f|g\rangle|. \tag{2.116}$$

In our toy Hilbert space $\mathbb{R}^2$, the Cauchy–Schwarz inequality follows simply from $|\cos\phi| \leq 1$ in Eq. (2.10).

The generalization of matrices to the complex field is straightforward. All the matrix formulas in Sect. 2.2 remain valid, with the real numbers replaced by complex numbers. A very important operation for complex $n \times n$ matrices, leading to the definition of the adjoint matrix, can defined as follows:

Definition 2.2 Given an operator A, represented in an orthonormal basis by the matrix A_{ij}, the operator whose matrix is obtained by taking the complex conjugate and interchanging rows and columns is called the *adjoint* of A, denoted by $A^\dagger$. Thus:

$$A_{ij}^\dagger = A_{ji}^* \quad (i, j = 1, 2, \dots, n). \tag{2.117}$$

The adjoint has the following properties:

$$\begin{aligned}
(AB)^\dagger &= B^\dagger A^\dagger, \\
(A^\dagger)^\dagger &= A, \\
(A + B)^\dagger &= A^\dagger + B^\dagger, \\
(cA)^\dagger &= c^* A^\dagger,
\end{aligned} \tag{2.118}$$

where c is a complex number. The *identity* is the operator I such that $I|x\rangle = |x\rangle$ for any $|x\rangle$. The matrix representing the identity has all elements along the main diagonal equal to 1, and 0 everywhere else (the same as for real matrices).

The adjoint operation allows us to move an operator from one side of a scalar product to the other, by virtue of the following Theorem:

Theorem 2.2 *For every pair of vectors* $|x\rangle$, $|y\rangle$ *and every operator A the following relations hold:*

$$\langle y|Ax\rangle = \langle A^\dagger y|\, x\rangle, \quad \langle Ay|x\rangle = \langle y|A^\dagger x\rangle. \tag{2.119}$$

The proof of the first equality follows from the sequence of operations:

$$\langle A^\dagger y | x \rangle = \langle x | A^\dagger y \rangle^* = \left(\sum_k x_k^* (A^\dagger y)_k \right)^* = \left(\sum_k x_k^* \sum_i A_{ki}^\dagger y_i \right)^* =$$

$$\left(\sum_k \sum_i x_k^* A_{ik}^* y_i \right)^* = \sum_{i,k} x_k A_{ik} y_i^* = \langle y | A x \rangle. \qquad (2.120)$$

The second equality follows from the first, since $(A^\dagger)^\dagger = A$. The relations (2.119) are easy to verify explicitly for the case $n = 2$.

Recall that rotations leave the lengths of real vectors invariant. What are the corresponding linear operators that leave the lengths of complex vectors in $\mathbb{C}^2$ or $\mathbb{C}^n$ invariant? The answer is *unitary* operators.

Definition 2.3 An operator U is called unitary if

$$U \, U^\dagger = U^\dagger U = I, \qquad (2.121)$$

where, as usual, I denotes the identity.

A unitary operation leaves invariant the scalar product of two vectors $|x\rangle$, $|y\rangle$, since, using (2.119), we have:

$$\langle Ux | Uy \rangle = \langle x | U^\dagger U y \rangle = \langle x | I y \rangle = \langle x | y \rangle. \qquad (2.122)$$

Thus the norm (or length) of a vector is left invariant by a unitary transformation. The correct generalization of ordinary rotations of vectors (which are generated by orthogonal matrices) are thus unitary operations on complex vectors. Our tentative postulate was that dynamical variables are represented by symmetric matrices; more generally, these should be *Hermitian* matrices. A matrix H_{ij} $(i, j = 1, 2 \ldots, n)$ is *Hermitian* (or *self-adjoint*) if $H = H^\dagger$, so that its elements are related by

$$H_{ij} = H_{ji}^* \ (i, j = 1, 2, \ldots, n). \qquad (2.123)$$

For example, if $n = 2$, H_{11} and H_{22} are real, and $H_{21} = H_{12}^*$. Thus, the matrix

$$\left\| \begin{array}{cc} 3 & 1+i \\ 1-i & 4 \end{array} \right\| \qquad (2.124)$$

is Hermitian. If H is both Hermitian and real, it is again simply a symmetric matrix.

Hermitian matrices admit an orthonormal basis of eigenvectors, just like symmetric matrices in the real case. Furthermore, the eigenvalues of an Hermitian matrix H are *real*; indeed, taking the scalar product of both sides of the equation

$$H |x\rangle = \lambda |x\rangle \qquad (2.125)$$

by the eigenvector $|x\rangle$ we obtain:

$$\langle x|H|x\rangle = \lambda\langle x|x\rangle, \quad \lambda = \frac{\langle x|H|x\rangle}{\langle x|x\rangle}. \tag{2.126}$$

and by (2.110), we have:

$$\langle x|H|x\rangle^* = \langle x|H^\dagger|x\rangle = \langle x|H|x\rangle. \tag{2.127}$$

Thus, both numerator and denominator of (2.126) are real, therefore the eigenvalue λ must be real. The matrix (2.124) has the eigenvectors $|u\rangle = \left(\frac{1+i}{\sqrt{3}}, -\frac{1}{\sqrt{3}}\right)$, $|v\rangle = \left(\frac{1+i}{2}, 1\right)$, thus

$$H|u\rangle = 2|u\rangle, \qquad H|v\rangle = 5|v\rangle, \qquad \langle u|v\rangle = 0. \tag{2.128}$$

Two Hermitian matrices A, B admit a common basis of eigenvectors, if and only if they *commute*, that is, if $[A, B] = 0$ or $A B = B A$. We omit the simple proof of these results for Hermitian matrices. We note that the fundamental interpretative postulate of QM (Postulate 1) is generalized with the simply substitution of "Hermitian" for "symmetric."

Observables are represented in QM by Hermitian matrices (more precisely, linear Hermitian operators), and states of a physical system by real or complex vectors. Of course, the case of two component vectors or spinors is the simplest (see Chap. 4). However, as noted above, the Hilbert space of realistic physical systems is usually infinite dimensional.

2.7 Coordinates and Momenta in Quantum Mechanics

We have seen in Chap. 1 that de Broglie's formula associates a "matter wave" with the rectilinear motion of a particle, with a wavelength $\lambda = h/mv = h/p$, where m is the mass, v the velocity and p the momentum of the particle. Accordingly, let us consider a very general instance of wave motion propagating in the x-direction. At a given instant of time, a periodic wave with wavelength λ might be represented by a function of the form

$$\psi(x) = f\left(\frac{2\pi x}{\lambda}\right), \tag{2.129}$$

where $f(\theta)$ is most often a sinusoidal function such as $\sin\theta$, $\cos\theta$, $e^{\pm i\theta}$, or some linear combination of these. Each of these is a periodic function, its value repeating every time its argument increases by 2π. This happens when x increases by one wavelength λ. The most useful form will turn out to be the complex exponential, which is related to the sine and cosine by Euler's formula $e^{i\theta} = \cos\theta + i\sin\theta$. We consider the wavefunction

$$\psi(x) = e^{i2\pi x/\lambda}, \tag{2.130}$$

apart from an arbitrary multiplicative constant. The wavelength λ of this complex-valued wavefunction can be replaced by h/p, where p is the particle momentum, in accordance with the de Broglie formula. Thus,

$$\psi(x) = e^{i2\pi px/h} = e^{ipx/\hbar}, \qquad (-\infty < x < \infty), \tag{2.131}$$

where $\hbar \equiv h/2\pi$. Since Planck's constant occurs in most formulas with the denominator 2π, this symbol, pronounced "aitch-bar," was introduced by Dirac in 1930.

Now that we have a mathematical representation of a matter wave, we should next try to find a "wave equation," a differential equation which the wavefunction satisfies. As a first step let us apply the operator $D = \frac{d}{dx}$ to Eq. (2.131). We find

$$\frac{d}{dx}\psi(x) = \frac{ip}{\hbar}\psi(x), \tag{2.132}$$

which can be rearranged to

$$-i\hbar\frac{d}{dx}\psi(x) = p\psi(x). \tag{2.133}$$

This can be recognized as an *eigenvalue equation* (see Eqs. 2.50 and 2.125) for the x-component of momentum p_x:

$$p_x\psi(x) = p\psi(x), \tag{2.134}$$

with the momentum operator evidently given by

$$p_x = -i\hbar\frac{d}{dx}. \tag{2.135}$$

This, incidentally, confirms our earlier speculation that the operator $D = \frac{d}{dx}$ is proportional to the velocity v (hence the momentum p) of a particle. In Dirac notation, the eigenvalue equation can be written

$$p_x|\psi\rangle = p\,|\psi\rangle. \tag{2.136}$$

Evidently, an eigenvalue $p = mv$ for a free particle, can be *any* real number: $-\infty < p < \infty$. This is a *continuous spectrum* of eigenvalues, in contrast to the energy levels of a bound atom or molecule, which was a distinguishing feature in the early development of quantum theory. Actually, highly excited states of atoms or molecules, in which ionization or dissociation has occurred, also show a continuum of energy eigenvalues. The momentum eigenfunctions $\psi(x)$ are complex-valued (except when $p = 0$). If $\psi_p(x) = e^{ipx/\hbar}$ and $\psi_{p'}(x) = e^{ip'x/\hbar}$ represent eigenstates with *different* eigenvalues, p and p', respectively, then the corresponding eigenfunctions are orthogonal. This can be shown by an intuitive (although not entirely mathematically rigorous) argument:

$$\langle p|p'\rangle = \int_{-\infty}^{+\infty} \psi_p(x)^* \psi_{p'}(x)\, dx = \int_{-\infty}^{+\infty} e^{i(p'-p)x/\hbar}\, dx =$$

$$\int_{-\infty}^{+\infty} \Big(\cos\big[(p'-p)x/\hbar\big] + i\, \sin\big[(p'-p)x/\hbar\big] \Big)\, dx = 0 \quad (p' \neq p). \quad (2.137)$$

In the last line, the infinite number of positive and negative contributions to the sine or the cosine integrals cancel each other out to give a result of zero.

The Hilbert space $\mathcal{L}^2(-\infty, +\infty)$ appropriate for QM is a set of *complex valued* functions $\psi(x)$ such that the following integral is finite:

$$\int_{-\infty}^{+\infty} \psi(x)^* \psi(x)\, dx = \int_{-\infty}^{+\infty} |\psi(x)|^2\, dx < \infty. \qquad (2.138)$$

An apparent disaster occurs when we try to evaluate Eq. (2.138) using a momentum eigenfunction (2.131). With $\psi(x) = e^{ipx/\hbar}$, the complex conjugate is $\psi(x)^* = e^{-ipx/\hbar}$. Thus $\psi(x)^* \psi(x) = |\psi(x)|^2 = 1$ and $\int_{-\infty}^{+\infty} 1\, dx = \infty$, violating the condition for a valid Hilbert space. There are several ways that we can talk our way out of this difficulty.

(1) We might limit our consideration to quantum systems with *bound states*, for which wavefunctions conforming to (2.138) can always be found. This could be done, for example, by replacing the infinite domain $-\infty < x < \infty$ by a finite interval $-a \leq x \leq a$. This excludes the free particle, Eq. (2.131), despite the fact that this system has been so fundamental in deriving some essential results in QM.

(2) We recognize that for a system in a momentum eigenstate, the probability density function $|\psi(x)|^2 = 1$ for *all* values of x, $-\infty < x < \infty$ (even beyond the bounds of the known Universe!). This is in accord with the uncertainty principle, since a precisely known momentum p implies a completely indefinite position x. More realistically, a free particle should be described by a *wavepacket*, which is a superposition of momentum eigenstates $\psi_p(x)$ of the form

$$\psi(x) = \int \phi(p)\psi_p(x)\, dp. \qquad (2.139)$$

Then the integral (2.138) converges, provided that $\int |\phi(p)|^2\, dp$ is finite.

(3) Hilbert space is redefined to accommodate continuous spectra and divergent integrals. Dirac himself was aware that "the bra and ket vectors that we now use form a more general space than a Hilbert space." A modern extension, known as

rigged Hilbert space[4] has the desired structure (the term "rigged" here implies "well-equipped and ready for action"). A "conventional" Hilbert space can accommodate a denumerably infinite number of basis vectors, labeled, for example, by $n = 1, 2, 3, \ldots$ But, in a rigged Hilbert space, the number of basis vectors can be *nondenumerably infinite*, labeled, perhaps by indices with a continuum of allowed values, such as ν, with $-\infty < \nu < \infty$.

Fortunately, we can carry on, using naive conventional Hilbert space, knowing that the mathematicians (however reluctantly) have us well covered regarding any inconsistencies or complications. As an illustrative example, consider the scalar product of eigenstates. For the discrete spectrum, we have

$$\langle m|n \rangle = \int_{-\infty}^{+\infty} \psi_m(x)^* \psi_n(x)\, dx = \delta_{m,n}, \tag{2.140}$$

where $\delta_{m,n}$ is the Kronecker delta, equal to 1 for $m = n$ and 0 for $m \neq n$. For eigenstates belonging to a continuous spectrum, we can write

$$\langle \mu|\nu \rangle = \int_{-\infty}^{+\infty} \psi_\mu(x)^* \psi_\nu(x)\, dx = \delta(\mu - \nu). \tag{2.141}$$

We have already seen that $\langle \mu|\nu \rangle = 0$ for $\mu \neq \nu$ (Eq. (2.137)). We have also found that $\langle \nu|\nu \rangle = \infty$ for $\mu = \nu$. But Dirac here introduced a special kind of infinity, as represented by the *delta function*, $\delta(\mu - \nu)$.

The Dirac delta function was intended as the continuum analog of the Kronecker delta. It is, however, *not* a true function in the mathematical sense, but rather a *generalized function* or *distribution*. The delta function was regarded with much disdain by mathematicians, until a rigorous theory was proposed by Laurent Schwartz in 1950.[5] It is defined as a hypothetical "function" such that $\delta(x - a) = 0$, if $x \neq a$, and $\delta(x - a) = \infty$, if $x = a$. However, the infinite value is somewhat special: an integral over that singular point is presumed to equal 1. Intuitively, the delta function can be pictured as the limit of a distribution with integrated area 1, of infinitesimal width but very large height, centered around the point $x = a$. The delta function is presumed to satisfy the integral relations: $\int_{-\infty}^{+\infty} \delta(x - a)\, dx = 1$ and $\int_{-\infty}^{+\infty} f(x)\delta(x - a)\, dx = f(a)$. To physicists, the delta function is invaluable for representing idealized objects such as point masses and point charges and for constructing Green's functions.

The eigenvalues and eigenvectors of the position operator q are rather tricky, but, fortunately, rarely needed. A particle *localized* around a point $x = a$ might be represented by a delta function: $\psi(x) = \delta(x-a)$. The relation $x\delta(x-a) = a\delta(x-a)$ can then be interpreted as an eigenvalue equation $x\psi(x) = a\psi(x)$. As a consequence

---

[4]R de la Madrid (2005), *The role of the rigged Hilbert space in Quantum Mechanics*, Eur J Phys 26:287–312.

[5]Schwartz L (1950–51) *Théorie des distributions*, Hermann, Paris. See also: Lighthill MJ (1958) *An Introduction to Fourier Analysis and Generalised Functions*. Cambridge University Press.

of the uncertainty principle, a particle with an exactly known value of x, has a totally undetermined value of p_x, the possible values being spread over $-\infty < p < \infty$.

The commutators of components of position and momentum are of central importance in the formalism of QM. Let us first evaluate $[x, p_x] = x\,p_x - p_x\,x$, where $p_x = -i\hbar\frac{d}{dx}$. These operators have meaning only when applied to a function of x, say $\phi(x)$. Now,

$$[x, p_x]\phi(x) = x(-i\hbar)\frac{d}{dx}\phi(x) - (-i\hbar)\frac{d}{dx}[x\phi(x)] = i\hbar\frac{dx}{dx}\phi(x) = i\hbar\phi(x),$$
(2.142)

in which we have taken the derivative of a product $x\psi(x)$ and simplified by cancellation. Since the function $\phi(x)$ is arbitrary, we can abstract the operator relation

$$[x, p_x] = i\hbar.$$
(2.143)

Obvious analogs of Eq. (2.135) for the y- and z-components of momentum are

$$p_y = -i\hbar\frac{d}{dy}, \qquad p_z = -i\hbar\frac{d}{dz}.$$
(2.144)

It is then simple to derive the analogous commutation relations:

$$[y, p_y] = i\hbar, \qquad [z, p_z] = i\hbar.$$
(2.145)

Since x commutes with p_z, etc., and the different components of position commute with one another, as do the different components of momentum, we can collect the entire set of commutation relations in the following compact form:

$$[q_i, q_j] = [p_i, p_j] = 0, \qquad [q_i, p_j] = i\hbar\delta_{i,j},$$
(2.146)

where $i, j = 1, 2, 3$ and $q_1 = x$, $p_1 = p_x$, etc.

2.8 Heisenberg Uncertainty Principle

The Heisenberg uncertainty principle is a fundamental consequence of the noncommutativity of position and momentum operators. Let us consider the one-dimensional case with x and p_x, which we write simply as p. The average, mean or *expectation variable* of a dynamical variable A in a quantum state $|\Psi\rangle$ is written

$$\overline{A} = \langle\Psi|A|\Psi\rangle.$$
(2.147)

The mean square deviation from the mean is then given by

$$(\Delta A)^2 = \langle \Psi | (A - \overline{A})^2 | \Psi \rangle, \tag{2.148}$$

where ΔA is the *root mean square*, which is designated as the *uncertainty* in A. Now define two functions

$$f = (x - \overline{x})\Psi, \quad \text{and} \quad g = i(p - \overline{p})\Psi. \tag{2.149}$$

We then find

$$\langle f | f \rangle = \langle (x - \overline{x})\Psi | (x - \overline{x})\Psi \rangle = \langle \Psi | (x - \overline{x})^2 | \Psi \rangle = (\Delta x)^2, \tag{2.150}$$

and, analogously,

$$\langle g | g \rangle = \langle i(p - \overline{p})\Psi | i(p - \overline{p})\Psi \rangle = \langle \Psi | (p - \overline{p})^2 | \Psi \rangle = (\Delta p)^2. \tag{2.151}$$

Next, let us evaluate:

$$\langle f | g \rangle + \langle g | f \rangle = \langle \Psi | (x - \overline{x})i(p - \overline{p}) - i(p - \overline{p})(x - \overline{x}) | \Psi \rangle. \tag{2.152}$$

After some cancelation, we find

$$\langle f | g \rangle + \langle g | f \rangle = i \langle \Psi | \left[x, p \right] | \Psi \rangle = -\hbar, \tag{2.153}$$

recalling the commutation relation $\left[x, p \right] = i\hbar$. From the Cauchy–Schwarz inequality, Eq. (2.116),

$$\sqrt{\langle f | f \rangle} \sqrt{\langle g | g \rangle} \geq \left| \langle f | g \rangle \right|. \tag{2.154}$$

Finally, we arrive at the Heisenberg uncertainty principle:

$$\Delta x \, \Delta p \geq \frac{\hbar}{2}. \tag{2.155}$$

This implies that exact values of a position variable and its conjugate momentum cannot be simultaneously known. Thus the trajectories of Bohr orbits are illusory: quantum behavior is not deterministic with regard to classical variables.

Chapter 3
The Schrödinger Equation

Abstract The Schrödinger equation is introduced and applied to some elementary problems: particle-in-a box, harmonic oscillator, angular momentum, and the hydrogen atom. The representation of eigenfunctions and eigenvalues is discussed, employing linear operators and matrices. The quantum theory of spin, as well as the Pauli exclusion principle, are described. These enable a theoretical understanding of atomic structure and the periodic table. The two disparate modes of time-dependence of a quantum system–unitary evolution and collapse of the wavefunction–are contrasted.

Keywords Schrödinger equation · spin · eigenfunctions and eigenvalues · atomic structure · evolution operator · collapse of the wavefunction

3.1 Heuristic Derivation

The Schrödinger equation, like Newton's equations of motion, expresses a fundamental law of nature and cannot, in the usual sense, be *derived*. The best we can do is to offer a heuristic argument to make the Schrödinger equation appear plausible. In Sect. 2.7, we sought a wave equation to describe de Broglie matter waves. We found there that a one-dimensional wave for a particle of wavelength λ, at a single instant of time, can be represented by a complex exponential $\psi(x) = \exp(2\pi i x/\lambda)$. We consider now the *time* dependence of the wave. At a fixed point x in space, the wave will also be sinusoidal in time, with a frequency ν. This can be represented by a sinusoidal function, which we again choose as a complex exponential, now, $\exp(-2\pi i \nu t)$. This is now periodic in the time $T = 1/2\pi\nu$, which is known as the *period* of the oscillation. The product of the x and t functions will evidently represent *both* the space and time variation of a matter wave[1]:

$$\Psi_0(x, t) = \exp\left(\frac{2\pi i x}{\lambda}\right) \times \exp(-2\pi i \nu t) = \exp\left[2\pi i \left(\frac{x}{\lambda} - \nu t\right)\right]. \tag{3.1}$$

---

[1] We choose the *negative* exponential for the time factor since the scalar product of the relativistic 4-vectors (ct, x_1, x_2, x_3) and $(E/c, p_1, p_2, p_3)$ gives $Et - \mathbf{p} \cdot \mathbf{x}$.

© Springer International Publishing AG 2017
G. Fano and S.M. Blinder, *Twenty-First Century Quantum Mechanics:*
Hilbert Space to Quantum Computers, UNITEXT for Physics,
DOI 10.1007/978-3-319-58732-5_3

The next step is to replace the wavelike variables λ and ν by particle variables, as given by the de Broglie and Planck–Einstein formulas: $\lambda = h/p$ and $\nu = E/h$, respectively. Introducing $\hbar = h/2\pi$ again, we obtain

$$\Psi_0(x, t) = \exp\left[\frac{i}{\hbar}(px - Et)\right], \tag{3.2}$$

as the wavefunction of a free particle (as indicated by the subscript 0). As we saw in the last chapter, taking the x-derivative of (3.2) gives

$$\frac{\partial}{\partial x}\Psi_0(x, t) = \frac{i}{\hbar}p \exp\left[\frac{i}{\hbar}(px - Et)\right] = \frac{i}{\hbar}p\,\Psi_0(x, t), \tag{3.3}$$

which suggests the definition of a momentum operator (p means p_x in one dimension)

$$p_x = -i\hbar\frac{\partial}{\partial x}. \tag{3.4}$$

Now that we have time dependence, we can also consider the time derivative

$$\frac{\partial}{\partial t}\Psi_0(x, t) = -\frac{i}{\hbar}E \exp\left[\frac{i}{\hbar}(px - Et)\right] = \frac{i}{\hbar}E\,\Psi_0(x, t). \tag{3.5}$$

This suggests an "energy operator"

$$E = i\hbar\frac{\partial}{\partial t}. \tag{3.6}$$

The energy of a particle of mass m in classical mechanics is equal to the sum of its kinetic and potential energies. In one dimension, as a function of position and velocity,

$$E = \frac{1}{2}mv^2 + V(x) \tag{3.7}$$

Energy expressed as a function of position and *momentum*, using $p = mv$, is more directly applicable in quantum mechanics. This function is called the *Hamiltonian*, $H(x, p)$. We can write

$$E = \frac{p^2}{2m} + V(x) = H(x, p). \tag{3.8}$$

Suppose we now promote (3.8) to an *operator* equation, replacing E and p (p_x) using (3.6) and (3.4), and apply it to a wavefunction $\Psi(x, t)$ (more general than a free particle). Note that the operator for the square of the momentum is given by:

$$p^2 = -i\hbar\frac{\partial}{\partial x}\left(-i\hbar\frac{\partial}{\partial x}\right) = -\hbar^2\frac{\partial^2}{\partial x^2}. \tag{3.9}$$

The potential energy function $V(x)$ can be treated as simply a *multiplicative operator*. Thereby we can transform (3.8) to a *wave equation*:

$$i\hbar\frac{\partial\Psi}{\partial t} = -\frac{\hbar^2}{2m}\frac{\partial^2\Psi}{\partial x^2} + V(x)\Psi. \tag{3.10}$$

We have thereby arrived at the one-dimensional *time-dependent Schrödinger equation* (TDSE).

The generalization for a particle in three dimensions is fairly straightforward. Note first that the momentum is a 3-component vector, so that $p^2 = p_x^2 + p_y^2 + p_z^2$ and the corresponding operator is given by

$$p^2 = -\hbar^2\left(\frac{\partial^2}{\partial x^2} + \frac{\partial^2}{\partial y^2} + \frac{\partial^2}{\partial z^2}\right) = -\hbar^2\nabla^2, \tag{3.11}$$

where ∇^2 is the *Laplacian operator*, also called "del-squared." The potential energy can now be a function of three variables, $V(x, y, z)$, written more compactly as $V(\mathbf{r})$. We can now extend the TDSE for a particle moving in three dimensions:

$$i\hbar\frac{\partial\Psi}{\partial t} = -\frac{\hbar^2}{2m}\nabla^2\Psi + V(\mathbf{r})\Psi. \tag{3.12}$$

In terms of the Hamiltonian operator H, the TDSE for a quantum system of any complexity can be expressed in the very compact form:

$$i\hbar\frac{\partial\Psi}{\partial t} = H\Psi. \tag{3.13}$$

The subsidiary conditions placed on the wavefunction Ψ are that it be finite, single-valued and continuous (collectively designated as *well behaved*). This famous and very fundamental equation is the starting point for most computational applications of quantum mechanics.

In many practical applications, including most of our subsequent development, we encounter a quantum system in a *stationary state*, meaning that it has no observable variation with time. If the potential energy V is independent of t, the Schrödinger equation (3.12) can be separated into two independent differential equations involving the space and time coordinates. This follows if the wavefunction can be written $\Psi(\mathbf{r}, t) = \psi(\mathbf{r})T(t)$. Putting this into (3.12), then dividing by $\psi(\mathbf{r})T(t)$, we obtain

$$\frac{i\hbar\frac{\partial T(t)}{\partial t}}{T(t)} = \frac{-\frac{\hbar^2}{2m}\nabla^2\psi(\mathbf{r}) + V(\mathbf{r})\psi(\mathbf{r})}{\psi(\mathbf{r})} = K. \tag{3.14}$$

The left-hand side of the equation is a function of t alone, while the right-hand side is a function only of $\mathbf{r}$. The only way for the equality to be valid is for both sides to be equal to a constant, say K. We now have two separate equations for the functions $T(t)$ and $\psi(\mathbf{r})$. The first of these can be written out as

$$i\hbar\frac{dT(t)}{dt} = K\,T(t). \tag{3.15}$$

This differential equation is readily solved to give $T(t) = \exp(-iKt/\hbar)$, which corresponds exactly to the time factor in $\Psi_0(x, t)$, Eq. (3.2), provided that the separation constant K is identified as the energy E. The equation for $\psi(\mathbf{r})$ in (3.14) then becomes the *time-independent Schrödinger equation* (TISE):

$$-\frac{\hbar^2}{2m}\nabla^2\psi(\mathbf{r}) + V(\mathbf{r})\psi(\mathbf{r}) = E\,\psi(\mathbf{r}). \tag{3.16}$$

The quantum mechanical *Hamiltonian operator* H, corresponding to the classical Hamiltonian function

$$H(\mathbf{r}, \mathbf{p}) = \frac{p^2}{2m} + V(\mathbf{r}) \tag{3.17}$$

(compare Eq. 3.8), can be written

$$H = -\frac{\hbar^2}{2m}\nabla^2 + V(\mathbf{r}). \tag{3.18}$$

Thus the TISE has the form of an *eigenvalue equation*

$$H\psi(\mathbf{r}) = E\,\psi(\mathbf{r}), \tag{3.19}$$

where the eigenvalues of the Hamiltonian evidently correspond to the allowed energies of the quantum system.

For a system in a stationary state, the complete wavefunction has the form $\Psi(\mathbf{r}, t) = \psi(\mathbf{r})e^{-iEt/\hbar}$, where the time dependence simply contributes a phase factor of modulus 1.

3.2   Particle in a Box

In this section and the next, we will consider solutions of the one-dimensional time-independent Schrödinger equation

$$-\frac{\hbar^2}{2m}\psi''(x) + V(x)\psi(x) = E\psi(x), \tag{3.20}$$

where, in the usual notation, $\psi'(x) = \frac{d\psi}{dx}$ and $\psi''(x) = \frac{d^2\psi}{dx^2}$. For a free particle, with $V(x) = 0$, the momentum eigenfunctions $\psi_0(x) = \exp(ipx/\hbar)$ are solutions to (3.20). The corresponding energy eigenvalues are $E = p^2/2m$, just as in classical mechanics. Since $-\infty < p < \infty$, the allowed energy values belong to a continuum, with $0 \leq E < \infty$.

The *particle in a box* is the simplest nontrivial application of the Schrödinger equation, but one which illustrates many of the fundamental concepts of quantum mechanics. Assume that a particle can move freely between two endpoints $x = 0$ and $x = a$, but cannot penetrate either end. This can be represented by a potential energy

$$V(x) = \begin{cases} 0 & \text{if} \quad 0 \le x \le a \\ \infty & \text{if} \quad x < 0 \text{ or } x > a \end{cases} \tag{3.21}$$

Infinite potential energy constitutes an impenetrable barrier. The particle is thus bound in a *potential well*, as represented by the dark lines in Fig. 3.1.

Since the particle cannot penetrate beyond $x = 0$ or $x = a$, we must have $\psi(x) = 0$ for $x < 0$ and $x > a$. By the requirement that the wavefunction be continuous, it must be true as well that

$$\psi(0) = 0 \quad \text{and} \quad \psi(a) = 0, \tag{3.22}$$

which constitutes a pair of boundary conditions on the wavefunction within the box. Inside the box, $V(x) = 0$, so the Schrödinger equation reduces to the free particle form. It is convenient to lump together the constants and write

$$\psi''(x) + k^2 \psi(x) = 0, \tag{3.23}$$

where

$$k^2 = \frac{2mE}{\hbar^2} \tag{3.24}$$

The general solution to Eq. (3.23) can be written

$$\psi(x) = A \sin kx + B \cos kx, \tag{3.25}$$

where A and B are constants to be determined by the boundary conditions (3.22). By the first condition, we find $\psi(0) = A \sin 0 + B \cos 0 = B = 0$, which reduces the solution to

$$\psi(x) = A \sin kx. \tag{3.26}$$

The second boundary condition at $x = a$ then implies $\psi(a) = A \sin ka = 0$. It is assumed that $A \ne 0$, for otherwise $\psi(x)$ would be zero everywhere and the particle would disappear. The condition that $\sin kx = 0$ implies that

$$ka = n\pi, \tag{3.27}$$

where n is a positive integer ($n = 0$ again gives $\psi(x) = 0$). Eliminating k between (3.27) and (3.24), we obtain

$$E_n = \frac{\hbar^2 k^2}{2m} = \frac{\hbar^2 \pi^2 n^2}{2ma^2} = \frac{h^2}{8ma^2} n^2 \tag{3.28}$$

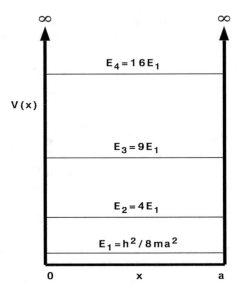

Fig. 3.1 Potential well and lowest energy levels for particle in a box

These are the only values of the energy which allow solution of the Schrödinger equation (3.23) consistent with the boundary conditions (3.22). The integer n, called a *quantum number*, is appended as a subscript on E to label the allowed energy levels. Figure 3.1 shows an energy-level diagram for the particle in a box. The occurrence of discrete or quantized energy levels is characteristic of a bound system, that is, one confined to a finite region in space. For the free particle, the absence of confinement gave rise to an energy continuum. Note that, in both cases, the number of energy levels is infinite—denumerably infinite for the particle in a box but nondenumerably infinite for the free particle.

The particle in a box assumes its lowest possible energy when $n = 1$, namely

$$E_1 = \frac{h^2}{8ma^2} \tag{3.29}$$

The state of lowest energy for a quantum system is termed its *ground state*. An interesting point is that $E_1 > 0$, whereas the corresponding classical system would have a minimum energy of zero. This is a recurrent phenomenon in quantum mechanics. The residual energy of the ground state, that is, the energy in excess of the classical minimum, is known as *zero point energy*. In effect, the kinetic energy, hence the momentum, of a bound particle cannot be reduced to zero. The minimum value of momentum is found by equating E_1 to $p^2/2m$, giving $p_{min} = \pm h/2a$. This can be expressed as an uncertainty in momentum given by $\Delta p \approx h/a$. Coupling this with the uncertainty in position, $\Delta x \approx a$, the size of the box, we can write $\Delta x \Delta p \approx h$, which is in qualitative accord with the Heisenberg uncertainty principle.

The particle-in-a-box eigenfunctions are given by Eq. (3.26), with $k = n\pi/a$, in accordance with (3.27). This gives

$$\psi_n(x) = A \sin\left(\frac{n\pi x}{a}\right), \qquad n = 1, 2, 3, \ldots \tag{3.30}$$

The eigenfunctions, like the eigenvalues, can be labeled by the quantum number n. The constant A, thus far arbitrary, can be adjusted so that $\psi_n(x)$ is normalized. The normalization condition is, in this case,

$$\int_0^a [\psi_n(x)]^2\, dx = 1, \tag{3.31}$$

the integration running over the domain of the particle, $0 \le x \le a$. Substituting (3.30) into (3.31) and integrating after the substitution $\theta = n\pi x/a$

$$A^2 \int_0^a \sin^2\left(\frac{n\pi x}{a}\right) dx = A^2 \frac{a}{n\pi} \int_0^{n\pi} \sin^2\theta\, d\theta = A^2 \frac{a}{2} = 1. \tag{3.32}$$

We have made the substitution $\theta = n\pi x/a$ and used the fact that the average value of $\sin^2\theta$ over an integral number of half wavelengths equals $\frac{1}{2}$. We can thus identify the normalization constant $A = (2/a)^{1/2}$, for all values of n.

Finally we can write the normalized eigenfunctions:

$$\psi_n(x) = \left(\frac{2}{a}\right)^{1/2} \sin\left(\frac{n\pi x}{a}\right), \qquad n = 1, 2, 3, \ldots \tag{3.33}$$

The first few eigenfunctions $\psi_n(x)$ and the corresponding probability distributions $\rho_n(x) = [\psi_n(x)]^2$ are plotted in Fig. 3.2. There is a close analogy between the states of this quantum system and the modes of vibration of a violin string. The patterns of standing waves on the string are, in fact, identical in form with particle-in-a-box wavefunctions, as shown on the left side of Fig. 3.2. A noteworthy feature of particle-in-a-box quantum states is the occurrence of *nodes*. These are points, other than the two end points (which are fixed by the boundary conditions), at which the wavefunction vanishes. At a node there is exactly zero probability of finding the particle. The nth quantum state has, in fact, $n - 1$ nodes. It is generally true that the number of nodes increases with the energy of a quantum state, which can be rationalized by the following qualitative argument. As the number of nodes increases, so does the number and steepness of the "wiggles" in the wavefunction. It is like skiing down a slalom course. Accordingly, the average curvature, given by the second derivative, must increase. But the second derivative is proportional to the kinetic energy operator. Therefore, the more nodes, the higher the energy. This will prove to be an invaluable guide in more complex quantum systems. Another important property is the mutual orthogonality of two different eigenfunctions. It is easy to see from Fig. 3.3 that the integral

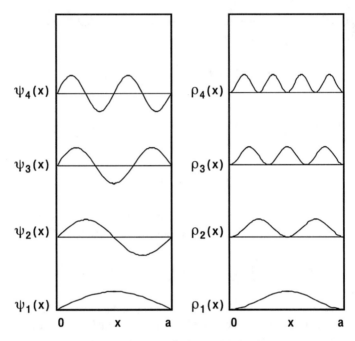

Fig. 3.2 Eigenfunctions and probability densities for particle in a box

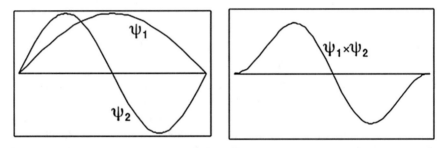

Fig. 3.3 Product of $n = 1$ and $n = 2$ eigenfunctions

$$\int_0^a \psi_2(x)\psi_1(x)\,dx = 0 \tag{3.34}$$

As we have shown in Chap. 2, this is a general result for quantum mechanical eigen-functions. The normalization, together with the orthogonality of the eigenfunctions, can be combined into a single relationship

$$\int_0^a \psi_m(x)\psi_n(x)\,dx = \delta_{mn}. \tag{3.35}$$

A set of functions $\{\psi_n\}$ which obeys Eq. (3.35) is called *orthonormal*.

3.3   The Harmonic Oscillator

Oscillatory motion is frequently encountered in everyday experience. Among numerous examples are pendulums, springs slightly displaced from equilibrium, strings of a musical instrument, etc. In all these cases there is a restoring force F proportional to the displacement x of the object; thus $F = -kx$, which is known as *Hooke's law*. Newton's law $F = ma$ becomes:

$$m\frac{d^2x}{dt^2} = -kx. \tag{3.36}$$

This has the general solution:

$$x(t) = A\cos\omega t + B\sin\omega t, \tag{3.37}$$

where $\omega = \sqrt{\frac{k}{m}}$. Classically, the harmonic oscillator has just this one natural frequency ω.

With a potential energy $V(x) = \frac{1}{2}kx^2$, the force is given by $F = -\frac{dV}{dx}$ and the total energy E of a classical oscillator is

$$E = \frac{1}{2m}p^2 + V(x) = \frac{1}{2m}p^2 + \frac{1}{2}kx^2 = \frac{1}{2m}\left(p^2 + m^2\omega^2x^2\right). \tag{3.38}$$

The corresponding quantum problem begins with a Hamiltonian operator of the same form:

$$H = \frac{1}{2m}\left(p^2 + m^2\omega^2x^2\right). \tag{3.39}$$

An elegant method to find the eigenfunctions and eigenvalues of H makes use of two new operators, a and a^+, defined by

$$a = \frac{1}{\sqrt{2m\hbar\omega}}(m\omega x + ip), \qquad a^+ = \frac{1}{\sqrt{2m\hbar\omega}}(m\omega x - ip). \tag{3.40}$$

These are called "raising" and "lowering" operators: ordering the eigenstates ψ_n, or $|n\rangle$ in Dirac notation, in order of increasing energy E_n, they transform $|n\rangle$ into $|n+1\rangle$ or $|n-1\rangle$ (up to a constant). In a different context they have more dramatic (magical?) names. a^+ is called a "creation operator" and a is called an "annihilation" operator. We will explain later the reason for these designations. The product of a^+ and a works out to

$$a^+a = \frac{1}{2m\hbar\omega}(m\omega x - ip)(m\omega x + ip) = \frac{1}{2m\hbar\omega}[m^2\omega^2x^2 + p^2 + im\omega(xp - px)]. \tag{3.41}$$

Using the commutation relation $[x, p] = xp - px = i\hbar$, we find,

$$a^+a = \frac{H}{\hbar\omega} + \frac{i(i\hbar)I}{2\hbar} = \frac{H}{\hbar\omega} - \frac{1}{2}I. \tag{3.42}$$

Thus we can write the Hamiltonian H, in terms of a, a^+, in the form

$$H = \hbar\omega \left(a^+a + \frac{1}{2}I \right). \tag{3.43}$$

Some relevant relations involving the raising and lowering operators are the following. The fundamental commutation relation for a and a^+:

$$[a, a^+] = aa^+ - a^+a = I. \tag{3.44}$$

The corresponding *anticommutation* relation is

$$\{a, a^+\} = aa^+ + a^+a = \frac{2H}{\hbar\omega}, \tag{3.45}$$

with an *anticommutator* defined by $\{F, G\} = FG + GF$. We obtain thereby:

$$[a, H] = aH - Ha = \hbar\omega(aa^+a - a^+aa) = \hbar\omega(aa^+ - a^+a)a = \hbar\omega a \tag{3.46}$$

and

$$[a^+, H] = a^+H - Ha^+ = \hbar\omega(a^+a^+a - a^+aa^+) = \hbar\omega a^+(a^+a - aa^+) = -\hbar\omega a^+. \tag{3.47}$$

The following theorem summarizes the structure of the eigenvalues and eigenfunctions of the quantum harmonic oscillator:

Theorem 3.1 *If $|n\rangle$ is an eigenfunction of H corresponding to the eigenvalue E_n, then $a|n\rangle$ either equals zero or else is an eigenfunction of H corresponding to the eigenvalue $E_n - \hbar\omega$. Conversely, $a^+|n\rangle$ is an eigenfunction of H corresponding to the eigenvalue $E_n + \hbar\omega$. The eigenvalues of H are the numbers $E_n = (n + \frac{1}{2})\hbar\omega$, $n = 0, 1, 2, 3, \ldots$.*

Proof From the eigenvalue equation:

$$H|n\rangle = E_n|n\rangle \tag{3.48}$$

and Eqs. (3.46), (3.47) we obtain:

$$\begin{aligned} (aH - Ha)|n\rangle = (aE_n - Ha)|n\rangle = \hbar\omega a|n\rangle, \\ (a^+H - Ha^+)|n\rangle = (a^+E_n - Ha^+)|n\rangle = -\hbar\omega a^+|n\rangle. \end{aligned} \tag{3.49}$$

Therefore:

$$Ha \, |n\rangle = (E_n - \hbar\omega)a \, |n\rangle, \qquad Ha^+|n\rangle = (E_n + \hbar\omega)a^+|n\rangle. \qquad (3.50)$$

Thus, if E_n is an eigenvalue, then $E_n - \hbar\omega$ and $E_n + \hbar\omega$ are also eigenvalues, unless one of the functions $a|n\rangle$ or $a^+|n\rangle$ vanishes. But $a^+|n\rangle$ never vanishes, since

$$|a^+|n\rangle|^2 = \langle a^+n|a^+n\rangle = \langle n|aa^+n\rangle = \langle n|(I + a^+a)|n\rangle = \langle n|n\rangle + \langle an|an\rangle \geq \langle n|n\rangle > 0.$$

By contrast, $a|n\rangle$ *can* equal the null vector. Assuming this is the case, let us solve the equation $a|\psi_0\rangle = 0$. Neglecting a numerical factor:

$$(m\omega x + ip)\psi_0(x) = m\omega x \psi_0(x) + \hbar\frac{d\psi_0}{dx} = 0 \qquad (3.51)$$

This simplifies to $\frac{d \log \psi_0}{dx} = -\frac{m\omega}{\hbar} x$, and integrating, $\log \psi_0(x) = -\frac{m\omega}{2\hbar} x^2 + \text{const}$. Thus

$$\psi_0(x) = A \exp\left(-\frac{m\omega}{2\hbar}x^2\right) \qquad (3.52)$$

where A is a normalization constant. Using the well-known Gaussian integral $\int_{-\infty}^{+\infty} e^{-\alpha x^2} dx = \left(\frac{\pi}{\alpha}\right)^{\frac{1}{2}}$, we can find A by imposing the condition $\int_{-\infty}^{+\infty} \psi_0(x)^2 \, dx = 1$. The result is $A = \left(\frac{2m\omega}{h}\right)^{\frac{1}{4}}$. Since $H = \hbar\omega(a^+a + \frac{1}{2}I)$, we obtain finally:

$$H \, |\psi_0\rangle = \frac{\hbar\omega}{2}|\psi_0\rangle, \qquad (3.53)$$

where ψ_0 is the *ground state* $|0\rangle$ and $E_0 = \frac{1}{2}\hbar\omega$ is the lowest energy level. Indeed, this must be the lowest possible energy, since, applying the lowering operator a, we obtain the null vector. The higher energy levels are $E_1 = \frac{3}{2}\hbar\omega$, $E_2 = \frac{5}{2}\hbar\omega$, ..., $E_n = (n + \frac{1}{2})\hbar\omega$ (see Fig. 3.4). The corresponding eigenfunctions $|n\rangle$ are easily obtained by repeatedly applying the operator a^+. Clearly then:

$$|n\rangle = A_n(a^+)^n|0\rangle, \qquad (3.54)$$

where A_n is a normalization constant. This completes the proof of the Theorem. Figure 3.5 shows a sketch of the ground-state wavefunction.

Remarkably, the quantization of the electromagnetic field in free space requires operators very similar to a and a^+. The underlying reason is that the Hamiltonian for electromagnetic radiation can also be written as a quadratic form in the operators $a_{\mathbf{k}}$, $a_{\mathbf{k}}^+$ where $\mathbf{k}$ denotes a three-dimensional momentum. We will not go into any more detail here, but note only that $a_{\mathbf{k}}^+$ sends a state $|\mathbf{k}_1\rangle$ into a state $|\mathbf{k}_2\rangle$, as a photon of momentum $\mathbf{k}$ is absorbed, and conversely, $a_{\mathbf{k}}$ sends $|\mathbf{k}_2\rangle$ back into $|\mathbf{k}_1\rangle$,

Fig. 3.4 Energy levels of the
quantum harmonic oscillator

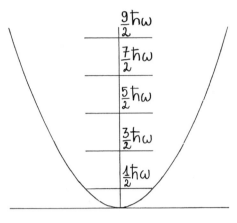

Fig. 3.5 Ground-state
wavefunction for the
harmonic oscillator

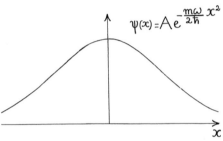

as a photon is emitted. Accordingly, $a_\mathbf{k}^+$, $a_\mathbf{k}$ are designated, respectively, as creation
and annihilation operators for light quanta.

3.4   Angular Momentum

3.4.1   *Particle in a Ring*

As a preliminary, let us consider first a *particle in a ring*, a variant of the one-
dimensional particle-in-a-box problem in which the x-axis is bent into a ring of
radius R. We can write the same Schrödinger equation

$$-\frac{\hbar^2}{2M}\frac{d^2\psi(x)}{dx^2} = E\psi(x), \tag{3.55}$$

where the particle mass is now designated M. There are no boundary conditions
in this case since the x-axis closes upon itself. A more appropriate independent
variable for this problem is the angular position on the ring given by, $\phi = x/R$. The
Schrödinger equation would then read

$$-\frac{\hbar^2}{2MR^2}\frac{d^2\psi(\phi)}{d\phi^2} = E\psi(\phi). \tag{3.56}$$

The kinetic energy of a body rotating in the xy-plane can be expressed as

$$E = \frac{L_z^2}{2I}, \tag{3.57}$$

where $I = MR^2$ is the moment of inertia and L_z, the z-component of angular momentum. (Since $\mathbf{L} = \mathbf{r} \times \mathbf{p}$, if $\mathbf{r}$ and $\mathbf{p}$ lie in the x–y plane, $\mathbf{L}$ points in the z-direction.) The structure of Eq. (3.56) suggests that this angular momentum operator is given by

$$L_z = -i\hbar\frac{d}{d\phi} \tag{3.58}$$

This result will follow from a more general derivation later. The Schrödinger equation (3.56) can now be written more compactly as

$$\psi''(\phi) + m^2\psi(\phi) = 0 \tag{3.59}$$

where

$$m^2 \equiv 2IE/\hbar^2 \tag{3.60}$$

(Please do *not* confuse this symbol m with mass!) Possible solutions to (3.56) are

$$\psi(\phi) = \text{const } e^{\pm im\phi} \tag{3.61}$$

In order for this wavefunction to be physically acceptable, it must be *single-valued*. Since ϕ increased by any multiple of 2π represents the same point on the ring, we must have $\psi(\phi + 2\pi) = \psi(\phi)$, and therefore

$$e^{im(\phi+2\pi)} = e^{im\phi}. \tag{3.62}$$

This requires that $e^{2\pi im} = 1$, which is true only is m is an integer: $m = 0$, ± 1, ± 2, ... Using (3.60), this gives the quantized energy values

$$E_m = \frac{\hbar^2}{2I}m^2. \tag{3.63}$$

Thus, in contrast to the particle-in-a-box, the eigenfunctions corresponding to $+m$ and $-m$ in Eq. (3.61) are linearly independent, so both must be included. Therefore all eigenvalues, except E_0, are twofold (doubly) degenerate. The eigenfunctions can therefore be written simply as const $e^{im\phi}$, with m running over *all* integer values. The normalized eigenfunctions are

$$\psi_m(\phi) = \frac{1}{\sqrt{2\pi}} \, e^{im\phi}, \qquad m = 0, \pm 1, \pm 2, \dots \tag{3.64}$$

and can be verified to satisfy the complex generalization of the normalization condition

$$\int_0^{2\pi} \psi_m(\phi)^* \, \psi_m(\phi) \, d\phi = 1 \tag{3.65}$$

with the complex conjugate $\psi_m(\phi)^* = (2\pi)^{-1/2} \, e^{-im\phi}$. It is easy to show that the functions $\psi_m(\phi)$ and $\psi_{m'}(\phi)$ are orthogonal for $m \neq m'$. We thus have the orthonormality relations

$$\int_0^{2\pi} \psi_m(\phi)^* \, \psi_{m'}(\phi) \, d\phi = \delta_{m,m'}. \tag{3.66}$$

The solutions (3.64) are also eigenfunctions of the angular momentum operator (3.58), with

$$L_z \psi_m(\phi) = m\hbar \, \psi_m(\phi), \quad m = 0, \pm 1, \pm 2, \dots \tag{3.67}$$

This is an instance of a fundamental result in quantum mechanics, that any measured component of orbital angular momentum is restricted to integral multiples of $\hbar$. The formulas of the Bohr theory of the hydrogen atom (Chap. 1) can be derived from this assumption alone.

3.4.2 Angular Momentum Operators

In classical mechanics a particle at point $\mathbf{r} = (x, y, z)$ with momentum $\mathbf{p} = (p_x, p_y, p_z)$ has an orbital angular momentum $\mathbf{L} = (L_x, L_y, L_z)$ with respect to the origin $(0, 0, 0)$ given by $\mathbf{L} = \mathbf{r} \times \mathbf{p}$. The Cartesian components of $\mathbf{L}$ can thereby be written

$$L_x = yp_z - zp_y, \qquad L_y = zp_x - xp_z, \qquad L_z = xp_y - yp_x. \tag{3.68}$$

Recall the linear momentum operators in quantum mechanics:

$$p_x = -i\hbar \frac{\partial}{\partial x}, \qquad p_y = -i\hbar \frac{\partial}{\partial y}, \qquad p_z = -i\hbar \frac{\partial}{\partial z}. \tag{3.69}$$

When applied to a wave function $\psi(x, y, z)$, the expression $yp_z\psi$ implies that we must first apply the operator p_z to ψ and then multiply the result by y, etc. We thereby obtain the operators for the 3 components of angular momentum

$$L_x = -i\hbar \left(y\frac{\partial}{\partial z} - z\frac{\partial}{\partial y} \right),$$
$$L_y = -i\hbar \left(z\frac{\partial}{\partial x} - x\frac{\partial}{\partial z} \right), \tag{3.70}$$
$$L_z = -i\hbar \left(x\frac{\partial}{\partial y} - y\frac{\partial}{\partial x} \right).$$

We have already found an alternative expression $L_z = -i\hbar\frac{\partial}{\partial\phi}$, namely, Eq. (3.58), in terms of the angle ϕ, which is actually one of the spherical polar coordinates (r, θ, ϕ) in which $\mathbf{r}$ can be expressed. The eigenvalues of L_z were found to be equal to $m\hbar$ $(m = 0, \pm 1, \pm 2, \dots)$. It would then be expected that the eigenvalues of L_x, L_y are likewise integer multiples of $\hbar$.

The components of the angular momentum obey the following commutation relations[2]

$$[L_y, L_z] = L_y L_z - L_z L_y = i\hbar L_x,$$
$$[L_z, L_x] = L_z L_x - L_x L_z = i\hbar L_y, \tag{3.71}$$
$$[L_x, L_y] = L_x L_y - L_y L_x = i\hbar L_z.$$

Since no two of the angular momentum components L_x, L_y, L_z commute, it follows that one cannot know the eigenvalues of more than one component (with one exception: when the total angular momentum equals 0). Proof of the commutation relations follows from the sequence of steps:

$$[L_x, L_y] = [yp_z - zp_y, zp_x - xp_z] = y[p_z, z]p_x + [z, p_z]p_y x =$$
$$-i\hbar[yp_x - xp_y] = i\hbar L_z, \tag{3.72}$$

with analogs for the other two commutators. By contrast, different components of coordinates or momenta have been found to be mutually commutative (for example $p_y x = xp_y$, $p_x p_y = p_y p_x$, etc.).

3.4.3 Eigenvalues and Eigenfunctions

Another important set of commutation relations involves the square of the total angular momentum, $L^2 = L_x^2 + L_y^2 + L_z^2$, for example,

$$[L^2, L_z] = [L_x^2, L_z] + [L_y^2, L_z] + [L_z^2, L_z] =$$
$$[L_x, L_z]L_x + L_x[L_x, L_z] + [L_y, L_z]L_y + L_y[L_y, L_z] = \tag{3.73}$$
$$i\hbar(-L_x L_y - L_y L_x + L_y L_x + L_x L_y) = 0,$$

---

[2] An interesting mnemonic is the symbolic vector formula: $\mathbf{L} \times \mathbf{L} = i\hbar\mathbf{L}$. The Cartesian components then give the three commutation relations.

and analogously for L_x and L_y. We have used the commutator identity $[X^2, Y] = [X, Y]X + X[X, Y]$. Thus, the square of the total angular momentum commutes with each of its components:

$$[L^2, L_x] = [L^2, L_y] = [L^2, L_z] = 0. \tag{3.74}$$

It follows, therefore, that there exist simultaneous eigenstates of L^2 and one component, generally chosen as L_z. Let us denote the corresponding eigenvectors by $|l, m\rangle$, and write the eigenvalue equations

$$L_z|l, m\rangle = m\hbar|l, m\rangle \quad \text{and} \quad L^2|l, m\rangle = \lambda\hbar^2|l, m\rangle \tag{3.75}$$

with λ remaining to be determined.

At this point we introduce the operators

$$L^+ = L_x + iL_y, \qquad L^- = L_x - iL_y, \tag{3.76}$$

which, we will see in a moment, are *raising* and *lowering* operators for the eigenvalues of L_z. It is easy to see that L^2 commutes with L^+, since it commutes with both L_x and L_y. Therefore

$$L^2L^+|l, m\rangle = L^+L^2|l, m\rangle = \lambda\hbar^2 L^+|l, m\rangle, \tag{3.77}$$

using the eigenvalue equation for L^2, Eq. (3.75). We can then write: $L^2(L^+|l, m\rangle) = \lambda\hbar^2(L^+|l, m\rangle)$, which shows that $L^+|l, m\rangle$ is an eigenvector of L^2 with the *same eigenvalue* $\lambda\hbar^2$ as $|l, m\rangle$. Thus, the operator L^+ applied to $|l, m\rangle$ does *not* change the *magnitude* of the angular momentum.

Consider now the commutator $[L_z, L^+]$

$$[L_z, L^+] = [L_z, L_x] + i[L_z, L_y] = i\hbar L_y + i(-i\hbar)L_x = \hbar(iL_y + L_x) = \hbar L^+. \tag{3.78}$$

Operating on the eigenvector $|l, m\rangle$:

$$\begin{aligned}
[L_z, L^+]|l, m\rangle &= L_zL^+|l, m\rangle - L^+L_z|l, m\rangle = \hbar L^+|l, m\rangle, \\
L_z(L^+|l, m\rangle) - m\hbar(L^+|l, m\rangle) &= \hbar(L^+|l, m\rangle), \\
L_z(L^+|l, m\rangle) &= (m + 1)\hbar(L^+|l, m\rangle).
\end{aligned} \tag{3.79}$$

Evidently, $L^+|l, m\rangle$ is an eigenvector of L_z with the eigenvalue $(m + 1)\hbar$, thus the designation of L^+ as a *raising operator*, meaning that

$$L^+|l, m\rangle = \text{const}|l, m + 1\rangle. \tag{3.80}$$

Analogously, L^- is a *lowering operator*, with

$$L^-|l, m\rangle = \text{const}|l, m - 1\rangle. \tag{3.81}$$

Suppose that the quantum number l is now defined as the *maximum* allowed value of m for a given eigenvalue $\lambda\hbar^2$ of L^2. (Correspondingly, $-l$ would be the *minimum* value of m.) Since there is no higher possible value of m, the raising operator on $|l, l\rangle$ should annihilate the vector:

$$L^+|l, l\rangle = 0. \tag{3.82}$$

Now here is an identity you would never ordinarily think of:

$$L^-L^+ = L_x^2 + L_y^2 + i(L_xL_y - L_yL_x) = L^2 - L_z^2 + i(i\hbar)L_z. \tag{3.83}$$

Thus

$$L^2 = L^-L^+ + L_z^2 + \hbar L_z. \tag{3.84}$$

Applying this to $|l, l\rangle$

$$L^2|l, l\rangle = L^-L^+|l, l\rangle + L_z^2|l, l\rangle + \hbar L_z|l, l\rangle, \tag{3.85}$$

giving

$$\lambda\hbar^2|l, l\rangle = 0 + l^2\hbar^2|l, l\rangle + l\hbar^2|l, l\rangle. \tag{3.86}$$

Finally, we can identify

$$\lambda = l^2 + l = l(l + 1). \tag{3.87}$$

The eigenvalues and eigenvectors for orbital angular momentum can now be summarized:

$$\begin{aligned} L^2|l, m\rangle &= l(l + 1)\hbar^2|l, m\rangle, \quad l = 0, 1, 2, \ldots \\ L_z|l, m\rangle &= m\hbar|l, m\rangle, \quad m = 0, \pm1, \pm2, \ldots, \pm l. \end{aligned} \tag{3.88}$$

Incidentally, l is usually denoted the *angular momentum quantum number*, since it determines the magnitude of the angular momentum, while m is called the *magnetic quantum number*, since it gives the component of angular momentum along an external magnetic field.

The magnitude of the total angular momentum $|\mathbf{L}| = \sqrt{l(l + 1)}\hbar$ is greater than its maximum observable component in any direction, namely $l\hbar$. The quantum mechanical behavior of the angular momentum and its components can be represented by a vector model, as shown in Fig. 3.6. The angular momentum vector $\mathbf{L}$ can be pictured as precessing about the z-axis, with its z-component L_z constant. The components L_x and L_y fluctuate in the course of precession, mirroring the fact that the system is not in an eigenstate of either, as implied by the commutation relations. There are $2l + 1$ possible values for L_z, with eigenvalues $m\hbar$ ($m = 0, \pm1, \pm2, \ldots, \pm l$), equally spaced between $+l\hbar$ and $-l\hbar$. This discreteness in the allowed directions of the angular momentum vector is called *space quantization*.

Fig. 3.6 Vector model for
angular momentum, showing
the case $l = 2$

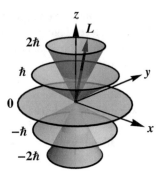

3.4.4   Matrix Representations

In Sect. 2.2, it was shown that operators and eigenvectors can be represented by
matrices. We will illustrate for the states of a system with $l = 1$. There are three
eigenstates, with eigenvalues $m = 1, 0, -1$. These can evidently be represented by
three-dimensional column vectors, with

$$[1, 1\rangle = \begin{Vmatrix} 1 \\ 0 \\ 0 \end{Vmatrix}, \qquad |1, 0\rangle = \begin{Vmatrix} 0 \\ 1 \\ 0 \end{Vmatrix}, \qquad |1, -1\rangle = \begin{Vmatrix} 0 \\ 0 \\ 1 \end{Vmatrix}. \qquad (3.89)$$

In a representation in which L_z is diagonal, the three angular momentum operators
can be represented by the matrices:

$$L_x = \hbar \begin{Vmatrix} 0 & \frac{1}{\sqrt{2}} & 0 \\ \frac{1}{\sqrt{2}} & 0 & \frac{1}{\sqrt{2}} \\ 0 & \frac{1}{\sqrt{2}} & 0 \end{Vmatrix}, \quad L_y = \hbar \begin{Vmatrix} 0 & -\frac{i}{\sqrt{2}} & 0 \\ \frac{i}{\sqrt{2}} & 0 & -\frac{i}{\sqrt{2}} \\ 0 & \frac{i}{\sqrt{2}} & 0 \end{Vmatrix}, \quad L_z = \hbar \begin{Vmatrix} 1 & 0 & 0 \\ 0 & 0 & 0 \\ 0 & 0 & -1 \end{Vmatrix},$$
$$(3.90)$$

and also

$$L^2 = \hbar^2 \begin{Vmatrix} 2 & 0 & 0 \\ 0 & 2 & 0 \\ 0 & 0 & 2 \end{Vmatrix}. \qquad (3.91)$$

You can verify that these matrices satisfy the commutation relations and the eigenvalue equations. Note also that all of these matrices are Hermitian ($L = L^\dagger$), with
$L_{i,j} = L^*_{j,i}$, as required by the postulates of quantum mechanics for all observables.

3.4.5   Electron Spin

We consider briefly the angular momentum associated with electron spin. This will
be indispensable in our later work on Bell's theorem and quantum computers. Many

atomic spectral lines appear, under sufficiently high resolution, to be closely spaced
doublets, for example, the $17.2\,cm^{-1}$ splitting of the yellow sodium D lines. Uhlen-
beck and Goudsmit suggested in 1925 that such doublets were due to an intrinsic angular momentum possessed by the electron (in addition to its orbital angular
momentum) that could be oriented in just two possible ways. This property, known as
spin, occurs as well in other elementary particles. Spin and orbital angular momenta
are roughly analogous to the daily and annual motions, respectively, of the Earth
around the Sun. To distinguish the spin angular momentum from the orbital, we designate the quantum numbers as s and m_s, in place of l and m. For the electron, the
quantum number s always has the value $\frac{1}{2}$, while m_s can have one of two values, $\pm\frac{1}{2}$.
The electron is said to be an elementary particle of spin $\frac{1}{2}$. The proton and neutron
also have spins $\frac{1}{2}$ and belong to the classification of particles called *fermions*, which
are governed by the Pauli exclusion principle. Other particles, including the photon,
have integer values of spin and are classified as *bosons*. These do *not* obey the Pauli
principle, so that an arbitrary number can occupy the same quantum state. A complete theory of spin requires relativistic quantum mechanics. For our purposes, it is
sufficient to recognize the two possible internal states of the electron, which can be
called *spin-up* and *spin-down*. Electron spin plays an essential role in determining
the possible electronic states of atoms and molecules.

The two spin states can be designated $\left|\frac{1}{2}, \frac{1}{2}\right\rangle$ and $\left|\frac{1}{2}, -\frac{1}{2}\right\rangle$. They obey the angular
momentum eigenvalue equations

$$S^2\left|\frac{1}{2}, \pm\frac{1}{2}\right\rangle = \frac{3}{4}\hbar^2\left|\frac{1}{2}, \pm\frac{1}{2}\right\rangle, \quad S_z\left|\frac{1}{2}, \pm\frac{1}{2}\right\rangle = \pm\frac{1}{2}\hbar\left|\frac{1}{2}, \pm\frac{1}{2}\right\rangle. \tag{3.92}$$

Note that $s(s+1) = \frac{1}{2}(\frac{1}{2}+1) = \frac{3}{4}$. Spin eigenvectors can alternatively be represented
by two-dimensional column vectors (actually *spinors*):

$$\left|\frac{1}{2}, \frac{1}{2}\right\rangle = \left\|\begin{matrix}1\\0\end{matrix}\right\|, \qquad \left|\frac{1}{2}, -\frac{1}{2}\right\rangle = \left\|\begin{matrix}0\\1\end{matrix}\right\|, \tag{3.93}$$

and spin operators by 2×2 matrices:

$$S_x = \frac{\hbar}{2}\left\|\begin{matrix}0 & 1\\1 & 0\end{matrix}\right\|, \quad S_y = \frac{\hbar}{2}\left\|\begin{matrix}0 & -i\\i & 0\end{matrix}\right\|, \quad S_z = \frac{\hbar}{2}\left\|\begin{matrix}1 & 0\\0 & -1\end{matrix}\right\|, \quad S^2 = \frac{3}{4}\hbar^2\left\|\begin{matrix}1 & 0\\0 & 1\end{matrix}\right\|.$$
$$\tag{3.94}$$

These are the famous *Pauli spin matrices*. These matrices (or operators) obey commutation relations completely analogous to Eqs. (3.71) and (3.74).

3.4.6 Abstract Theory of Angular Momentum

A completely general theory of angular momentum can be based on a set of commutation relations, not necessarily connected to either orbital or spin angular momentum.

Suppose we begin with a set of operators J_1, J_2, J_3, J^2 obeying the commutation relations:

$$[J^2, J_i] = 0, \qquad [J_1, J_2] = i\hbar J_3, \; et \; cyc, \qquad i = 1, 2, 3, \qquad (3.95)$$

where *et cyc* means all cyclic permutations of $(1, 2, 3)$. A more elegant way to write these commutations makes use of the *permutation symbol* or *Levi-Civita symbol*, ε_{ijk}, which is equal to $+1$ if (i, j, k) is an even permutation of $(1, 2, 3)$, -1 if it is an odd permutation, and 0 otherwise. More explicitly,

$$\varepsilon_{ijk} = \begin{cases} +1 & \text{if } (i, j, k) \text{ is } (1, 2, 3), (2, 3, 1) \text{ or } (3, 1, 2), \\ -1 & \text{if } (i, j, k) \text{ is } (3, 2, 1), (1, 3, 2) \text{ or } (2, 1, 3), \\ 0 & \text{if } i = j \text{ or } j = k \text{ or } k = i \end{cases} \qquad (3.96)$$

The commutation relations are now:

$$[J_i, J_j] = i\hbar\varepsilon_{ijk}J_k. \qquad (3.97)$$

The eigenvalue equations can now be derived by the same sequence of manipulations leading to Eqs. (3.88). The one generalization is that the quantum numbers j and m can now be odd half-integers, as well as integers, just so the sequence $-j$ to j can be spanned by a finite number of integer jumps. We find then:

$$\begin{aligned} J^2|j, m\rangle &= j(j+1)\hbar^2|j, m\rangle, & j &= 0, \tfrac{1}{2}, 1, \tfrac{3}{2}, 2, \ldots \\ J_3|j, m\rangle &= m\hbar|j, m\rangle, & m &= -j, -j+1, \ldots, j. \end{aligned} \qquad (3.98)$$

3.5   The Hydrogen Atom

Historically, the hydrogen atom has played a major role in the development of the quantum theory, as described in Chap. 1. With successive refinements of our theories, through nonrelativistic, then relativistic quantum mechanics, quantum electrodynamics up to the current standard model based on quantum field theory, the hydrogen atom has provided a definitive test of each new theory. In several ways, the hydrogen atom is fundamental to our understanding of the Universe. Hydrogen comprises over 90% of the atoms in the visible matter of the Universe (not counting the overwhelming prevalence of mysterious dark matter).

The time-independent Schrödinger equation for an electron, of mass m and charge $-e$, in a hydrogenlike atom takes the form:

$$-\frac{\hbar^2}{2m}\nabla^2\psi(\mathbf{r}) - \frac{Ze^2}{r}\psi(\mathbf{r}) = E\,\psi(\mathbf{r}), \qquad (3.99)$$

where Z is the atomic number of the nucleus. It is simple to generalize the problem from the hydrogen atom ($Z = 1$) to all one-electron (hydrogenlike) ions. Thus $Z = 2$ for He^+, $Z = 3$ for Li^{2+}, and so on. The potential energy for the attractive Coulomb interaction between an electron of charge $-e$ and a nucleus of charge $+Ze$ is given by $V(r) = (Ze)(-e)/r$ where r is the interparticle distance. (We use the Gaussian system of units, since, in SI units, the additional factor $1/4\pi \varepsilon_0$, is extraneous in applications to atomic phenomena.) A force dependent only on r, the magnitude of $\mathbf{r}$, is known as a *central force*. In Cartesian coordinates, the potential energy would take the form $V(x, y, z) = -Ze^2/\sqrt{x^2 + y^2 + z^2}$, which would make the Schrödinger equation rather difficult to solve. It is much more convenient to use *spherical coordinates*, (r, θ, ϕ), in which the potential energy takes the simplest form $V(r) = -Ze^2/r$. It should be noted that the nuclear kinetic energy has not been taken into account. For simplicity, the nuclear mass M, which is at least 1836 times the electron mass, can be approximated as infinite, thus implying zero nuclear kinetic energy. (A correction for finite nuclear mass can be made by using the *reduced mass* of the electron, $\mu = mM/(m + M)$, which is of the order of 1 part per 1000 smaller than m.)

3.5.1 Spherical Polar Coordinates

Following is a brief review of this coordinate system, sufficient for our present purposes. The three Cartesian coordinates can be expressed in terms of spherical coordinates as follows:

$$x = r \sin \theta \cos \phi, \qquad y = r \sin \theta \sin \phi, \qquad z = r \cos \theta, \qquad (3.100)$$

which can be readily be deduced from Fig. 3.7. Spherical coordinates are closely analogous to the geographical coordinate system, which locates points by latitude, longitude and altitude. Referring to the familiar globe of the world, r represents the radius of the globe, with the range $0 \leq r < \infty$. The *azimuthal angle* θ is the angle between the vector $\mathbf{r}$ and the z-axis or "north pole," with the range $0 \leq \theta \leq \pi$. Thus

Fig. 3.7 Spherical coordinates (r, θ, ϕ); r is the length of the vector $\mathbf{r}$, while θ and ϕ are the central angles of the shaded sectors

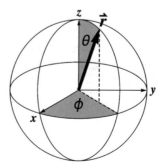

$\theta = 0$ points to the "north pole," $\theta = \pi$, to the "south pole" and $\theta = \pi/2$ runs around the "equator." The circles of constant θ on the surface of a sphere are analogous to the parallels of latitude on the globe (although the geographic conventions are different, with the equator at 0° latitude, while the poles are at 90° N and S latitude). The *polar angle* ϕ measures the rotation of the vector **r** around the z-axis, with $0 \leq \phi < 2\pi$, counterclockwise from the x-axis. The loci of constant ϕ on the surface of a sphere are great circles through both poles. These clearly correspond to *meridians* in the geographic specification of *longitude* (measured in degrees, 0° to 180° E and W of the Greenwich Meridian).

Integration of a function over all space in spherical coordinates takes the form:

$$\int_0^\infty \int_0^\pi \int_0^{2\pi} F(r, \theta, \phi)\, r^2 \sin\theta\, dr\, d\theta\, d\phi. \tag{3.101}$$

For a spherically symmetric function $F(r)$, independent of θ or ϕ, this integral simplifies to

$$\int_0^\infty F(r)\, 4\pi r^2\, dr, \tag{3.102}$$

as implied by the division of space into spherical shells of area $4\pi r^2$ and thickness dr.

3.5.2 Solution of the Schrödinger Equation

The Laplacian operator in spherical coordinates is given by

$$\nabla^2 = \frac{1}{r^2}\frac{\partial}{\partial r}r^2\frac{\partial}{\partial r} + \frac{1}{r^2\sin\theta}\frac{\partial}{\partial\theta}\sin\theta\frac{\partial}{\partial\theta} + \frac{1}{r^2\sin^2\theta}\frac{\partial^2}{\partial\phi^2} \tag{3.103}$$

Thus the hydrogenic Schrödinger equation, in all its glory, looks like this:

$$-\frac{\hbar^2}{2m}\left(\frac{1}{r^2}\frac{\partial}{\partial r}r^2\frac{\partial}{\partial r} + \frac{1}{r^2\sin\theta}\frac{\partial}{\partial\theta}\sin\theta\frac{\partial}{\partial\theta} + \frac{1}{r^2\sin^2\theta}\frac{\partial^2}{\partial\phi^2}\right)\psi(r, \theta, \phi)$$
$$-\frac{Ze^2}{r}\psi(r, \theta.\phi) = E\psi(r, \theta, \phi). \tag{3.104}$$

The long expression within the brackets following $-\hbar^2/2m$ represents the total kinetic energy of the electron. This corresponds, in classical mechanics, to the kinetic energy part of the Hamiltonian in spherical coordinates:

$$E_{\text{kin}} = \frac{p_r^2}{2m} + \frac{L^2}{2mr^2}, \tag{3.105}$$

where p_r is the momentum in the radial direction, while L is the angular momentum. The rotational part of the electron's kinetic energy can be expressed $E_{rot} = L^2/2I$ (see Eq. 3.57), where the relevant moment of inertia is given by $I = mr^2$. Clearly, the first term in the Laplacian (containing r alone) must correspond to the radial part of the kinetic energy:

$$H_{rad} = -\frac{\hbar^2}{2m} \frac{1}{r^2} \frac{\partial}{\partial r} r^2 \frac{\partial}{\partial r} = -\frac{\hbar^2}{2m} \left(\frac{\partial^2}{\partial r^2} + \frac{2}{r} \frac{\partial}{\partial r} \right), \tag{3.106}$$

while the remaining two terms represent angular or rotational motion:

$$H_{rot} = -\frac{\hbar^2}{2m} \left(\frac{1}{r^2 \sin \theta} \frac{\partial}{\partial \theta} \sin \theta \frac{\partial}{\partial \theta} + \frac{1}{r^2 \sin^2 \theta} \frac{\partial^2}{\partial \phi^2} \right). \tag{3.107}$$

The relation $H_{rot} = L^2/2mr^2$ identifies the angular momentum operator, in spherical coordinates:

$$L^2 = -\hbar^2 \left(\frac{1}{\sin \theta} \frac{\partial}{\partial \theta} \sin \theta \frac{\partial}{\partial \theta} + \frac{1}{\sin^2 \theta} \frac{\partial^2}{\partial \phi^2} \right). \tag{3.108}$$

The eigenvalue equation for angular momentum (3.88) in Dirac notation,

$$L^2 |l, m\rangle = l(l+1)\hbar^2 |l, m\rangle, \tag{3.109}$$

can be expressed in terms of wavefunctions and differential operators as

$$-\hbar^2 \left(\frac{1}{\sin \theta} \frac{\partial}{\partial \theta} \sin \theta \frac{\partial}{\partial \theta} + \frac{1}{\sin^2 \theta} \frac{\partial^2}{\partial \phi^2} \right) Y_{lm}(\theta, \phi) = \hbar^2 l(l+1) Y_{lm}(\theta, \phi). \tag{3.110}$$

The eigenfunctions $Y_{lm}(\theta, \phi)$ are known as *spherical harmonics*, which also occur in many other applications of mathematical physics. The quantum number l is often specified by a code that originated in early atomic physics: $l = 0, 1, 2, 3, \ldots$ are designated $s, p, d, f, \ldots$ states, respectively. For an s-state, $Y_{00}(\theta, \phi) = 1/\sqrt{4\pi}$; the wavefunction has no dependence on the angles θ or ϕ, only on r. Thus an s-state is *spherically symmetrical*.

The hydrogenic Schrödinger equation (3.104) can evidently be simplified by separation of variables, with

$$\psi(r, \theta, \phi) = R_{nl}(r) Y_{lm}(\theta, \phi), \tag{3.111}$$

where n is the *principal quantum number*, $n = 1, 2, 3, \ldots$, with the same meaning as in the Bohr theory. For a given n, the angular momentum quantum number l has the possible values $l = 0, 1, 2, \ldots, n-1$. As a short cut, we can use the known energies from the Bohr theory: $E_n = -Z^2 e^4 m/2n^2 \hbar^2 = -Z^2 e^2/2n^2 a_0$ The ground state, with $n = 1$, can admit only $l = 0$. It is designated the 1s-state, with a spherically symmetrical wavefunction $\psi(r)$. The ground-state energy equals $E_1 = E_{1s} = -Z^2 e^2 / 2a_0$.

After some algebraic reduction, we obtain an ordinary differential equation for the *radial function* $R_{nl}(r)$:

$$-\frac{\hbar^2}{2m}\left(\frac{d^2 R_{nl}}{dr^2} + \frac{2}{r}\frac{dR_{nl}}{dr} - \frac{l(l+1)}{r^2}R_{nl}\right) - \frac{Ze^2}{r}R_{nl} = E_n R_{nl}. \tag{3.112}$$

It is convenient, in atomic and molecular applications, to introduce *atomic units*, obtained by setting $\hbar = m = e = 1$. The unit of length is the *bohr*, equal to the Bohr radius $a_0 = \hbar^2/me^2 \approx 0.529 \times 10^{-10}$ m. The unit of energy is the *hartree*, equal to $e^2/a_0 \approx 27.211$ eV. Thus, for the ground-state energy $E_{1s} = -Z^2/2$ hartrees. Setting $l = 0$ for an s-state, the radial equation for the ground state wavefunction now simplifies to:

$$\frac{1}{2}R''(r) + \frac{1}{r}R'(r) + \frac{Z}{r}R(r) = \frac{Z^2}{2}R(r), \tag{3.113}$$

It is often useful, as a first step, to obtain an asymptotic solution of a differential equation, in this case for large values of r. Thus, neglecting the two terms containing $\frac{1}{r}$, the equation reduces to

$$R''(r) \approx Z^2 R(r) \tag{3.114}$$

A solution is $R(r) \approx e^{-Zr}$. An alternative solution e^{+Zr} is rejected since it increases without limit as $r \to \infty$. We are very fortunate to find that this asymptotic solution is also an exact solution to Eq. (3.113): $R_{1s}(r) = Ae^{-Zr}$. To determine the normalization constant A, such that the total probability of finding the electron somewhere is equal to 1, we calculate

$$\int_0^\infty [R_{1s}(r)]^2 r^2 dr = A^2 \int_0^\infty e^{-2Zr} r^2 dr = \frac{A^2}{4Z^3} = 1. \tag{3.115}$$

This gives $A = 2Z^{3/2}$ and

$$R_{1s}(r) = 2Z^{3/2} e^{-Zr}, \tag{3.116}$$

or, since $Y_{00} = 1/\sqrt{4\pi}$,

$$\psi_{1s}(r) = \frac{Z^{3/2}}{\sqrt{\pi}} e^{-Zr}. \tag{3.117}$$

We have made use of the definite integral:

$$\int_0^\infty r^n e^{-\alpha r} dr = \frac{n!}{\alpha^{n+1}}. \tag{3.118}$$

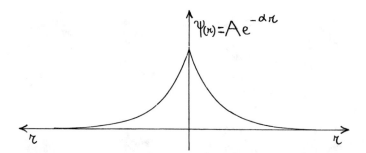

$\psi(r) = A e^{-\alpha r}$

Fig. 3.8 Ground-state wavefunction of the hydrogen atom. In order to simulate the cusp at the origin in three dimension, we have shown the behavior of $\psi_{1s}(r)$ both on the positive and "negative" r axes

Fig. 3.9 Density plot of the hydrogen $1s$ wavefunction. The nucleus is shown as a black dot in the center

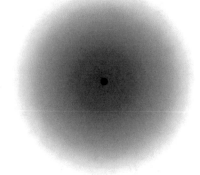

A one-dimensional projection of $\psi_{1s}(r)$ is shown in Fig. 3.8 and a three-dimensional density plot of the wavefunction are simulated in Fig. 3.9. Note the exponential decrease from the maximum value at the nucleus. The $1s$-wavefunction has no nodes (loci where the function equals zero), in common with the ground states of the particle in a box and harmonic oscillator. As we have seen, higher-energy states are associated with an increasing number of nodes. The $2s, 3s, \ldots$ states have *radial nodes*, spheres on which $\psi = 0$, while $p, d, f, \ldots$ states have the angular nodes of the corresponding spherical harmonics Y_{lm}, which are surfaces through the origin.

3.5.3 Atomic Structure

In many-electron atoms, to a reasonable level of approximation, each individual electron can be pictured as occupying an *atomic orbital* (AO).[3] This provides a

---

[3]Technically, since electrons are *indistinguishable* particles, every electron is equally associated with every occupied atomic orbital. Expressed another way, the electrons in an atom or molecule have a *cardinality* but no *ordinality*.

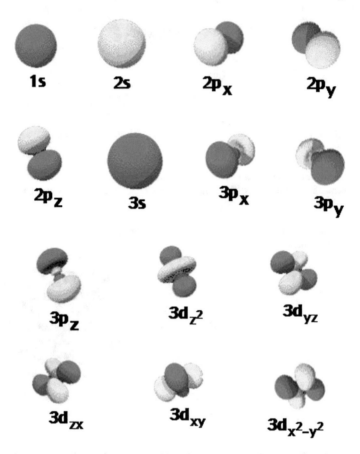

Fig. 3.10 Schematic images of some hydrogenic atomic orbitals. The wavefunctions are positive in the *blue areas* and negative in the *yellow areas*

foundation for much of quantum chemistry. (The term "orbital" is an adaptation from "orbit" in the Bohr theory, recognizing that an electron is described by a probability distribution rather than a classical trajectory.) The AOs have the same designations as hydrogenic functions, $1s$, $2s$, $2p$, ..., are have very similar shapes. Some of the lower energy AOs are pictured in Fig. 3.10. These are highly schematic representations since the actual wavefunctions fall off exponentially. These figures convey a general idea of the shape of the orbitals and of their nodal surfaces, boundaries between the positive (blue) and negative (yellow) regions of the wavefunction. Each AO is described by a unique set of four quantum numbers: n, l, m_l and m_s. The first three belong to the wavefunction $\psi_{nlm}(r, \theta, \phi)$; the fourth is the electron spin orientation $m_s = \pm\frac{1}{2}$. According to the *Pauli exclusion principle*, no two electrons can share the same set of four quantum numbers. According to the *Aufbau principle*, the ground state of an atom is constructed conceptually by successively filling orbitals in the order of increasing energy. Given the multiplicity of values of $m_l = 0, \pm1, \pm2, ..., \pm l$ and of

$m_s = \pm\frac{1}{2}$, a set of AOs with quantum numbers n, l constitute a *subshell*, containing, at most, $2(2l + 1)$ electrons: 2, 6, 10, ..., respectively, for $ns, np, nd, \ldots$. The Pauli principle and the Aufbau principle provide a rational understanding of the structure of the Periodic Table. As an example, potassium (K) with atomic number $Z = 19$ has a ground-state electronic configuration written $1s^2 2s^2 2p^6 3s^2 3p^6 4s^1$. Note that $4s$ subshell is open (not complete), which accounts for the relative ease with which this electron can be removed, to leave a K^+ ion. The remaining 18 electrons form a *closed shell* configuration, which tends to bestow enhanced chemical stability.

3.6 Time Evolution and Collapse of the Wavefunction

3.6.1 The Evolution Operator

The time-dependent Schrödinger equation (TDSE), Eq. (3.13), shows how the Hamiltonian H governs the time evolution of a quantum system. In Dirac notation, the TDSE can be written

$$i\hbar \frac{d|\Psi(t)\rangle}{dt} = H|\Psi(t)\rangle. \tag{3.119}$$

We now introduce an operator $U(t)$ which maps (or "rotates") $|\Psi(0)\rangle$ into $|\Psi(t)\rangle$, such that

$$|\Psi(t)\rangle = U(t)|\Psi(0)\rangle. \tag{3.120}$$

Substituting this into the TDSE, we obtain

$$i\hbar \frac{d}{dt} U(t)|\Psi(0)\rangle = HU(t)|\Psi(0)\rangle. \tag{3.121}$$

Since this applies to an arbitrary function $|\Psi(0)\rangle$, the relation can be reduced to an operator equation

$$i\hbar \frac{d}{dt} U(t) = HU(t), \tag{3.122}$$

with the initial condition $U(0) = I$ (the identity operator). If this were a differential equation involving ordinary functions, the solution would then be:

$$U(t) = e^{-iHt/\hbar}. \tag{3.123}$$

This can also be considered the *formal* solution to the corresponding operator equation. The exponential implicitly represents the power series operator expansion:

$$e^{-iHt/\hbar} = I + \left(-\frac{i}{\hbar}Ht\right) + \frac{1}{2!}\left(-\frac{i}{\hbar}Ht\right)^2 + \frac{1}{3!}\left(-\frac{i}{\hbar}Ht\right)^3 + \ldots \tag{3.124}$$

The adjoint operator $U(t)^\dagger$ is equal to $e^{+iH^\dagger t/\hbar}$. Since the Hamiltonian is Hermitian, $H^\dagger = H$, it follows that $U(t)^\dagger = U(t)^{-1}$. Thus $U(t)$ is a *unitary* operator, which is known as the *evolution operator* or *propagator*, since it governs the development of the quantum system with time. Applying the evolution operator to an eigenfunction of H, denoted ψ_n or $|n\rangle$, we find

$$U(t)\psi_n = e^{-iHt/\hbar}\psi_n = e^{-iE_n t/\hbar}\psi_n. \tag{3.125}$$

Since $H^k \psi_n = E_n^k \psi_n$, the exponential series for H, Eq. (3.124) turns into the corresponding series for E_n. This agrees with the form of the wavefunction for a stationary state: $\Psi = \psi\, e^{-iEt/\hbar}$, which we obtained in Sect. 3.1.

Let us now apply this formalism to the simplest possible case, a two-dimensional Hilbert space. The Hamiltonian can then be represented by a 2×2 matrix in the orthonormal basis of the eigenvectors $|\psi_1\rangle$, $|\psi_2\rangle$, with the corresponding eigenvalues E_1, E_2. Since $H_{ij} = \langle\psi_i|H|\psi_j\rangle = E_j\langle\psi_i|\psi_j\rangle$, H, in this basis, is represented by the simple diagonal matrix:

$$H = \left\|\begin{matrix} E_1 & 0 \\ 0 & E_2 \end{matrix}\right\|. \tag{3.126}$$

All powers H^2, H^3, etc. are then also diagonal matrices:

$$H^2 = \left\|\begin{matrix} E_1^2 & 0 \\ 0 & E_2^2 \end{matrix}\right\|, \qquad H^3 = \left\|\begin{matrix} E_1^3 & 0 \\ 0 & E_2^3 \end{matrix}\right\|, \qquad \dots \tag{3.127}$$

It follows that the evolution operator is represented by the matrix:

$$U(t) = e^{-iHt/\hbar} = \left\|\begin{matrix} e^{-iE_1 t/\hbar} & 0 \\ 0 & e^{-iE_2 t/\hbar} \end{matrix}\right\|. \tag{3.128}$$

This result can be generalized to a Hilbert space of any dimension.

The derivative of an $n \times n$ matrix A with respect to a parameter t can be denoted by a matrix B, whose elements are the derivatives of the corresponding matrix elements of A. Thus

$$B_{i,k} = \frac{dA_{ik}}{dt} \quad i, k = 1, 2, \dots, n \tag{3.129}$$

In the case of the matrix $U(t)$ we have then:

$$i\hbar\frac{dU}{dt} = \left\|\begin{matrix} E_1\, e^{-iE_1 t/\hbar} & 0 \\ 0 & E_2\, e^{-iE_2 t/\hbar} \end{matrix}\right\| \tag{3.130}$$

This is evidently the 2×2 matrix form of the operator relation (3.122)

$$i\hbar\frac{dU}{dt} = HU. \tag{3.131}$$

When applied to $|\psi(0)\rangle$, the state at $t = 0$, this produces $|\psi(t)\rangle$, a solution of the TDSE. From an operational point of view, the Hamiltonian H can thereby be pictured as the generator of time evolution of a quantum system.

3.6.2 Schrödinger and Heisenberg Pictures

Thus far, we have followed a formulation of QM referred to as the *Schrödinger picture*. Its distinguishing features are state vectors $|\Psi(t)\rangle$ depending on time and satisfying the TDSE, but operators which are fixed in time (apart from cases in which there is explicit time dependence). The time evolution of a quantum system is then carried entirely by the state vectors $|\Psi(t)\rangle$. In abstract geometrical terms, Schrödinger picture can be associated with a coordinate system in Hilbert space in which dynamical variables are fixed but state vectors are moving. Possibilities for alternative formulations exist, however, since both operators and state vectors are abstract quantities not directly accessible to measurement. Quantities which *do* have objective reality are scalars in Hilbert space: eigenvalues, probability densities, expectation values and transition probabilities.

Consider now a Dirac bracket expression such as $f(t) = \langle\Phi(t)|A|\Psi(t)\rangle$. Now introduce the evolution operator, so that $|\Psi(t)\rangle = U(t)|\Psi(0)\rangle$ and $\langle\Phi(t)| = \langle\Phi(0)|U^{\dagger}(t)$. The bracket expression transforms as follows

$$f(t) = \langle\Phi(t)|A|\Psi(t)\rangle = \langle\Phi(0)|U^{\dagger}(t)AU(t)|\Psi(0)\rangle = \langle\Phi(0)|A(t)|\Psi(0)\rangle,$$

(3.132)

where

$$A(t) = U^{\dagger}(t)AU(t) = e^{iHt/\hbar}Ae^{-iHt/\hbar}.$$

(3.133)

Here, the time-independent operator A in the Schrödinger picture is transformed to a time-dependent operator $A(t)$ in what is known as the *Heisenberg picture*. Often these alternative representations of an operator are distinguished by the notation A_S and A_H. In the Heisenberg picture, it is the *operators* that now evolve with time, while the state vectors remain as a fixed, time-independent basis. The focus on operators in Hilbert space, and their matrix representations, closely resembles the formulation of matrix mechanics. And, in addition, like classical mechanics, deals with equations of motion for dynamical variables.

Noting that H and $U(t)$ commute for a Hamiltonian which has no explicit time dependence ($[H, U(t)] = 0$), it follows that the Hamiltonian is the same in both Schrödinger and Heisenberg pictures: $H_H = U^{\dagger}(t)H_S U(t) = H_S$.

The time dependence of an observable, represented by an operator in Heisenberg picture, such as $A(t)$ in Eq. (3.133), follows from Eq. (3.131) and its conjugate, in the following sequence of steps:

$$\frac{d}{dt}A(t) = \frac{d}{dt}\left(U^{\dagger}(t)AU(t)\right) = \left(\frac{d}{dt}U^{+}(t)\right)AU(t) + U^{\dagger}(t)A\frac{d}{dt}U(t) =$$

$$- (i\hbar)^{-1}HU^{\dagger}(t)AU(t) + U(t)^{\dagger}A(i\hbar)^{-1}HU(t) = (i\hbar)^{-1}\left(A(t)H - HA(t)\right).$$

$$(3.134)$$

We obtain thereby the *Heisenberg equation of motion*:

$$i\hbar\frac{d}{dt}A(t) = \left[A(t), H\right], \tag{3.135}$$

showing more explicitly how the Hamiltonian is the generator of the time evolution of a quantum system. An observable which commutes with the Hamiltonian, so that $\left[A(t), H\right] = 0$, must be constant in time: $dA/dt = 0$. Such a quantity is known as a *constant of the motion*.

A very interesting consequence of the Heisenberg equation of motion is *Ehrenfest's theorem*, which shows that quantum mechanical expectation values obey Newton's classical equations of motion. Let us illustrate this by considering a particle in one dimension with Hamiltonian $H = p^2/2m + V(x)$. The Heisenberg equations for $x(t)$ and $p(t)$ are given by

$$i\hbar\frac{dx}{dt} = [x, H], \qquad i\hbar\frac{dp}{dt} = [p, H]. \tag{3.136}$$

The commutators can be evaluated in the following steps:

$$[x, H] = \frac{1}{2m}[x, p^2] = \frac{1}{2m}\left(p[x, p] + [x, p]p\right) = i\hbar\frac{p}{m}, \tag{3.137}$$

$$[p, H] = pV(x) - V(x)p = -i\hbar\frac{dV}{dx}. \tag{3.138}$$

We find then

$$\frac{dx}{dt} = \frac{p}{m} \qquad \text{and} \qquad \frac{dp}{dt} = -\frac{dV}{dx}. \tag{3.139}$$

Analogous relations can be derived for the *expectation values* $\overline{x} = \langle x \rangle$ and $\overline{p} = \langle p \rangle$:

$$\frac{d\langle x \rangle}{dt} = \frac{\langle p \rangle}{m} \qquad \text{and} \qquad \frac{d\langle p \rangle}{dt} = \left\langle -\frac{dV}{dx}\right\rangle = F, \tag{3.140}$$

where F can be identified with the classical force. The result is a quantum generalization of Newton's second law:

$$F = m\frac{d^2\langle x \rangle}{dt^2}. \tag{3.141}$$

The TDSE describes the evolution of a quantum system, represented by the state vector $|\Psi(t)\rangle$ in Hilbert space, in a perfectly deterministic way, as represented by the unitary evolution operator $U(t)$. This deterministic behavior is quite analogous to the way Newton's laws govern the motion of a classical system subject to internal or external forces. Ehrenfest's theorem above, in fact, gives an explicit connection between classical and quantum behavior. In general, when physical objects approach macroscopic dimensions, QM reduces to classical mechanics (apart from certain low-temperature phenomena, including superconductivity and Bose–Einstein condensation).

3.6.3 Collapse of the Wavefunction

In contrast to the unitary deterministic evolution of a quantum system, there is the phenomenon called *collapse (or reduction) of the wavefunction*. This is what happens when a measurement is made on the system, or when the system is subjected to a perturbation which suddenly changes the Hamiltonian. A collapse can also be associated with an unpredictable random event, such as the radioactive decay of a nucleus. Before collapse, a state vector might consist of a superposition of a set of basis vectors in Hilbert space, something like

$$|\Psi\rangle = \sum_n c_n |\Phi_n\rangle, \tag{3.142}$$

where the $|\Phi_n\rangle$ are eigenfunctions of some dynamical variable Λ, such that

$$\Lambda|\Psi\rangle = \lambda_n|\Phi_n\rangle. \tag{3.143}$$

If $|\Psi\rangle$ is normalized and $\{|\Phi_n\rangle\}$ is an orthonormal set:

$$\langle\Psi|\Psi\rangle = \left(\sum_m \langle\Phi_m|c_m^*\right)\left(\sum_n c_n|\Phi_n\rangle\right) = \sum_{n,m} c_m^* c_n \langle\Phi_m|\Phi_n\rangle =$$

$$\sum_{n,m} c_m^* c_n \delta_{m,n} = \sum_n |c_n|^2 = 1. \tag{3.144}$$

The expectation value of Λ in the state $|\Psi\rangle$, meaning the average over a large number of independent measurements, is given by

$$\langle\Psi|\Lambda|\Psi\rangle = \sum_n |c_n|^2 \lambda_n. \tag{3.145}$$

The probability of observing the result λ_n is evidently given by $P_n = |c_n|^2$. And the sum of probabilities, as it should, adds up to 1: $\sum_n P_n = 1$. The situation can be interpreted as follows. Suppose a system in a state $|\Psi\rangle$ is happily evolving in accordance with the TDSE. But then someone decides to make a measurement of the observable Λ. This causes the state vector $|\Psi\rangle$ to *collapse* into one of the eigenstates of Λ, say $|\Phi_n\rangle$, while the measurement registers a value λ_n. It cannot be predicted *which* eigenstate results from the collapse, but the probability of observing λ_n is equal to $|c_n|^2$. It is fairly obvious that there is no way in which continuous unitary time evolution can produce an instantaneous transition from a superposition to one of its components. Collapse is, in fact, an *irreversible* process, akin to the interaction of a thermodynamic system with its environment.

It should be mentioned that there have been some attempts to include the measurement apparatus in an extended quantum system. Quoting (Griffiths 2002, pp. 246): "...after all, what is special about a quantum measurement? Any real measurement apparatus is constructed out of aggregates of particles to which the laws of QM apply, so the apparatus ought to be described by those laws, and not used to provide an excuse for their breakdown."

Collapse of the wavefunction, the second possible mode of evolution of a quantum system, has created controversy since the inception of quantum mechanics. In contrast to the TDSE, the mechanism of collapse had to be added to the theory as an ad hoc postulate. The replacement of deterministic outcomes by random events, following only a probabilistic distribution, has long been one of the most controversial epistemological aspects of quantum mechanics. Einstein described the time evolution of a state in QM as resting on two "legs," one very solid, but the other much weaker. Despite the unquestionable experimental success of QM, two aspects—the irreducible probabilistic character and the existence of two types of time evolution— have caused much intense opposition to the traditional (Copenhagen) interpretation of QM, as originally developed by Bohr, Heisenberg, and Dirac.

Fig. 3.11 Does God play dice with the Universe?

Fig. 3.12 The 1927 Solvay Congress on the quantum theory. Colorized version from the American Physical Society historical collection

To reiterate, in QM it is not possible to obtain more detailed knowledge that can replace probability with certainty. This interpretation of the meaning of the Schrödinger wavefunction provides the foundation of the *Copenhagen interpretation* of QM, which suffices for most practicing scientists nowadays. The most eminent critic of the probabilistic interpretation of QM was none other than Albert Einstein, as epitomized in his famous phrase: "God does not play dice with the Universe!" (see Fig. 3.11). By some accounts, Niels Bohr replied with the rejoinder: "Stop telling God what to do!"

Figure 3.12 is a portrait of the participants in the Fifth Solvay International Conference in 1927, including all of the founding fathers of quantum mechanics. They came together to contemplate the foundations of the newly formulated theory. Here the long-running dialog between Niels Bohr and Albert Einstein first began.

3.7   Philosophical Issues in Quantum Mechanics

The philosophical inclinations of a number of scientists has shaped attitudes regarding the orthodox interpretation of quantum mechanics. Three alternative philosophical viewpoints have been espoused by physicists who worry about such things: these are often classified as Neopositivism, Realism and Idealism:

(1) Neopositivism: The Copenhagen interpretation of QM is in complete agreement with the ideas of Neopositivism, which considers metaphysically meaningless the question of what actually is the individual trajectory of an electron or a photon. QM avoids talking about particles "in themselves," but provides only ways of computing the probabilities that a measurement will give a certain result. The only things we can talk about are the results of our measurements. An indication of the influ-

ence of positivism on the orthodox interpretation is the word "observable" chosen to denote any dynamical variable.

(2) Realism: The strongest adversaries of the orthodox interpretation of QM have been the realists. Among them, there are renowned physicists, including Einstein, Bohm, and de Broglie. As Einstein once asked Abraham Pais: "Do you really believe that the Moon is not there when nobody looks?"

(3) Idealism: (Von Neumann 1932; Stapp 2001) The idealists are more willing to accept the Copenhagen interpretation, albeit with an important stipulation. For them the state Ψ of a physical system is a mental construct. Therefore they do not find it strange at all that the state makes a "sudden jump" when we carry out a measurement, because at the moment in which we become aware of the result of that measurement, our knowledge of the physical system abruptly changes. Quantum state reduction is therefore a conscious act. Even their point of view is, under certain circumstances, paradoxical. Regarding this we will describe the well-known paradox of "Schrödinger's cat."

Let us imagine a cat enclosed in a sealed box, along with a vial of cyanide and a mechanism, driven by a Geiger counter, that can break the vial when the decay of an atom in a radioactive sample is detected. An atom's decay is a random event, so that its precise time cannot be predicted. Only its half-life is known. Assume that the experiment is run for a length of time such that a decay will occur with a probability of 50%. At the end of this time, is the cat alive or dead? If the "reduction" of the state Ψ of the cat happens only when an external observer opens the container (when the observer becomes aware of the result of the experiment) the cat remains in a superposition state—it is half alive and half dead (see Fig. 3.13) until the container is opened. Probably the cat would greatly object to this interpretation, since he knows if he is alive (and of course he would object even more strongly to participating in the experiment).

Some physicists (including de Broglie and Bohm) tried to eliminate the probabilistic aspect of QM introducing some type of "hidden variables" that would determine the result of a measurement. This would be analogous to statistical mechanics for atoms and molecules. We cannot see the detailed motions of the atoms with our eyes, but their statistical behavior explains the laws of thermodynamics. However, these theories have not been met with general acceptance. For the reader who has the patience to read Chap. 5 we will introduce Bell's inequality (see, e.g., Bell 1987), which would be valid for a hidden variable theory satisfying relativity, but would be violated by QM. Since experiments (Aspect et al. 1981) have confirmed the existence of quantum entanglement that violates Bell's inequality, the acceptance of the orthodox interpretation is further supported. To again quote Griffiths (2002), pp. 334: "In summary, the basic lesson to be learned from the Bell inequalities is that it is difficult to construct a plausible hidden variable theory which will mimic the sorts of correlations predicted by quantum theory and confirmed by experiment". Such a theory must either exhibit peculiar nonlocality which violates relativity, or else incorporate influences which travel backwards in time, counter to everyday experience. This seems a rather high price to pay just to have a theory which is more "classical" than ordinary QM.

Fig. 3.13 Schrödinger's cat

Of course, most physicists do not believe that the current interpretation of QM will survive forever. Among other things, QM relies strongly upon linearity, while modern physics has uncovered a preponderance of nonlinear phenomena in nature. Also, it is recognized that quantum mechanics remains incomplete until it can be coherently incorporated into a "Theory of Everything," which also contains general relativity. In fact, Penrose has proposed that *quantum gravity* is involved in the loss of coherence of a quantum state, which can purportedly occur whenever space-time curvature increases to a critical level (Penrose 1996). Probably the most popular current viewpoint is the hypothesis of *decoherence*, which we shall discuss further in Chap. 5. Among other things, decoherence explains why a superposition state of macroscopic objects is generally extremely unstable: this helps to clarify the relation between classical and quantum mechanics. Some progress has recently been made by the consistent histories formalism of Gell-Mann, Griffiths, Hartle, and Omnes, (see Griffiths 2002, Omnes 1994, Zurek 1991), but it seems that the situation is still very fluid and far from being satisfactorily resolved.

We might also note that a number of even more exotic interpretations of quantum mechanics have been suggested. Everett and Wheeler once suggested the *many worlds* interpretation of quantum mechanics, in which each random event causes the splitting of the entire Universe into disconnected parallel Universes, in each of which, one possible outcome becomes the reality. David Deutsch (1997) believes that the implicit large-scale parallel processing by a quantum computer can be explained only by the many worlds interpretation of QM. Needless to say, not many people are willing to accept such a metaphysically unwieldy view of reality. Most scientists are content to apply the highly successful computational mechanisms of quantum theory to their work, without worrying unduly about its philosophical underpinnings. As Feynman put it, "Shut up and calculate!" Much like most of us happily using our computers without acquiring a detailed knowledge of either semiconductor technology or operating system programming.

The dichotomy of appearance and reality is, in fact, not limited to the quantum theory, but pervades much of modern science. It applies notably to neuroscience, regarding the origin of consciousness. In our daily lives, we intuitively assume that our perceptions, visual, auditory, tactile, etc., actually represent the "real world." This simulation has evidently been a reasonably successful one for the human race. Otherwise natural selection would have weeded us out by now. But philosophers, going back to Plato/Aristotle, have recognized that our perceptions provide merely a *model* of external reality (recall Plato's *Parable of the Cave*). The most elegant formulation of the dichotomy was given by Immanuel Kant in his *Critique of Pure Reason*. Kant distinguished the separate realms of *phenomena* and *noumena*. Phenomena are the experiences through our senses; noumena are the underlying things which actually constitute reality. Since the thing in itself (*Ding an sich*) is, by supposition, entirely beyond our capabilities, we must be completely unaware of the noumenal realm. This should provide a healthy dose of modesty for the quantum theorist! Possibly the best summary of these ideas was expressed in some scribbled graffiti one of us (SMB) once encountered: "Reality is a crutch."

We end this chapter with a few thoughts of Feynman (1982), on the prospects of QM becoming part of a consistent and aesthetically beautiful structure. Einstein thought that quantum mechanics was incomplete. But Einstein was wrong, because "spooky action at a distance" has now been shown to be real. Where does Feynman enter the story? Partly because he earned a Nobel Prize for his contributions to quantum mechanics, but also because he expressed some of his thoughts with a sublimely poetic perspective, such as the following[4]:

> *We have always had a great deal of difficulty*
> *understanding the world view*
> *that quantum mechanics represents.*
> *At least I do,*
> *because I'm an old enough man*
> *that I haven't got to the point*
> *that this stuff is obvious to me.*
> *Okay, I still get nervous with it....*
> *You know how it always is,*
> *every new idea,*
> *it takes a generation or two*
> *until it becomes obvious*
> *that there's no real problem.*
> *I cannot define the real problem,*
> *therefore I suspect there's no real problem,*
> *but I'm not sure*
> *there's no real problem.*

---

[4]See also Mermin, Physics Today, April 1985, pp. 38–47.

Chapter 4
New Adventures: Isotropic Vectors, Rotations, Spinors, and Groups

Abstract Some more specialized mathematical topics are introduced, including isotropic vectors, rotations, spinors, and Lie groups. The concept of invariance in the objective world is discussed. The stereographic projection is introduced to describe the behavior of spinors. The Lie groups SO(3) and SU(2) are studied in detail.

Keywords Isotropic vector · Spinor · Stereographic projection · Lie groups · SO(3) · SU(2).

According to many scientific thinkers, notably Henri Poincaré, intuitive pictures are essential for the deeper understanding of scientific principles. This is particularly true for the subjects covered in this chapter, in which geometrical realizations play an essential role. We refer to Felix Klein's classic *Erlangen program* (Klein 1872), which characterizes the properties of different geometries. The ideas of Einstein, Dirac, and Nozick, are then extended to provide a more explicit meaning of the concept of the "objective world." We will then introduce *isotropic vectors*, vectors which are orthogonal to the rotation axis yet remain invariant under rotation. Intuitively, an apparent contradiction, but resolved when complex vector components are admitted. Finally, the spin of elementary particles is connected to isotropic vectors by application of stereographic projections. The correspondence between "rotations" of the spinors associated with the SU(2) group and rotations of the physical objects by the SO(3) group is analyzed, with particular attention to infinitesimal rotations.

In this chapter, we will use symbols of both types: $\overrightarrow{OP}$ and $\mathbf{v}$ to denote vectors in physical space. Since spinors are quantum mechanical attributes, we will use Dirac notation to represent spin states of particles.

4.1 Invariance and the Objective World

In this section, before resuming the physical and mathematical treatment of our subject, we will discuss some important epistemological considerations connected with the concept of invariance. Important contributions were made by Klein, Dirac, Einstein, and Nozick.

© Springer International Publishing AG 2017
G. Fano and S.M. Blinder, *Twenty-First Century Quantum Mechanics:*
Hilbert Space to Quantum Computers, UNITEXT for Physics,
DOI 10.1007/978-3-319-58732-5_4

The geometrical properties of a figure are never properties of a *single* object, but can be considered to belong to a *class* of objects. For example, the property of a square, wherein the length of a diagonal is equal to the length of a side multiplied by $\sqrt{2}$, does not change if we translate or rotate the square, or if we shrink or enlarge it. The same is true for the circle, and in general for all the figures in elementary geometry. The remarkable intuition of Felix Klein, encapsulated in his famous Erlangen program, is based on classification and characterization of a geometry by a group G of transformations (translations, rotations, and changes of scale for the case of elementary geometry), and a study of the properties of the figures that are *invariant* with respect to these transformations. Only such properties are meaningful, while all others are irrelevant for that geometry.

Given a *group* G of transformations, we can say that two figures (for example two triangles or two polygons) A and B are *equivalent* if there exists a transformation of G that maps A into B; we write $A \sim B$. In general, such equivalence is

(1) Reflexive: $A \sim A$.
(2) Symmetric: If $A \sim B$, $B \sim A$.
(3) Transitive: If $A \sim B$ and $B \sim C$, then $A \sim C$.

From property (1) it follows that the group G contains the *identity* transformation I, which leaves any figure unchanged: $IA = A$ for any A. From property (2) it follows that for any element $g \in G$ that sends A into B, there exists a corresponding element denoted by $g^{-1} \in G$ which sends B into A (see Fig. 4.1); g^{-1} is called *inverse* of g. Finally, from (3) it follows that if $B = g_1 A$ and $C = g_2 B$, there exists a transformation denoted by $g_2 g_1$ that sends directly A into C. Thus $g_2 g_1 A = C$; $g_2 g_1$ is called *product* of g_2 and g_1 (see Fig. 4.2). The existence of the identity I, the inverse g^{-1}, and the closure of product $g_2 g_1$ (meaning that it is also a member of the group) characterize the mathematical properties of a group. (Technically, another property is needed: the associative property: $g_1(g_2 g_3) = (g_1 g_2) g_3$). In general the *commutative* property does not hold: it is not necessarily true that $g_1 g_2 = g_2 g_1$.

An elementary group used in physics is the group of *translations*. Denoting by $g_\mathbf{v}$ the operation of translating a physical object in the $\mathbf{v}$ direction a distance $|\mathbf{v}|$ we

Fig. 4.1 The group element
g sends A into B; the inverse
g^{-1} sends B into A

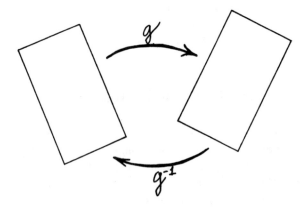

Fig. 4.2 g_1 sends A in B; g_2 sends B in C. The product $g_2 g_1$ sends A in C

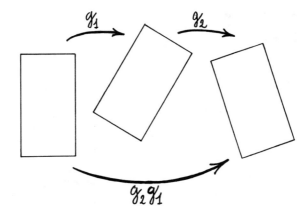

Fig. 4.3 The translation vectors **v** and **w** adds according to the parallelogram's rule

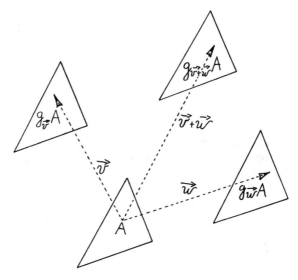

have (see Fig. 4.3) $g_{\mathbf{v}} g_{\mathbf{w}} = g_{\mathbf{v+w}}$; the identity of the group is $g_{\mathbf{0}}$, and the inverse of $g_{\mathbf{v}}$ is $g_{\mathbf{-v}}$. All quantities used in elementary geometry, such as lengths, angles, areas, volumes, are *invariants* with respect to the group of all translations and rotations; this "big" group is called the *Euclidean* group.

In our daily lives, we encounter other groups: for example, let us project a figure F lying in a plane π from a point P (not belonging to π) to another plane π'. This is called a *perspective projection*. It results in a new figure F', shown in Fig. 4.4. This procedure can be iterated by repeating the projection from different points to different planes. A new group of transformations is obtained, from the initial to the final plane. This is called the *projective group*. Lengths, angles, areas are *not* invariant for operations in this group; triangles remain triangles, but equilateral triangles do not remain equilateral; squares become quadrilaterals; circles are transformed in other conic sections. In fact, conics always transform to conics. Projections are common

Fig. 4.4 Perspective
projection of a figure from
the plane π to the plane π'

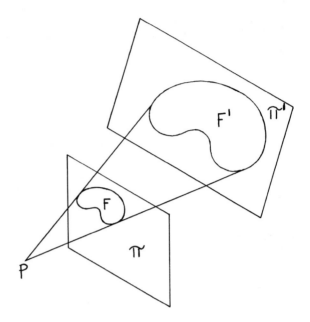

in everyday life. If we walk along a long corridor, it appears from our viewpoint
to be a parallelepiped, with the four lines of intersection of the walls with the floor
and the ceiling appearing to converge to a point. There is some kind of projectivity
from the floor and the ceiling onto our retinas. A classic result which applies to
both a given figure and its projection is *Brianchon's theorem*: if a hexagon ABCDEF
circumscribes a conic, the lines AD, BE, CF connecting opposite vertices intersect
at a single point P, as shown in Fig. 4.5; this remains valid even when projecting the
hexagon onto another plane.

Henri Poincaré wrote: "La geometrie est avant tout l'etude analytique d'un
groupe" (Geometry is, from the most fundamental point of view, the analytic study
of a group). Changing the group varies the geometrical properties which come into
play. In this way, we can have metric, linear, projective, and topological properties

Fig. 4.5 Brianchon's
theorem: in the hexagon
ABCDEF circumscribed
about a conic, the diagonals
AD, BE, CF intersect at a
single point P

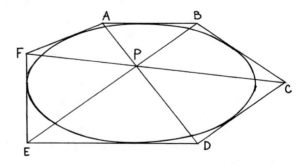

of geometrical objects. Another very important aspect concerns our observations of the physical world. Consider a cube-shaped object, and suppose that we move it three meters to the east or, what amounts to the same thing, the observer moves three meters to the west; in the first case we have performed an active transformation, in the second a passive transformation, in which it is the reference frame which moves. In either case, the transformation belongs to the translation group. Distances between any two vertices of the cube remain invariant. The same happens if we *rotate* the cube, or, equivalently, if we turn our head. Many other properties, including mass, size, and color, also remain unchanged under translational or rotational transformations.

This need for the observer to "look for invariants" in the observed world is of such generality that, in our opinion, it is inherent also in the brain of animals. Let us imagine a gazelle who sees two lions, motionless but ready to pounce. The gazelle quickly glances at the first lion, and then at the second; in doing so, she turns her head. But there must be something in her brain that enables her to realize that even if she turns her head (so that her picture of the world changes), the first lion is still there, somehow remaining *invariant*. If we wanted to build a robot with visual sensors, able to distinguish objects even when his "eyes" are moving, probably an efficient way would be to teach him to apply to the input of every visual observation the appropriate mathematical transformations; he will thus be able to compare images obtained at different instants, keeping in mind some invariant (or nearly invariant) aspects of the object. "Smart missiles" operate this way, but we doubt that Nature does things the same way. However, there must be something "invariant" in relationships among the various images. These considerations bring us to propose two fundamental concepts

(1) Variations of images.
(2) Existence of common elements, or *invariants*, as emphasized in Klein's formulation of possible geometries.

According to the philosopher Robert Nozick (2001), the deep root of objectivity is the *invariance* under different transformations. He says that an objective fact must be accessible from different points of view, and there is an intersubjective agreement among them. Before Nozick, Dirac wrote: "The important things in the world appear as the invariants ... of these transformations." In our opinion, Nozick was perfectly right. For example, when we sleep, our dreams lack this type of "invariance." Nozick gave an appealingly convincing meaning to the concept of the "objective world." It is interesting to recall that, about 2500 years ago, two great Greek philosophers, Heraclitus and Parmenides, affirmed the importance of these two conceptual elements: the former said *panta rhei*, "everything flows," while the latter stressed the "being, timeless, uniform, and unchanging nature of things." After these currents of thought, Plato/Aristotle tried to reconcile being with becoming by means of the notion of *eidos*, essence. The world would be formed by stable structures in the midst of continuous unpredictable change. But only in the modern formulation of geometry (Klein) and the physical laws (Einstein, Dirac), has a most fruitful and consistent synthesis of the two elements emerged.

4.2 Isotropic Vectors

We begin this section by considering a somewhat counterintuitive mathematical object. We have seen that points (x, y) in the Cartesian plane can be considered as "endpoints" of real vectors $OP = (x, y)$. Now consider the equation

$$x^2 + y^2 = 0, \tag{4.1}$$

which, in the real number field, has only the trivial solution $x = 0$, $y = 0$. But using complex numbers, (4.1) can be factored using

$$x^2 + y^2 = x^2 + y^2 - ixy + ixy = (x + iy)(x - iy) = 0. \tag{4.2}$$

It follows that all points on the complex straight lines

$$y = \pm ix \tag{4.3}$$

belong to the circle of zero radius (4.1)! Of course, the "magic" enters with the use of complex coordinates; still it is amazing that an infinitesimal circle can split into two straight lines.

Let us now consider two vectors z_1, z_2, belonging, respectively, to the two lines (4.3)

$$z_1 = (1, i), \qquad z_2 = (1, -i). \tag{4.4}$$

(We do not denote these vectors by $\mathbf{z}_1$, $\mathbf{z}_2$ since, being complex, they do not belong to our physical space). Note that each $z^2 = 0$ (z^2 not to be confused with z^*z). These complex vectors have the very interesting property of being *invariant under rotation*. All vectors can be rotated, but z_1, z_2 remain fixed. Applying the rotation matrix (see Chap. 2) to z_1, we find:

$$R z_1 = \left\| \begin{matrix} \cos\theta & -\sin\theta \\ \sin\theta & \cos\theta \end{matrix} \right\| \left\| \begin{matrix} 1 \\ i \end{matrix} \right\| = \left\| \begin{matrix} \cos\theta - i\sin\theta \\ \sin\theta + i\cos\theta \end{matrix} \right\| =$$

$$(\cos\theta - i\sin\theta) \left\| \begin{matrix} 1 \\ i \end{matrix} \right\| = e^{-i\theta} z_1. \tag{4.5}$$

We see that z_1 and z_2 are *eigenvectors* of the rotation matrix for any value of θ, with corresponding eigenvalues $e^{\mp i\theta}$, which are just phase factors of magnitude 1; z_1 and z_2 are called *isotropic vectors*.[1,2]

---

[1] A useful account of isotropic vectors and spinors, by their inventor, is found in Cartan (1966).

[2] Isotropic vectors can be pictured as having a direction, but zero magnitude. In four-dimensional Minkowski spacetime of special relativity, there occur *null* or *lightlike* 4-vectors. For example, $x^2 + y^2 + z^2 - c^2t^2 = 0$ describes the path of a light ray. It might be surmised that a light ray has a direction, but zero magnitude.

If we remain in the real plane $\mathbb{R}^2$, or in its complex generalization $\mathbb{C}^2$ (two-component complex vectors), there are no other invariant vectors with respect to the rotation matrix. But if we consider rotations in physical three-dimensional space $\mathbb{R}^3$, there is another invariant direction, that of the rotation axis. For example, considering the rotation R_z in the x–y plane

$$R_z = \begin{Vmatrix} \cos\theta & -\sin\theta & 0 \\ \sin\theta & \cos\theta & 0 \\ 0 & 0 & 1 \end{Vmatrix}. \tag{4.6}$$

The rotation is about the z axis, and the isotropic vectors Z_1, Z_2 are (we use capital letters to indicate that these are now three-dimensional):

$$Z_1 = (1, i, 0), \qquad Z_2 = (1, -i, 0). \tag{4.7}$$

In the following we will see that isotropic vectors are closely related to the representation of spin for elementary particles.

A random philosophical observation: humans existed for hundreds of thousands of years before finding a technical application for a rotation axis, namely, the wheel; mathematicians understood the importance of the isotropic vectors Z_1, Z_2 only a century ago. But Nature knew all about them since the beginning of time, implicitly using them in the rotations of stars and planets and in the spin of particles.

4.3   The Stereographic Projection

An elegant way to introduce the quantum mechanical spin of particles is to establish a correspondence between the points P with coordinates (x, y, z) on a sphere of radius 1, satisfying the equation

$$x^2 + y^2 + z^2 = 1, \tag{4.8}$$

and points in the complex plane. We will also see that there is a close correspondence between the two isotropic directions, Z_1, Z_2 in Eq. (4.7) and the representation of spin states, which show analogous behavior under rotation. Also, as we shall see, the mathematical description of spin will necessarily involve complex variables.

Let $O(0, 0, 0)$ be the center of a sphere, $N(0, 0, 1)$ the north pole and $S(0, 0, -1)$ the south pole (see Fig. 4.6). Let $P'(x', y', 0)$ be the intersection of the line SP with the equatorial plane $z = 0$, and $Q(0, 0, z)$ the projection of $P(x, y, z)$ on the z axis. P' is called the *stereographic projection*[3] of P. From the similar triangles SOP', SQP we find:

---

[3]In complex analysis, the inverse transformation is considered: the extended complex plane (including the point at infinity) can be projected onto the *Riemann sphere*.

Fig. 4.6 Stereographic projection of the unit sphere from the south pole S onto the plane $z = 0$; it sends the northern hemisphere onto the region inside the unit circle, the southern hemisphere onto the region outside the unit circle; the equator coincides with the circle. The closer P is to S, the more distant is the image P' from O in the *shaded plane*

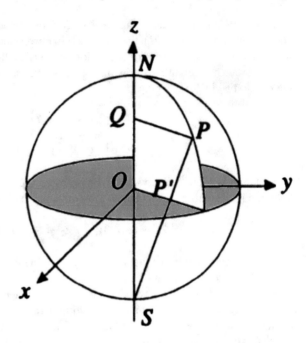

$$\frac{x'}{x} = \frac{y'}{y} = \frac{SO}{SQ} = \frac{1}{1+z}.$$

(4.9)

Now introduce, in the plane $z = 0$, the complex variable ζ

$$\zeta = x' + iy' = \frac{x + iy}{1+z},$$

(4.10)

It is useful to write ζ as the ratio of two complex numbers

$$\zeta = \frac{\phi}{\psi}, \qquad \phi = \alpha(x + iy), \qquad \psi = \alpha(1 + z),$$

(4.11)

where α is a constant to be determined. We might perhaps anticipate that a vector (ψ, ϕ) might represent the quantum mechanical spin state associated with the fundamental particles of matter: electrons, protons, neutrons, etc. The term *spin* implies rotation, since the pioneers of quantum theory originally imagined the electron as a tiny rotating sphere. In our case, the rotation axis is parallel to $P(x, y, z)$. For example, if the rotation is about the z-axis, one isotropic vector can be $\left\| \begin{matrix} 1 \\ 0 \end{matrix} \right\|$, corresponding to the *spin-up* state. The spin state is represented by a *spinor* $\left\| \begin{matrix} \psi \\ \phi \end{matrix} \right\|$, belonging to $\mathbb{C}^2$, the two-dimensional complex vector space. Two vectors $|u\rangle$, $|v\rangle$ form an orthonormal basis in $\mathbb{C}^2$ if they are of unit length and orthogonal:

$$\langle \mathbf{u}|\mathbf{u}\rangle = \langle \mathbf{v}|\mathbf{v}\rangle = 1, \qquad \langle \mathbf{u}|\mathbf{v}\rangle = 0. \tag{4.12}$$

Since quantum states are represented by vectors of unit length, the spinor $|\Psi\rangle = \left\| \begin{array}{c} \psi \\ \phi \end{array} \right\|$ must fulfill the condition

$$\langle \Psi|\Psi\rangle = \psi^*\psi + \phi^*\phi = |\psi|^2 + |\phi|^2 = 1. \tag{4.13}$$

Using (4.8) and (4.11), we find

$$|\psi|^2 + |\phi|^2 = |\alpha|^2\left[x^2 + y^2 + (1+z)^2\right] = |\alpha|^2(2+2z) = 1, \tag{4.14}$$

which determines the constant α

$$|\alpha| = \frac{1}{\sqrt{2(1+z)}}. \tag{4.15}$$

to within a phase factor $e^{i\varphi}$. Taking the complex conjugate of (4.11) we have

$$\psi^* = \alpha^*(1+z), \qquad \phi^* = \alpha^*(x - iy). \tag{4.16}$$

From Eqs. (4.11) and (4.14), the following relations between the spinor $|\Psi\rangle = \left\| \begin{array}{c} \psi \\ \phi \end{array} \right\|$ and the vector $\overrightarrow{OP} = (x, y, z)$ can be derived[4]:

$$x = \psi\phi^* + \psi^*\phi, \qquad y = i(\psi\phi^* - \psi^*\phi), \qquad z = \psi\psi^* - \phi\phi^*. \tag{4.17}$$

Note that while the spinor (ψ, ϕ) uniquely determines $\overrightarrow{OP}(x, y, z)$, the converse is not true, since α is determined only up to a phase factor $e^{i\varphi}$. In particular, the spinor $|\Psi\rangle = (\psi, \phi)$ and the spinor $-|\Psi\rangle = (-\psi, -\phi)$ correspond to the *same* vector $\overrightarrow{OP}(x, y, z)$.

At the beginning of this section we showed that, to the point $P(x, y, z)$ of the sphere of radius 1, there corresponds (up to a phase factor $e^{i\varphi}$) a complex spinor $|\Psi\rangle = \left\| \begin{array}{c} \psi \\ \phi \end{array} \right\|$ of norm 1. It is easy to verify that $|\Psi\rangle$ is an eigenvector of the Hermitian matrix

---

[4]By elementary calculations, using (4.8):

(1) $|\alpha|^2\left[(1+z)(x-iy) + (1+z)(x+iy)\right] = 2|\alpha|^2(1+z)x = x$;

(2) $i|\alpha|^2\left[(1+z)(x-iy) - (1+z)(x+iy)\right] = 2|\alpha|^2(1+z)y = y$;

(3) $|\alpha|^2\left[(1+z)^2 - x^2 - y^2\right] = |\alpha|^2\left[(1+z^2+2z-x^2-y^2\right] = |\alpha|^2\left[x^2+y^2+z^2+z^2+2z-x^2-y^2\right] = 2|\alpha|^2(1+z)z = z$.

$$H = H(\overrightarrow{OP}) = \left\| \begin{matrix} z & x - iy \\ x + iy & -z \end{matrix} \right\|, \tag{4.18}$$

with the eigenvalue $+1$:

$$H|\Psi\rangle = \left\| \begin{matrix} z & x - iy \\ x + iy & -z \end{matrix} \right\| \left\| \begin{matrix} \psi \\ \phi \end{matrix} \right\| = \left\| \begin{matrix} z\psi + (x - iy)\phi \\ (x + iy)\psi - z\phi \end{matrix} \right\| =$$

$$\alpha \left\| \begin{matrix} z(1 + z) + x^2 + y^2 \\ (x + iy)(1 + z) - z(x + iy) \end{matrix} \right\| = \alpha \left\| \begin{matrix} 1 + z \\ x + iy \end{matrix} \right\| = \left\| \begin{matrix} \psi \\ \phi \end{matrix} \right\| = |\Psi\rangle. \tag{4.19}$$

Let us denote by $|\Phi\rangle = (X, Y)$ the second normalized eigenvector of H; $|\Phi\rangle$ must fulfill the conditions

$$\langle \Phi | \Phi \rangle = 1, \qquad \langle \Psi | \Phi \rangle = 0. \tag{4.20}$$

Thus $X\psi^* + Y\phi^* = 0$, $X^*X + Y^*Y = 1$, and we can set:

$$X = \phi^*, \qquad Y = -\psi^*, \qquad \text{and thus} \qquad |\Phi\rangle = \left\| \begin{matrix} \phi^* \\ -\psi^* \end{matrix} \right\|. \tag{4.21}$$

Clearly, $H|\Phi\rangle = -|\Phi\rangle$; so that the eigenvalue of $|\Phi\rangle$ equals -1.

The matrix H represents a physical observable, the projection of the spin in the direction $\overrightarrow{OP} = (x, y, z)$. The possible results of a measurement are $+1$ or -1. In the former case, after the measurement the spin state is Ψ, we can say that the spin is *parallel* to $\overrightarrow{OP}$; in the latter case, the spin state is $|\Phi\rangle$, and the spin is *antiparallel* to $\overrightarrow{OP}$. This interpretation is consistent with the following: changing $\overrightarrow{OP}$ in $-\overrightarrow{OP}$ we have:

$$H(-\overrightarrow{OP}) = \left\| \begin{matrix} -z & -x + iy \\ -x - iy & z \end{matrix} \right\| = -H(\overrightarrow{OP}), \tag{4.22}$$

thus the eigenvalues are reversed

$$H(-\overrightarrow{OP})|\Psi\rangle = -|\Psi\rangle, \qquad H(-\overrightarrow{OP})|\Phi\rangle = |\Phi\rangle. \tag{4.23}$$

If $x = y = 0$, $z = 1$, $\overrightarrow{OP}$ coincides with the north pole $\overrightarrow{ON}$, and H is diagonal

$$H(\overrightarrow{ON}) = \left\| \begin{matrix} 1 & 0 \\ 0 & -1 \end{matrix} \right\|. \tag{4.24}$$

The eigenvectors are $e_1 = (1, 0)$ and $e_2 = (0, 1)$. Suppose now we measure $H(\overrightarrow{ON})$ on an arbitrary spinor $|\Psi\rangle = (\psi, \phi)$. The probabilities p_1, p_2 of obtaining the results $+1$ and -1 are respectively

$$|\langle e_1 | \Psi \rangle|^2 = |\psi|^2, \qquad |\langle e_2 | \Psi \rangle|^2 = |\phi|^2. \tag{4.25}$$

From (4.11) and (4.15), we see that the probabilities p_1 and p_2 depend only on z

$$p_1 = |\psi|^2 = \frac{1+z}{2}, \quad p_2 = |\phi|^2 = \frac{1-z}{2}. \tag{4.26}$$

This result is reasonable: if $\overrightarrow{OP} = \overrightarrow{ON}$, $z = 1$, so that $p_1 = 1$ and $p_2 = 0$; if $\overrightarrow{OP} = \overrightarrow{OS}$ (south pole), $z = -1$, $p_1 = 0$ and $p_2 = 1$. Finally, if $\overrightarrow{OP}$ lies somewhere on the equatorial circle, $z = 0$ and the two probabilities are equal: $p_1 = p_2 = \frac{1}{2}$.

As mentioned at the beginning of this section, we can prove that there is a correspondence between isotropic vectors and spinors. First, let us generalize the concept of isotropic vectors to an arbitrary plane. The equation

$$x^2 + y^2 + z^2 = 0 \tag{4.27}$$

admits only the solution $x = y = z = 0$ in the real field. However in the complex field there is an infinite number of solutions. In the plane $z = 0$, we find the invariant vectors Z_1, Z_2, from Eq. (4.7). For example, $x = 3, y = 4, z = 5i$, etc. All these vectors determine an *isotropic* direction. Let us now consider a plane π through the origin, and let $\mathbf{X}_1$, $\mathbf{X}_2$ be two orthonormal vectors belonging to π. Let $\mathbf{n}$ be a unit vector orthogonal to π (see Fig. 4.7). Therefore $\{\mathbf{X}_1, \mathbf{X}_2, \mathbf{n}\}$ forms an orthonormal basis. Since both sets $\{\mathbf{e}_1, \mathbf{e}_2, \mathbf{e}_3\}$ and $\{\mathbf{X}_1, \mathbf{X}_2, \mathbf{X}_3\}$ are orthonormal bases, we can use the following correspondence between vectors related to the plane $z = 0$ and vectors belonging to π (Table 4.1).

Fig. 4.7 The axis $\mathbf{n}$ of a rotation in the plane π, and two orthonormal vectors $\mathbf{x}_1$, $\mathbf{x}_2$ in the plane

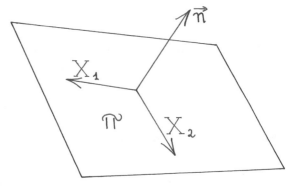

Table 4.1 Relationship between vectors

Plane $z = 0$ normal $\mathbf{e}_3 = (0, 0, 1)$	Plane π normal $\mathbf{n}$
$\mathbf{e}_1 = (1, 0, 0)$, $\mathbf{e}_2 (0, 1, 0)$	$\mathbf{X}_1$, $\mathbf{X}_2$
$Z_1 = \mathbf{e}_1 + i\mathbf{e}_2$; $Z_2 = \mathbf{e}_1 - i\mathbf{e}_2$	$Z_1 = \mathbf{X}_1 + i\mathbf{X}_2$; $Z_2 = \mathbf{X}_1 - i\mathbf{X}_2$

$Z_1 = \mathbf{X}_1 + i\mathbf{X}_2$ and $Z_2 = \mathbf{X}_1 - i\mathbf{X}_2$ belong to π since $\mathbf{X}_1$ and $\mathbf{X}_2$ do, while Z_1 and Z_2 are *isotropic* directions in the plane π. Indeed we can write, using the ordinary (not Hermitian) scalar product

$$(Z_1, Z_1) = (\mathbf{X}_1 + i\mathbf{X}_2, \mathbf{X}_1 + i\mathbf{X}_2) = \\ (\mathbf{X}_1, \mathbf{X}_1) + i(\mathbf{X}_1, \mathbf{X}_2) + i(\mathbf{X}_2, \mathbf{X}_1) - (\mathbf{X}_2, \mathbf{X}_2) = +1 - 1 = 0, \tag{4.28}$$

and analogously $(Z_2, Z_2) = 0$. As an example:, let $\mathbf{n}$ be the vector $\frac{1}{\sqrt{3}}(1, 1, 1)$, and $\mathbf{R} = (x, y, z)$ an arbitrary vector. The condition that $\mathbf{R}$ belongs to π is $(\mathbf{n}, \mathbf{R}) = 0$, or

$$x + y + z = 0. \tag{4.29}$$

We can choose:

$$\mathbf{X}_1 = \frac{1}{\sqrt{2}}(-1, 1, 0), \qquad \mathbf{X}_2 = \frac{1}{\sqrt{6}}(1, 1, -2). \tag{4.30}$$

Thus the isotropic directions of π can be chosen as

$$Z_{1,2} = \left(-\frac{1}{\sqrt{2}} \pm \frac{i}{\sqrt{6}}, \frac{1}{\sqrt{2}} \pm \frac{i}{\sqrt{6}}, \mp 2i\right) \tag{4.31}$$

Returning to the general case, we identify the unit vector $\mathbf{n} = (n_1, n_2, n_3)$ with the vector $\overrightarrow{OP}$ to the sphere of radius 1 in the stereographic projection (see Fig. 4.6). We still denote by π the plane through the origin orthogonal to $\mathbf{n}$, and by $Z_1 = (a, b, c)$, $Z_2 = (a^*, b^*, c^*)$ the two isotropic vectors of π. We prove the following:

Theorem 4.1 *Suppose the spin of a particle is parallel to the direction* $\mathbf{n} = (n_1, n_2, n_3)$ *of physical space. Then the two components* ψ, ϕ *of the corresponding spinor are related to the isotropic directions of* π *by the following simple formulas:*

$$\psi^2 = \frac{1}{2}(a - ib), \qquad \phi^2 = -\frac{1}{2}(a + ib), \qquad -2\psi\phi = c \tag{4.32}$$

Proof Clearly (4.32) is equivalent to

$$a = \psi^2 - \phi^2, \qquad b = i(\psi^2 + \phi^2), \qquad c = -2\psi\phi. \tag{4.33}$$

The theorem can be proved in two steps

(1) Show that the vector (a, b, c), as defined by Eq. (4.33), is isotropic;
(2) Show that it belongs to π.

Step (1) follows easily, since

$$a^2 + b^2 + c^2 = (\psi^2 - \phi^2)^2 - (\psi^2 + \phi^2)^2 + 4\psi^2\phi^2 = 0 \tag{4.34}$$

In order to prove Step (2), we compute the scalar product of (a, b, c) with (n_1, n_2, n_3); but first let us slightly modify (4.11), using $|\alpha| = \frac{1}{\sqrt{2(1+z)}}$. Since now $(x, y, z) = (n_1, n_2, n_3)$, we have

$$\psi = \frac{1}{\sqrt{2(1 + n_3)}}(1 + n_3), \qquad \phi = \frac{1}{\sqrt{2(1 + n_3)}}(n_1 + in_2). \tag{4.35}$$

From Eqs. (4.35) and (4.33) we have

$$a = \psi^2 - \phi^2 = \frac{1}{2(1 + n_3)}\left[(1 + n_3)^2 - (n_1^2 - n_2^2 + 2in_1n_2)\right], \tag{4.36}$$

$$b = i(\psi^2 + \phi^2) = \frac{i}{2(1 + n_3)}\left[(1 + n_3)^2 + (n_1^2 - n_2^2 + 2in_1n_2)\right], \tag{4.37}$$

$$c = -2\psi\phi = \frac{1}{2(1 + n_3)}\left[-2(1 + n_3)(n_1 + in_2)\right]. \tag{4.38}$$

Neglecting the common factor $\frac{1}{2(1+n_3)}$, it is tedious but elementary to multiply Eqs. (4.36)–(4.38) by n_1, n_2, n_3, respectively, add the three results, and prove that both the real and the imaginary part of $(\mathbf{n}, Z_1) = an_1 + bn_2 + cn_3$ vanish. Thus (a, b, c) is an isotropic vector belonging to π. The Theorem is proved. Since, in formulas (4.32) and (4.33), the squares ψ^2, ϕ^2 are related linearly with a, b, c, it is sometimes surmised that "spinors are square roots of vectors."[5]

Let us next consider what happens to the spinor (ψ, ϕ) when we perform a *rotation* R in physical space. In the simplest case is $R = R_z$ (Eq. (4.6)), a rotation by an angle θ in the xy-plane

$$R_z = \left\|\begin{matrix} \cos\theta & -\sin\theta & 0 \\ \sin\theta & \cos\theta & 0 \\ 0 & 0 & 1 \end{matrix}\right\|. \tag{4.39}$$

From (4.7) we know that:

$$Z_1 = \left\|\begin{matrix} 1 \\ i \\ 0 \end{matrix}\right\| = \left\|\begin{matrix} a \\ b \\ c \end{matrix}\right\|, \qquad Z_2 = \left\|\begin{matrix} 1 \\ -i \\ 0 \end{matrix}\right\| = \left\|\begin{matrix} a^* \\ b^* \\ c^* \end{matrix}\right\|, \tag{4.40}$$

and from (4.32) we find the vectors corresponding to the spinors $\left\|\begin{matrix} \psi_1 \\ \phi_1 \end{matrix}\right\|, \left\|\begin{matrix} \psi_2 \\ \phi_2 \end{matrix}\right\|$:

$$\left\|\begin{matrix} \psi_1^2 \\ \phi_1^2 \end{matrix}\right\| = \left\|\begin{matrix} \frac{a-ib}{2} \\ -\frac{a+ib}{2} \end{matrix}\right\| = \left\|\begin{matrix} 1 \\ 0 \end{matrix}\right\|, \qquad \left\|\begin{matrix} \psi_2^2 \\ \phi_2^2 \end{matrix}\right\| = \left\|\begin{matrix} \frac{a^*-ib^*}{2} \\ -\frac{a^*+ib^*}{2} \end{matrix}\right\| = \left\|\begin{matrix} 0 \\ -1 \end{matrix}\right\|. \tag{4.41}$$

---

[5]Note that a two-dimensional vector can be represented by a complex number $z = x + iy = re^{i\theta}$. The square root is $\sqrt{z} = \sqrt{x + iy} = \sqrt{r}e^{i\theta/2}$, containing the half-angle, characteristic of spinors.

Since Z_1 and Z_2 are invariant under rotation, we have (see (4.5)):

$$R_z Z_1 = e^{-i\theta} Z_1, \qquad R_z Z_2 = e^{i\theta} Z_2. \tag{4.42}$$

Thus the effect of R_z is to multiply $\left\| \begin{matrix} \psi_1^2 \\ \phi_1^2 \end{matrix} \right\|$ and $\left\| \begin{matrix} \psi_2^2 \\ \phi_2^2 \end{matrix} \right\|$ by $e^{-i\theta}$ and $e^{i\theta}$ respectively; consequently the spinors $\left\| \begin{matrix} \psi_1 \\ \phi_1 \end{matrix} \right\|$ and $\left\| \begin{matrix} \psi_2 \\ \phi_2 \end{matrix} \right\|$ will be multiplied by $e^{\mp i\theta/2}$. This result is general: A rotation of angle θ in physical space, corresponds a "rotation" of $\theta/2$ in spinor space. In order to "prove" this statement, let us do some "experimental mathematics" by examining other relevant particular cases (which will be useful in the sequel).

Let R_x, R_y represent rotations by angle θ in the y–z and z–y planes, respectively

$$R_x = \left\| \begin{matrix} 1 & 0 & 0 \\ 0 & \cos\theta & -\sin\theta \\ 0 & \sin\theta & \cos\theta \end{matrix} \right\|, \qquad R_y = \left\| \begin{matrix} \cos\theta & 0 & \sin\theta \\ 0 & 1 & 0 \\ -\sin\theta & 0 & \cos\theta \end{matrix} \right\| \tag{4.43}$$

These relations can be summarized in a table

Table 4.2, it is very simple to check the eigenvalue equations (4.5) for $R = R_x$ and $R = R_y$; the eigenvalues are always $e^{\mp i\theta}$, and the eigenvectors are the isotropic vectors. The values of ψ^2 and ϕ^2 shown in the Table follow immediately. Since ψ^2 and ϕ^2 depend linearly on a and b, if we perform a rotation they will be multiplied by $e^{i\theta}$ and everything works as in the case of R_z. The spin-up spinor $\left\| \begin{matrix} \psi \\ \phi \end{matrix} \right\|$ turns out to be multiplied by $e^{-i\theta/2}$ and the spin-down spinor $\left\| \begin{matrix} \phi^* \\ -\psi^* \end{matrix} \right\|$ by $e^{i\theta/2}$. In the last column of Table 4.2, three particular cases of the observable H appear. We recall that H represents the spin components in the $\mathbf{n} = (n_1, n_2, n_3)$ direction. For the three cases: $\mathbf{n} = (1, 0, 0)$, $\mathbf{n} = (0, 1, 0)$, $\mathbf{n} = (0, 0, 1)$, the corresponding H matrices are the famous *Pauli spin matrices* $\sigma_x, \sigma_y, \sigma_z$, respectively

$$\sigma_x = \left\| \begin{matrix} 0 & 1 \\ 1 & 0 \end{matrix} \right\|, \qquad \sigma_y = \left\| \begin{matrix} 0 & -i \\ i & 0 \end{matrix} \right\|, \qquad \sigma_z = \left\| \begin{matrix} 1 & 0 \\ 0 & -1 \end{matrix} \right\|. \tag{4.44}$$

Considering $\sigma_x, \sigma_y, \sigma_z$ as the components of a vector $\boldsymbol{\sigma}$, we can write formally: $H = \mathbf{n} \cdot \boldsymbol{\sigma} = n_1 \sigma_x + n_1 \sigma_y + n_3 \sigma_z$. Thus $\sigma_x, \sigma_y, \sigma_z$ represent the spin components in the directions of the three axes, and H represents the spin component in the $\mathbf{n}$ direction. Of course these components are not numbers, but *matrices*. The eigenvalues of the spin matrices are ± 1 (see Eq. (4.19)) and the eigenvectors are the corresponding spinors for spin-up and spin-down states.

It is instructive to compare the representation of *orthogonal* states in real three-dimensional space to that in Hilbert space. The two opposite spatial orientations of a spin-$\frac{1}{2}$ particle, up ($\uparrow$) and down ($\downarrow$), are $180°$ apart. But in Hilbert space,

Table 4.2 Rotations and spinors

Rotation axis **n**	Rotation matrix	Isotropic vector Z_1	$\psi^2 = \frac{a-ib}{2}$	$\phi^2 = \frac{-a-ib}{2}$	Spin-up spinor $\lvert\Phi_1\rangle = (\psi, \phi)$	Spin-down spinor $\lvert\Phi_2\rangle = (\phi^*, -\psi^*)$	matrix H $H = \begin{Vmatrix} n_3 & n_1 - in_2 \\ n_1 + in_2 & -n_3 \end{Vmatrix}$
$\begin{Vmatrix} 1 \\ 0 \\ 0 \end{Vmatrix}$	R_x	$\begin{Vmatrix} 0 \\ 1 \\ i \end{Vmatrix}$	$-\frac{i}{2}$	$-\frac{i}{2}$	$\frac{1}{2}\begin{Vmatrix} 1-i \\ 1-i \end{Vmatrix}$	$\frac{1}{2}\begin{Vmatrix} 1+i \\ -1-i \end{Vmatrix}$	$\sigma_x = \begin{Vmatrix} 0 & 1 \\ 1 & 0 \end{Vmatrix}$
$\begin{Vmatrix} 0 \\ 1 \\ 0 \end{Vmatrix}$	R_y	$\begin{Vmatrix} -i \\ 0 \\ 1 \end{Vmatrix}$	$-\frac{i}{2}$	$+\frac{i}{2}$	$\frac{1}{2}\begin{Vmatrix} 1-i \\ 1+i \end{Vmatrix}$	$\frac{1}{2}\begin{Vmatrix} 1-i \\ -1-i \end{Vmatrix}$	$\sigma_y = \begin{Vmatrix} 0 & -i \\ i & 0 \end{Vmatrix}$
$\begin{Vmatrix} 0 \\ 0 \\ 1 \end{Vmatrix}$	R_z	$\begin{Vmatrix} 1 \\ i \\ 0 \end{Vmatrix}$	1	0	$\begin{Vmatrix} 1 \\ 0 \end{Vmatrix}$	$\begin{Vmatrix} 0 \\ -1 \end{Vmatrix}$	$\sigma_z = \begin{Vmatrix} 1 & 0 \\ 0 & -1 \end{Vmatrix}$

orthogonal vectors or spinors are oriented perpendicularly, 90° apart. This accounts for the occurrence of half-angles (such as $\theta/2$) in formulas involving spinors.

4.3.1   Spinors in Spherical Coordinates

An instructive alternative approach to the preceding results can be found by expressing $\mathbf{n} = (n_1, n_2, n_3)$ in spherical polar coordinates. Since $\mathbf{n}$ lies on the unit sphere, we find, for its components, $n_1 = \sin\theta\cos\varphi$, $n_2 = \sin\theta\sin\varphi$, $n_3 = \cos\theta$. Note also the combinations: $n_1 \pm in_2 = \sin\theta\, e^{\pm i\varphi}$. The Hermitian operator H in Eq. (4.18) can be expressed in spherical coordinates as

$$H = \mathbf{n}\cdot\boldsymbol{\sigma} = \left\| \begin{matrix} n_3 & n_1 - in_2 \\ n_1 + in_2 & -n_3 \end{matrix} \right\| = \left\| \begin{matrix} \cos\theta & \sin\theta\, e^{-i\varphi} \\ \sin\theta\, e^{i\varphi} & -\cos\theta \end{matrix} \right\|. \tag{4.45}$$

The eigenvalues of the spin, in any direction, have previously found to be ± 1, which we also call spin-up ($\uparrow$) and spin-down ($\downarrow$), respectively. Now let $|\Psi(\theta)\,\uparrow\rangle = \left\| \begin{matrix} \psi \\ \phi \end{matrix} \right\|$ be an eigenvector of H with the eigenvalue $+1$, so that

$$\left\| \begin{matrix} \cos\theta & \sin\theta\, e^{-i\varphi} \\ \sin\theta\, e^{i\varphi} & -\cos\theta \end{matrix} \right\| \left\| \begin{matrix} \psi \\ \phi \end{matrix} \right\| = \left\| \begin{matrix} \psi \\ \phi \end{matrix} \right\|. \tag{4.46}$$

Expanding out the matrix equation, we find

$$(\cos\theta)\psi + (\sin\theta\, e^{-i\varphi})\phi = \psi, \qquad (\sin\theta\, e^{i\varphi})\psi - (\cos\theta)\phi = \phi. \tag{4.47}$$

The second equation can be rearranged to

$$(\sin\theta\, e^{i\varphi})\psi - (1 + \cos\theta)\phi = 0. \tag{4.48}$$

An obvious solution is $\psi = \text{const}\,(1 + \cos\theta)$, $\phi = \text{const}\,\sin\theta\, e^{i\varphi}$. The constant can be determined by the normalization condition $|\psi|^2 + |\phi|^2 = 1$, which leads to $\text{const} = \pm 1/\sqrt{2 + 2\cos\theta}$. Choosing the $+$ sign, we can write

$$\psi = \frac{1 + \cos\theta}{\sqrt{2 + 2\cos\theta}} \quad \text{and} \quad \phi = \frac{\sin\theta\, e^{i\varphi}}{\sqrt{2 + 2\cos\theta}}. \tag{4.49}$$

We now make use of two half-angle trigonometric identities

$$\sqrt{\frac{1 + \cos\theta}{2}} = \cos\frac{\theta}{2} \quad \text{and} \quad \sin\theta = 2\sin\frac{\theta}{2}\cos\frac{\theta}{2}, \tag{4.50}$$

which simplifies Eq. (4.49) to

$$\psi = \cos\frac{\theta}{2}, \qquad \phi = \sin\frac{\theta}{2}e^{i\varphi}. \tag{4.51}$$

Finally, the spinor for the eigenvalue $+1$ (spin-up $\uparrow$) can be written:

$$|\Psi(\theta)\uparrow\rangle = \left\|\begin{array}{c} \cos(\theta/2) \\ \sin(\theta/2)e^{i\varphi} \end{array}\right\|. \tag{4.52}$$

By an analogous analysis for the eigenvalue -1 (spin-down $\downarrow$), it can be shown that

$$|\Psi(\theta)\downarrow\rangle = \left\|\begin{array}{c} \sin(\theta/2) \\ -\cos(\theta/2)e^{i\varphi} \end{array}\right\|. \tag{4.53}$$

It is very clear that the spinors (4.52) and (4.53) are mutually orthogonal.

For this Hilbert space, the basis spinors are the states with spin-up $|\uparrow\rangle$ and spin-down $|\downarrow\rangle$ along the axis with $\theta = 0$

$$|\uparrow\rangle = \left\|\begin{array}{c} 1 \\ 0 \end{array}\right\| \quad \text{and} \quad |\downarrow\rangle = \left\|\begin{array}{c} 0 \\ 1 \end{array}\right\|. \tag{4.54}$$

The spinors (4.52) and (4.53) can thereby be expressed as the linear combinations

$$|\Psi(\theta)\uparrow\rangle = \cos\frac{\theta}{2}|\uparrow\rangle + \sin\frac{\theta}{2}e^{i\varphi}|\downarrow\rangle, \quad |\Psi(\theta)\downarrow\rangle = \sin\frac{\theta}{2}|\uparrow\rangle - \cos\frac{\theta}{2}e^{i\varphi}|\downarrow\rangle. \tag{4.55}$$

In a measurement of the z spin component, the probabilities of observing spin-up and spin-down in the state $|\Psi(\theta)\uparrow\rangle$ are, respectively,

$$p(\uparrow) = |\langle\uparrow|\Psi(\theta)\rangle|^2 = \cos^2\frac{\theta}{2} \quad \text{and} \quad p(\downarrow) = |\langle\downarrow|\Psi(\theta)\rangle|^2 = \sin^2\frac{\theta}{2}. \tag{4.56}$$

It should be noted that these relations apply to a single electron spin. In the following chapter, we will be dealing with analogous measurements in a two-electron system.

4.4 Lie Groups: SO(3) and Vector Rotation, SU(2) and Spinor Rotation

We know that 3×3 rotation matrices leave invariant the scalar product of vectors in the physical space $\mathbb{R}^3$; this set of matrices comprise the group SO(3). S stands for *special* meaning that the matrices have determinant $+1$. As a consequence, the group does not include any transformations that map an object into its mirror image in any plane. Similarly, the 2×2 complex matrices that leave invariant the Hermitian scalar product between spinors (normalized vectors in $\mathbb{C}^2$) constitute a group called SU(2). The norms (lengths) of both types of vectors is left invariant by their respective transformations. A simple and instructive way to study the relation between rotations in physical space and "rotations" in spinor space is to consider infinitesimal rotations: rotations that differ infinitesimally from the identity. Recall that the identity matrix I is defined, in any number of dimensions, by $I\mathbf{v} = \mathbf{v}$ for any $\mathbf{v}$. The 2×2 and 3×3 identity matrices are:

$$I_2 = \begin{Vmatrix} 1 & 0 \\ 0 & 1 \end{Vmatrix}, \qquad I_3 = \begin{Vmatrix} 1 & 0 & 0 \\ 0 & 1 & 0 \\ 0 & 0 & 1 \end{Vmatrix}. \tag{4.57}$$

Let us consider first the two-dimensional case. Let $R(\theta)$ denote the rotation matrix

$$R(\theta) = \begin{Vmatrix} \cos\theta & -\sin\theta \\ \sin\theta & \cos\theta \end{Vmatrix}. \tag{4.58}$$

Recall the series expansions:

$$\cos\theta = 1 - \frac{\theta^2}{2!} + \frac{\theta^4}{4!} - \dots, \qquad \sin\theta = \theta - \frac{\theta^3}{3!} + \frac{\theta^5}{5!} - \dots \tag{4.59}$$

As $\theta \to 0$, $\cos\theta \to 1$ and $\sin\theta \to 0$. We can safely use the approximations $\cos\delta\theta \approx 1$ and $\sin\delta\theta \approx 0$ as long as $\delta\theta$ is infinitesimal. Thus, given a vector $\mathbf{v} = \begin{Vmatrix} x \\ y \end{Vmatrix}$ in the plane, we can write

$$R\,\mathbf{v} \simeq \begin{Vmatrix} 1 & -\delta\theta \\ \delta\theta & 1 \end{Vmatrix} \begin{Vmatrix} x \\ y \end{Vmatrix} = \begin{Vmatrix} 1 & 0 \\ 0 & 1 \end{Vmatrix} \begin{Vmatrix} x \\ y \end{Vmatrix} + \delta\theta \begin{Vmatrix} 0 & -1 \\ 1 & 0 \end{Vmatrix} \begin{Vmatrix} x \\ y \end{Vmatrix} \simeq (I_2 + \Lambda_0 \delta\theta)\mathbf{v}, \tag{4.60}$$

where Λ_0 is the derivative $\frac{dR}{d\theta}$ computed at $\theta = 0$

$$\Lambda_0 = \begin{Vmatrix} 0 & -1 \\ 1 & 0 \end{Vmatrix}, \tag{4.61}$$

Note that $\Lambda_0\mathbf{v} = \begin{Vmatrix} -y \\ x \end{Vmatrix}$, thus $\langle \mathbf{v} | \Lambda_0 \mathbf{v} \rangle = 0$, so that $\Lambda_0\mathbf{v}$ is orthogonal to $\mathbf{v}$.

Let us now examine the geometrical meaning of these formulas. For any value of θ, the vector $R(\theta)\mathbf{v}$ is obtained by a rotation of the vector $\mathbf{v}$ by an angle θ. But for infinitesimal $\delta\theta$, $R(\delta\theta)\mathbf{v} - \mathbf{v}$ is tangent to the circle of radius $|\mathbf{v}|$ at the point $\mathbf{v}$ (see Fig. 4.8). This is completely analogous to the behavior of the velocity vector for a particle in circular motion: tangent to the circle and normal to the radius. The case of an infinitesimal rotation $R_z(\delta\theta)$ around the z axis is entirely analogous. With $\mathbf{v} = (x, y, z)$, we have

Fig. 4.8 For small $\delta\theta$, $R(\delta\theta)\mathbf{v} - \mathbf{v}$ is "almost" tangent to the circle of radius $|\mathbf{v}|$ at the point $\mathbf{v}$

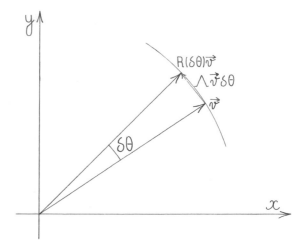

$$R_z(\delta\theta)\,\mathbf{v} \simeq \left\| \begin{matrix} 1 & -\delta\theta & 0 \\ \delta\theta & 1 & 0 \\ 0 & 0 & 1 \end{matrix} \right\| \left\| \begin{matrix} x \\ y \\ z \end{matrix} \right\| = (I_3 + \Lambda_3\delta\theta)\mathbf{v}, \tag{4.62}$$

where

$$\Lambda_3 = \left\| \begin{matrix} 0 & -1 & 0 \\ 1 & 0 & 0 \\ 0 & 0 & 0 \end{matrix} \right\|. \tag{4.63}$$

The vector $\Lambda_3\mathbf{v} = (-y, x, 0)$ is orthogonal to $\mathbf{v} = (x, y, z)$ and can be imagined to be tangent to the circle traced out by $R(\theta)\mathbf{v}$ as θ varies between 0 and 2π (see Fig. 4.9). We call Λ_3 the *infinitesimal rotation operator* (or *generator*) about the z axis. We can repeat the same arguments for the cases of the x and y axis (see Figs. 4.10 and 4.11). In this way we find the infinitesimal operators Λ_1, Λ_2:

$$R_x(\delta\theta)\,\mathbf{v} \simeq \left\| \begin{matrix} 1 & 0 & 0 \\ 0 & 1 & -\delta\theta \\ 0 & \delta\theta & 1 \end{matrix} \right\| \mathbf{v} = (I_3 + \Lambda_1\delta\theta)\mathbf{v}, \qquad \Lambda_1 = \left\| \begin{matrix} 0 & 0 & 0 \\ 0 & 0 & -1 \\ 0 & 1 & 0 \end{matrix} \right\|. \tag{4.64}$$

$$R_y(\delta\theta)\,\mathbf{v} \simeq \left\| \begin{matrix} 1 & 0 & \delta\theta \\ 0 & 1 & 0 \\ -\delta\theta & 0 & 1 \end{matrix} \right\| \mathbf{v} = (I_3 + \Lambda_2\delta\theta)\mathbf{v}. \qquad \Lambda_2 = \left\| \begin{matrix} 0 & 0 & 1 \\ 0 & 0 & 0 \\ -1 & 0 & 0 \end{matrix} \right\| \tag{4.65}$$

Let us now consider the general case of a rotation about an axis directed along the

Fig. 4.9 The vector $\Lambda_3 \mathbf{v}$ for
a rotation R_z around the z
axis

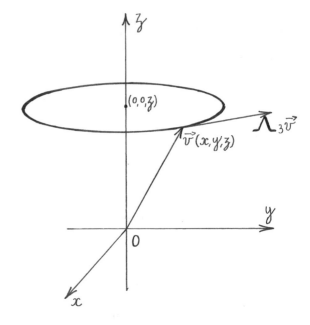

Fig. 4.10 The vector $\Lambda_1 \mathbf{v}$
for a rotation R_x around the x
axis

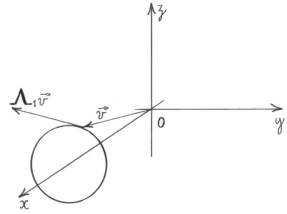

unit vector $\mathbf{n} = (n_1, n_2, n_3)$. We assume again that $\delta\theta$ is infinitesimal. The following
theorem assures that it is "reasonable" to write $R(\delta\theta) \simeq I_3 + \Lambda\delta\theta$, where

$$\Lambda = \begin{Vmatrix} 0 & -n_3 & n_2 \\ n_3 & 0 & -n_1 \\ -n_2 & n_1 & 0 \end{Vmatrix} = n_1\Lambda_1 + n_2\Lambda_2 + n_3\Lambda_3 \qquad (4.66)$$

Theorem 4.2 *For any* $\mathbf{v} = (x, y, z)$, *the vector* $\Lambda\mathbf{v}$ *is orthogonal both to* $\mathbf{v}$ *and the
rotation axis* $\mathbf{n}$. *The rotation from* $\mathbf{v}$ *to* $\Lambda\mathbf{v}$ *is "seen" as counterclockwise from* $\mathbf{n}$.

Fig. 4.11 The vector $\Lambda_2 \mathbf{v}$ for a rotation R_y around the y axis

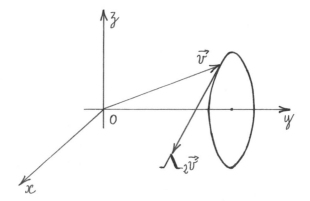

Proof We have

$$\Lambda \mathbf{v} = \begin{Vmatrix} 0 & -n_3 & n_2 \\ n_3 & 0 & -n_1 \\ -n_2 & n_1 & 0 \end{Vmatrix} \begin{Vmatrix} x \\ y \\ z \end{Vmatrix} = \begin{Vmatrix} -n_3 y + n_2 z \\ n_3 x - n_1 z \\ -n_2 x + n_1 y \end{Vmatrix}. \tag{4.67}$$

By the definition of scalar product, we have

$$\mathbf{v} \cdot \Lambda \mathbf{v} = x(-n_3 y + n_2 z) + y(n_3 x - n_1 z) + z(-n_2 x + n_1 y) = 0, \tag{4.68}$$

$$\mathbf{n} \cdot \Lambda \mathbf{v} = n_1(-n_3 y + n_2 z) + n_2(n_3 x - n_1 z) + n_3(-n_2 x + n_1 y) = 0. \tag{4.69}$$

With regard to the last part of the Theorem, it is sufficient to note that the general case $\mathbf{n} = (n_1, n_2, n_3)$ can be transformed continuously to $\mathbf{n} = (0, 0, 1)$, where the property holds (see also Figs. 4.9, 4.10 and 4.11). Remark: The Theorem can be shown more directly by noting that $\Lambda \mathbf{v} = \mathbf{n} \times \mathbf{v}$. We have however chosen to show the elementary calculations in detail.

For all the possible axis $\mathbf{n}$, the relative positions of $\mathbf{v}$, $\Lambda \mathbf{v}$ and $\mathbf{n}$ are the same as in the three particular cases $\mathbf{n} = (1, 0, 0)$, $\mathbf{n} = (0, 1, 0)$, $\mathbf{n} = (0, 0, 1)$. The reader might wonder if an infinitesimal rotation $R(\delta\theta)$ about $\mathbf{n}$ can always be written as

$$R(\delta\theta) \simeq I_3 + \Lambda\delta\theta = I_3 + \delta\theta(n_1 \Lambda_1 + n_2 \Lambda_2 + n_3 \Lambda_3) \tag{4.70}$$

This conjecture turns out to be true, although we will not give a rigorous proof.

Now that we have learned about infinitesimal rotations in three dimensions, we turn to the analogous case for two complex dimensions. Since normalized two-dimensional vectors of $\mathbb{C}^2$ represent physical states, we will use Dirac notation. Let us consider a general 2×2 complex matrix

$$U = \begin{Vmatrix} a & b \\ c & d \end{Vmatrix}. \tag{4.71}$$

U maps the two orthonormal basis vectors $|i\rangle = \left\|\begin{matrix}1\\0\end{matrix}\right\|$, $|j\rangle = \left\|\begin{matrix}0\\1\end{matrix}\right\|$, to the vectors $U|i\rangle = \left\|\begin{matrix}a\\c\end{matrix}\right\|$, $U|j\rangle = \left\|\begin{matrix}b\\d\end{matrix}\right\|$. Let us require that, under U, the Hermitian scalar product remains invariant; since $\langle i|i\rangle = \langle j|j\rangle = 1$, $\langle i|j\rangle = 0$, the same relations must hold for $|Ui\rangle$, $|Uj\rangle$

$$\langle Ui|Ui\rangle = a^*a + c^*c = 1,$$
$$\langle Uj|Uj\rangle = b^*b + d^*d = 1,$$
$$\langle Uj|Ui\rangle = b^*a + d^*c = 0. \tag{4.72}$$

From these equations it follows that a possible solution for the matrix A is given by[6]

$$U = \left\|\begin{matrix} a & b \\ -b^* & a^* \end{matrix}\right\| \tag{4.73}$$

with the condition:

$$|a|^2 + |b|^2 = 1. \tag{4.74}$$

Let us denote by SU(2) the set of 2×2 complex matrices of the form (4.73) obeying the condition (4.74). This leads to the following:

Lemma 4.1 *A matrix U belonging to SU(2) leaves invariant the Hermitian norm of any vector.*

Proof Let $|r\rangle = \left\|\begin{matrix}x\\y\end{matrix}\right\|$, then $U|r\rangle = \left\|\begin{matrix} ax + by \\ -b^*x + a^*y \end{matrix}\right\|$, and

$$\langle Ur|Ur\rangle = (ax + by)(a^*x^* + b^*y^*) + (-b^*x + a^*y)(-bx^* + ay^*) =$$
$$|x|^2(aa^* + bb^*) + |y|^2(bb^* + aa^*)+ \tag{4.75}$$
$$xy^*(ab^* - b^*a) + yx^*(ba^* - a^*b) = |x|^2 + |y|^2.$$

In order to prove that SU(2) is a group, one can verify, by explicit calculation, that:

(1) If $U_1 \in$ SU(2), $U_2 \in$ SU(2), then $U_1 U_2 \in$ SU(2).
(2) The identity $I_2 \in$ SU(2).
(3) The inverse $U^{-1} \in$ SU(2), etc.

However we prefer to focus on the invariance of the norm, using the following identities:

$$|U_2(U_1\mathbf{r})| = |U_1\mathbf{r}| = |\mathbf{r}|, \qquad |I_2\mathbf{r}| = |\mathbf{r}|, \tag{4.76}$$

---

[6]Setting $A = |a|^2$, $B = |b|^2$, $C = |c|^2$ and $D = |d|^2$, Eq. (4.72) can be written as $A = 1 - C$, $B = 1 - D$; multiplying these two equations we have $AB = 1 - D - C + DC$; but $BA = DC$, thus $D + C = 1$. Finally, adding $A = 1 - C$ and $B = 1 - D$ we find $A + B = 2 - (D + C) = 1$, and (4.74) is proved. It is then simple to verify that if $c = -b^*$ and $d = a^*$, Eq. (4.72) are satisfied.

etc. Defining SU(2) as the set of 2×2 complex matrices that leave the Hermitian norm of vectors invariant, it follows from (4.76) that SU(2) is a group. Note that the same reasoning can be followed using the concept of Hermitian scalar product of two vectors. Invariance is the key idea connected with the group concept.

And now a nice geometrical result: The set of matrices SU(2) can be represented as a 3-sphere of unit radius in four-dimensional space $\mathbb{R}^4$ (a 3-sphere, also called a *hypersphere*, is the four-dimensional analog of a sphere; sometimes we will skip the label "3"). In fact the matrix U (4.73) depends upon the two complex numbers $a = a_1 + ia_2, b = b_1 + ib_2$; condition (4.74) in terms of the four numbers a_1, a_2, b_1, b_2 becomes:

$$a_1^2 + a_2^2 + b_1^2 + b_2^2 = 1 \tag{4.77}$$

which is the equation of the 3-sphere with unit radius and center at the point $O = (0, 0, 0, 0)$. Note that U depends on just three parameters, since the four real numbers must satisfy one additional condition. To any point on the sphere there corresponds a group element; conversely to any element of SU(2) there corresponds a point on the sphere. For example, the identity I_2 corresponds to $a = 1, b = 0$, so that the point $a_1 = 1, a_2 = 0, b_1 = 0, b_2 = 0$ still corresponds to I_2). As we have anticipated, it is useful to consider very small displacements from I_2. Geometrically, this corresponds to a very small neighborhood of points on the sphere (4.77) around $I_2 = (1, 0, 0, 0)$.

In the following, we denote by $\varepsilon_1, \varepsilon_2, \varepsilon_3$, infinitesimally small real numbers. Let us consider, instead of the sphere (4.77) in four-dimensional space, which we are not able to draw, the more accessible sphere $x^2 + y^2 + z^2 = 1$, in three-dimensional space. Denote by $\overrightarrow{ON} = (0, 0, 1)$ the vector from the origin to the north pole. Vectors like $(\varepsilon_1, \varepsilon_2, 1) = (0, 0, 1) + (\varepsilon_1, \varepsilon_2, 0)$ have their endpoints on the sphere in a neighborhood of N, since the vector $(\varepsilon_1, \varepsilon_2, 0)$ is orthogonal to the z axis (see Fig. 4.12). By analogy, it is clear that four-dimensional vectors $(1, \varepsilon_3, \varepsilon_2, \varepsilon_1)$ belong to the sphere (4.77) in a neighborhood of the point $(1, 0, 0, 0)$. In fact, $1 + \varepsilon_1^2 + \varepsilon_2^2 + \varepsilon_3^2$ differs from 1 by second-order infinitesimals. Since for these vectors $a_1 = 1, a_2 = \varepsilon_3, b_1 = \varepsilon_2, b_2 = \varepsilon_1$, the corresponding matrix $U \in$ SU(2) is:

$$U = \left\| \begin{matrix} 1 + i\varepsilon_3 & \varepsilon_2 + i\varepsilon_1 \\ -\varepsilon_2 + i\varepsilon_1 & 1 - i\varepsilon_3 \end{matrix} \right\| = \left\| \begin{matrix} 1 & 0 \\ 0 & 1 \end{matrix} \right\| + i \left\| \begin{matrix} \varepsilon_3 & \varepsilon_1 - i\varepsilon_2 \\ \varepsilon_1 + i\varepsilon_2 & -\varepsilon_3 \end{matrix} \right\| = I_2 + iE \tag{4.78}$$

Setting:

$$|\varepsilon| = \sqrt{\varepsilon_1^2 + \varepsilon_2^2 + \varepsilon_3^2}, \quad x = \frac{\varepsilon_1}{|\varepsilon|}, \quad y = \frac{\varepsilon_2}{|\varepsilon|}, \quad z = \frac{\varepsilon_3}{|\varepsilon|}, \tag{4.79}$$

the matrix E becomes equal to $|\varepsilon|H$ (see (4.18))

$$E = |\varepsilon| \left\| \begin{matrix} z & x - iy \\ x + iy & -z \end{matrix} \right\| = |\varepsilon|H \tag{4.80}$$

Fig. 4.12 A neighborhood of the north pole N. The point $(\varepsilon_1, \varepsilon_2, 1)$ belongs to the plane tangent to the sphere at the north pole

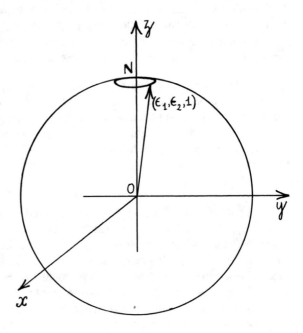

and $x^2 + y^2 + z^2 = 1$ as it should. Note also that if $\overrightarrow{OP} = (x, y, z)$ coincides with $(1, 0, 0)$, $(0, 1, 0)$, $(0, 0, 1)$, respectively, while H becomes one of the Pauli spin matrices:

$$\left\| \begin{matrix} 0 & 1 \\ 1 & 0 \end{matrix} \right\|, \quad \left\| \begin{matrix} 0 & -i \\ i & 0 \end{matrix} \right\|, \quad \left\| \begin{matrix} 1 & 0 \\ 0 & -1 \end{matrix} \right\| \tag{4.81}$$

which represent the spin component in the three-axis direction. It follows that $H = x\sigma_x + y\sigma_y + z\sigma_z$. The matrix U (4.78) in the neighborhood of the identity becomes:

$$U \simeq I_2 + i|\varepsilon|H = I_2 + i|\varepsilon|(x\sigma_x + y\sigma_y + z\sigma_z). \tag{4.82}$$

Equations (4.70) and (4.82) look quite similar $\delta\theta$ and ε are infinitesimals, $x^2 + y^2 + z^2 = 1$ and $n_1^2 + n_2^2 + n_3^2 = 1$; but we are about to discover another remarkable similarity. Recall that the *commutator* of two matrices A, B is given by $[A, B] = AB - BA$. Let us compute the commutator of the infinitesimal operators Λ_1, Λ_2, Λ_3 of the rotation group. From Eqs. (4.63)–(4.65), we easily find

$$\begin{aligned} [\Lambda_1, \Lambda_2] &= \Lambda_1 \Lambda_2 - \Lambda_2 \Lambda_1 = \Lambda_3, \\ [\Lambda_2, \Lambda_3] &= \Lambda_2 \Lambda_3 - \Lambda_3 \Lambda_2 = \Lambda_1, \\ [\Lambda_3, \Lambda_1] &= \Lambda_3 \Lambda_1 - \Lambda_1 \Lambda_3 = \Lambda_2. \end{aligned} \tag{4.83}$$

A condensed way of writing these three commutation relations is

$$\left[\Lambda_1, \Lambda_2\right] = \Lambda_3, \; et \; cyc, \tag{4.84}$$

where *et cyc* means that the relation holds when 1, 2, 3 are cyclically permuted (to 3, 1, 2 and 2, 3, 1). In Sect. 3.4, the analogous commutation relations for the components of angular momentum were shown to be

$$\left[J_1, J_2\right] = i\hbar J_3, \; et \; cyc, \tag{4.85}$$

looking quite similar to those for Λ_i. The main difference is in the domain of the operators: for the operators Λ_i here, it is the physical space $\mathbb{R}^3$, while for the components of angular momentum, it is the Hilbert space of states of the system. Moreover, the Λ_i operators are dimensionless, while the angular momentum operators J_i have the same dimension as $\hbar$. Comparing the commutation relations, we can identify

$$J_i = i\hbar\Lambda_i. \tag{4.86}$$

The commutation relations for the Pauli spin matrices σ_x, σ_y, σ_z of the complex group SU(2) (see Eq. 4.82) are given by

$$\left[\sigma_x, \sigma_y\right] = \sigma_x\sigma_y - \sigma_y\sigma_x = 2i\sigma_z, \; et \; cyc \tag{4.87}$$

These evidently are related to angular momentum operators by

$$J_1 = \frac{\hbar}{2}\sigma_x, \quad J_2 = \frac{\hbar}{2}\sigma_y, \quad J_3 = \frac{\hbar}{2}\sigma_z. \tag{4.88}$$

The factors $\hbar/2$ evidently reflect the fact that these represent spin-$\frac{1}{2}$ particles. From now on, for compactness in notation, we will be using units with $\hbar = 1$.

In summary, we find that in the neighborhood of the identity an element of the 3×3 rotation group can be written

$$R(\delta\theta) \simeq I_3 + \delta\theta(n_1\Lambda_1 + n_2\Lambda_2 + n_3\Lambda_3) = I_3 - i\delta\theta(n_1J_1 + n_2J_2 + n_3J_3), \tag{4.89}$$

and an element U of SU(2) can be written

$$U \simeq I_2 - i\frac{\delta\theta}{2}(n_1\sigma_x + n_2\sigma_y + n_3\sigma_z). \tag{4.90}$$

The infinitesimal rotations (4.89) in $\mathbb{R}^3$ and (4.90) in $\mathbb{C}^2$ show a one-to-one correspondence, provided we remain close to the identity. However, we will see later that this is no longer globally true for the full groups SO(3) and SU(2).

Accordingly, we can state

Theorem 4.3 *In the neighborhood of the identity, the operators $U(R)$ preserve the group structure.*

Proof We need to prove that $U(R)$ sends products into products, the identity I_3 to the identity I_2, the inverse to the inverse. Consider two infinitesimal rotations $R_1(\delta\alpha)$, $R_2(\delta\beta)$, and let $\mathbf{n} = (n_1, n_2, n_3)$, $\mathbf{m} = (m_1, m_2, m_3)$ denote the two unit vectors along the rotation axis. We have:

$$R_1 \simeq I_3 - i\delta\alpha(n_1 J_1 + n_2 J_2 + n_3 J_3), \quad R_2 \simeq I_3 - I\delta\beta(m_1 J_1 + m_2 J_2 + m_3 J_3). \quad (4.91)$$

Since $\delta\alpha$, $\delta\beta$ are infinitesimal, we can safely neglect higher order terms containing $\delta\alpha\,\delta\beta$. Furthermore $I_3 I_3 = I_3$. Thus

$$R_1 R_2 \simeq I_3 - i\Big[(\delta\alpha\,n_1 + \delta\beta\,m_1)J_1 + (\delta\alpha\,n_2 + \delta\beta\,m_2)J_2 + (\delta\alpha\,n_3 + \delta\beta\,m_3)J_3\Big]. \quad (4.92)$$

By Eq. (4.90) we have

$$U(R_1) \simeq I_2 - i\frac{\delta\alpha}{2}(n_1\sigma_1 + n_2\sigma_2 + n_3\sigma_3),$$

$$U(R_2) \simeq I_2 - i\frac{\delta\beta}{2}(m_1\sigma_1 + m_2\sigma_2 + m_3\sigma_3). \quad (4.93)$$

Again, neglecting terms containing the product $\delta\alpha\,\delta\beta$, we find

$$U(R_1)U(R_2) \simeq I_2 - \tfrac{i}{2}\Big[(\delta\alpha\,n_1 + \delta\beta\,m_1)\sigma_1 +$$

$$(\delta\alpha\,n_2 + \delta\beta\,m_2)\sigma_2 + (\delta\alpha\,n_3 + \delta\beta\,m_3)\sigma_3\Big] = U(R_1 R_2). \quad (4.94)$$

Therefore U maps "products into products". If the angle $\delta\theta$ is zero, $R = I_3$ and $U(R) = I_2$ (recall that I_2, I_3 are the identity operators). From (4.92) we see that for $\mathbf{n} = \mathbf{m}$ and $\delta\beta = -\delta\alpha$, $R_1 R_2 = I_3$, thus $R_2 = R_1^{-1}$. By the same argument $U(R_1)U(R_2) = I_2$, so that $U(R_2) = U(R_1)^{-1}$ and the Theorem is proved.

If the correspondence $U(R)$ were to preserve the group properties for the entire rotation group SO(3), we would say that the matrices $U(R)$ form a *representation* of SO(3). Actually we have proven the Theorem only in the neighborhood of the identity. A natural question: is (4.94) valid for *all* elements of the rotation group? The answer is affirmative, but given the heuristic nature of our approach, we will just show that it is plausible that $U(R_1)U(R_2) = U(R_1 R_2)$ for all R_1, R_2. We know that the rotation $R(\theta)$ about the z axis is (see Eq. (4.39))

$$R_z = \begin{Vmatrix} \cos\theta & -\sin\theta & 0 \\ \sin\theta & \cos\theta & 0 \\ 0 & 0 & 1 \end{Vmatrix} \quad (4.95)$$

Fig. 4.13 Geometrical proof that the sum of two rotations of $3\pi/2$ is equivalent to a rotation of π

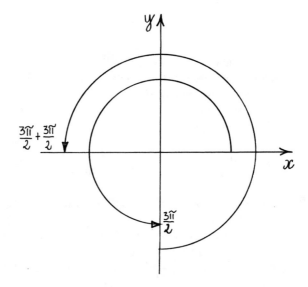

Note that θ can be any real number; however $R_z(\theta \pm 2\pi) = R_z(\theta \pm 4\pi) = \cdots = R_z(\theta)$, and in general $R_z(\theta \pm 2k\pi) = R_z(\theta)$ for all integer k. The matrices $R_z(\theta)$ form a group that we will call G_z. Of course,

$$R_z(\phi)\, R_z(\theta) = R_z(\phi + \theta) \tag{4.96}$$

since by performing two successive rotations about the same axis the rotation angles are additive. Different matrices $R_z(\theta)$ are obtained only for $0 \le \theta < 2\pi$. For example, $R_z(\frac{3\pi}{2})R_z(\frac{3\pi}{2}) = R_z(3\pi) = R_z(\pi)$ (see Fig. 4.13). Let us now consider two different representations of the group G_z: the first is simply the set of matrices

$$D(\theta) = \left\| \begin{matrix} \cos\theta & -\sin\theta \\ \sin\theta & \cos\theta \end{matrix} \right\| \tag{4.97}$$

and the second is obtained by associating with $R_z(\theta)$ the matrix:

$$F(\theta) = \left\| \begin{matrix} \exp(-i\theta/2) & 0 \\ 0 & \exp(i\theta/2) \end{matrix} \right\| \tag{4.98}$$

Both set of matrices D and F satisfy the relations

$$\begin{matrix} D(\theta)D(\phi) = D(\theta + \phi), & D(\theta)^{-1} = D(-\theta), & D(0) = I \\ F(\theta)F(\phi) = F(\theta + \phi), & F(\theta)^{-1} = F(-\theta), & F(0) = I \end{matrix} \tag{4.99}$$

For very small $\delta\theta$, $F(\delta\theta)$ becomes

$$F(\delta\theta) = \left\| \begin{matrix} 1 - i\delta\theta/2 & 0 \\ 0 & 1 + i\delta\theta/2 \end{matrix} \right\| \tag{4.100}$$

which coincides with $U(R_z) \in SU(2)$, as it should (see Eq. (4.90)). In the first case the mapping between $R_z(\theta)$ and $D(\theta)$ is one to one. We say that the representation D is *faithful* (D is isomorphic with R_z). By contrast, in the second case (which is what happens for the group $SU(2)$) to one matrix $R_z(\theta)$ there correspond two matrices, $\pm F(\theta)$. For example, $R_z(0) = R_z(2\pi) = I_3$, while:

$$F(0) = \left\| \begin{matrix} 1 & 0 \\ 0 & 1 \end{matrix} \right\| = I_2, \qquad F(2\pi) = \left\| \begin{matrix} e^{-i\pi} & 0 \\ 0 & e^{i\pi} \end{matrix} \right\| = \left\| \begin{matrix} -1 & 0 \\ 0 & -1 \end{matrix} \right\| = -I_2 \quad (4.101)$$

Let us now prove that the set F_z of matrices $F(\theta)$, with $0 \le \theta \le 4\pi$ is the subgroup (a subset which is also a group) of $SU(2)$ and it corresponds to the group G_z of rotations about the z axis. For a simple proof, we ask the reader to use his imagination, and try to answer the question: where are the points on the sphere (see Eq. (4.77)) $a_1^2 + a_2^2 + b_1^2 + b_2^2 = 1$ which correspond to the set F_z? These points are, in fact, characterized by the equations

$$b_1 = b_2 = 0 \quad (\text{since } b = b_1 + i b_2 = 0) \tag{4.102}$$

$$a_1 = \cos\frac{\theta}{2}, \qquad a_2 = -\sin\frac{\theta}{2} \quad \left(\text{since } e^{-i\theta/2} = \cos\frac{\theta}{2} - i\sin\frac{\theta}{2}\right) \tag{4.103}$$

Fig. 4.14 The reader is asked to "imagine" the figure in four dimensions! The equations of the circle Γ are: $b_1 = b_2 = 0$, $a_1^2 + a_2^2 = 1$. The *arrows* indicate the directions of increasing θ

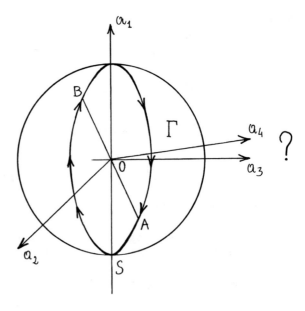

When θ varies from 0 to 4π, the point $(a_1, a_2, b_1, b_2) = (\cos\frac{\theta}{2}, \sin\frac{\theta}{2}, 0, 0)$ moves along the circle Γ of equation $a_1^2 + a_2^2 = 1$ (and of course $b_1 = b_2 = 0$). An attempt to draw the circle is shown in Fig. 4.14. We also note that

$$F(0) = F(4\pi) = \left\| \begin{matrix} 1 & 0 \\ 0 & 1 \end{matrix} \right\| = I_2 \qquad (4.104)$$

and the corresponding point on the circle Γ is N, the "north pole" of the sphere. However

$$F(2\pi) = \left\| \begin{matrix} e^{-i\pi} & 0 \\ 0 & e^{i\pi} \end{matrix} \right\| = \left\| \begin{matrix} -1 & 0 \\ 0 & -1 \end{matrix} \right\| = -I_2 \qquad (4.105)$$

and the corresponding point is S, the "south pole", since $a_1 = -1$, $a_2 = 0$. Therefore when θ varies from 0 to 2π, the rotation $R_z(\theta)$ runs over the *whole* subgroup G_z, while in four dimensions the corresponding point describes only half of the circle Γ, from N to S. When θ varies from 2π to 4π, also the second semicircle is covered.

It is easy to prove that not only N and S, but *any* pair of endpoints of a diameter of the circle Γ (*antipodes*) correspond to the same element $R_z(\theta)$ of G_z (see the points A and B of Fig. 4.15). Clearly,

$$F(\theta + 2\pi) = \left\| \begin{matrix} e^{-i(\theta+2\pi)/2} & 0 \\ 0 & e^{i(\theta+2\pi)/2} \end{matrix} \right\| = F(\theta) \left\| \begin{matrix} -1 & 0 \\ 0 & -1 \end{matrix} \right\| = -F(\theta), \quad (4.106)$$

while $R_z(\theta + 2\pi) = R_z(\theta)$. This result is completely general: no matter which direction we choose for the rotation axis, adding 2π to the rotation angle does not change the matrix R but merely changes the sign of the matrix $U(R)$. In formulas

$$R(\theta + 2\pi) = R(\theta), \qquad U[R(\theta + 2\pi)] = -U[R(\theta)]. \qquad (4.107)$$

Therefore, to *two* matrices $\pm U(R)$, there corresponds *one* rotation R.

The behavior under rotation is the characteristic feature distinguishing spinors from vectors. A vector will return to its initial orientation after a rotation by 2π, while a spinor will be transformed to its opposite orientation. We need a rotation of 4π to return a spinor to its initial state. There is an amusing analogy for a particle moving on a Möbius band. After one circuit (2π) the particle will find itself on the opposite side of the band from where it started. Only after a second circuit (4π) does the particle return to its starting point.

To "visualize" the whole rotation group SO(3) is much more difficult than visualizing the four-dimensional sphere that represents SU(2). To explain, consider a solid sphere S of radius π in our physical space R^3. A point $P(x, y, z)$ belongs to S if $x^2 + y^2 + z^2 \leq \pi^2$, where the origin O is at $(0, 0, 0)$. To each vector $\overrightarrow{OP}$, we associate an element of SO(3) with rotation angle θ equal to the distance $\pm|\overrightarrow{OP}| = \pm\sqrt{x^2 + y^2 + z^2}$, and rotation axis along $\overrightarrow{OP}$. Hence if $\overrightarrow{OP} = (x, y, z)$ corresponds to a rotation of angle $\theta \leq \pi$, $-\overrightarrow{OP} = (-x, -y, -z)$ corresponds to a

Fig. 4.15 The point A
corresponds to a rotation of
π around the axis OA, while
the point B corresponds to a
rotation of $-\pi$ around
$OB = -OA$. Thus the two
rotations coincide

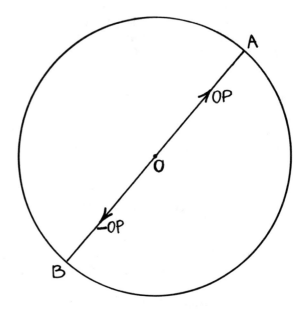

rotation of angle $-\theta$. Since rotations of angle $+\pi$ and $-\pi$ are identical, we identify
the endpoints A, B of a diameter of length 2π (see Fig. 4.15). A and B are called
antipodal points. With this convention, points of the solid ball S are in one-to-one
correspondence with the elements of the rotation group SO(3). However S is connected but not *simply connected*; this means that there are closed loops that cannot
be shrunk to a point (similar to what can happen for circles on a torus). For instance,
consider the path going from points A, B of Fig. 4.15, which is a *closed* path since
$A \equiv B$. If one makes the same identification for *all* the diameters of S one obtains a
bizarre set, with an infinite number of such loops. Of course each loop corresponds
to a subgroup of rotations of a given axis and angles $-\pi \le \theta \le +\pi$.

To summarize, an n-dimensional *Lie group* G "lives" on a differentiable manifold,
a continuous and smooth surface in some topological space. The elements g of G
depend upon n parameters $x_1, x_2, \ldots, x_n$. If $g_1 \in G$, $g_2 \in G$, the parameters of
the product $g_1 g_2$ must be smooth functions of the parameters of g_1, g_2; the same
must hold for the inverse. Usually one adopts the convention that $(0, 0, \ldots, 0)$ are
the parameters of the identity I. Near the identity a point g of the manifold can be
expressed in the form, analogous to (4.89)

$$g = I + x_1 T_1 + x_2 T_2 + \cdots + x_n T_n \tag{4.108}$$

where the $T_1, T_2, \ldots, T_n$ are the *infinitesimal generators* of the group, as we have
seen in the cases of SO(3) and SU(2). These "vectors" span the "plane" tangent to the
manifold at the point I. They can be multiplied, and the commutators $T_i T_j - T_j T_i$ must
remain in this "plane," so that they can be written as a *linear combination* of the T_i[7]:

---

[7]The T_i are said to belong to a *Lie algebra*.

$$\left[T_i, T_j\right] = \sum_{k}^{n} c_{ijk} T_k. \tag{4.109}$$

The angular momentum components provide a simple example of such commutation relations. The *structure constants* c_{ijk} determine the *local* structure of the group, but the *global* behavior of a Lie group G is not determined by its local structure. G can even be a *multiply-connected* set, such that, given two points $g_1 \in G$, $g_2 \in G$, there is no continuous path[8] connecting g_1 and g_2. One such case is the Lorentz group, which contains four connected components.

We note that there are other Lie groups of importance in theoretical physics. For example, the group SU(3), complex matrices which leave invariant the Hermitian scalar product in three dimensions and do not change the sign of volumes, is the basis of quantum chromodynamics. This underlies the theory of strong interactions, part of the standard model of elementary particles. Another example is the Lie group SO(3,1), the Lorentz group, which governs the behavior of relativistic transformations.

---

[8] A *path* or *curve* $g = g(\tau), 0 \le \tau \le 1$ on a Lie group is a mapping from the set $0 \le \tau \le 1$ and the subset $\{g(\tau)\}$ of G. The path is continuous if the parameters of $g(\tau)$ are continuous functions of τ.

Chapter 5
Quantum Entanglement and Bell's Theorem

Abstract A description of quantum entanglement and its applications. Bell's inequalities and Bell's theorem are described, along with their implications for local reality and hidden variables. Other topics: applications using electron spin and photon polarization, Aspect's experiments, decoherence of quantum states.

Keywords Entanglement · EPR experiment · Bell's theorem · Local reality · Aspect's experiments · Decoherence

5.1 Product States in Hilbert Space

Let us consider a composite system of two electrons, a and b. The state of electron a is represented by the vector $|A\rangle$ belonging to the Hilbert space H_A, and the state of electron b is represented by the vector $|B\rangle$ in the Hilbert space H_B. How can we construct the Hilbert space H of the "global states" representing both electrons? For simplicity we neglect the positions of the two electrons, and consider only their spins. Then $|A\rangle$ and $|B\rangle$ can be represented by the two-component complex vectors:

$$|A\rangle = \left\| \begin{matrix} a_1 \\ a_2 \end{matrix} \right\| \in \mathbb{C}^2, \qquad |B\rangle = \left\| \begin{matrix} b_1 \\ b_2 \end{matrix} \right\| \in \mathbb{C}^2, \qquad H_A = H_B = \mathbb{C}^2. \qquad (5.1)$$

A description of the composite system involves the set of all possible pairs of the two states $|A\rangle$, $|B\rangle$. Such a global state can be represented by a *tensor product* (or simply *product*) of $|A\rangle$, $|B\rangle$, denoted by $|A\rangle \otimes |B\rangle$. Given two product states:

$$|\Psi_1\rangle = |A_1\rangle \otimes |B_1\rangle, \qquad |\Psi_2\rangle = |A_2\rangle \otimes |B_2\rangle, \qquad (5.2)$$

we can define the scalar product $\langle \Psi_2 | \Psi_1 \rangle$ by

$$\langle \Psi_2 | \Psi_1 \rangle = \langle A_2 | A_1 \rangle \langle B_2 | B_1 \rangle. \qquad (5.3)$$

© Springer International Publishing AG 2017
G. Fano and S.M. Blinder, *Twenty-First Century Quantum Mechanics:*
Hilbert Space to Quantum Computers, UNITEXT for Physics,
DOI 10.1007/978-3-319-58732-5_5

Since H_A and H_B are two copies of $\mathbb{C}^2$, we can use the same basis vectors in both spaces. Let us denote by $|A\uparrow\rangle = \left\|\begin{matrix}1\\0\end{matrix}\right\|, |A\downarrow\rangle = \left\|\begin{matrix}0\\1\end{matrix}\right\|$ the basis vectors in H_A, and analogously by $|B\uparrow\rangle = \left\|\begin{matrix}1\\0\end{matrix}\right\|, |B\downarrow\rangle = \left\|\begin{matrix}0\\1\end{matrix}\right\|$ the basis vectors in H_B. It is easy to verify that the following product states

$$
\begin{aligned}
|\uparrow\uparrow\rangle &= |A\uparrow\rangle \otimes |B\uparrow\rangle, & |\uparrow\downarrow\rangle &= |A\uparrow\rangle \otimes |B\downarrow\rangle, \\
|\downarrow\uparrow\rangle &= |A\downarrow\rangle \otimes |B\uparrow\rangle, & |\downarrow\downarrow\rangle &= |A\downarrow\rangle \otimes |B\downarrow\rangle,
\end{aligned}
\tag{5.4}
$$

are orthonormal. Indeed using the definition (5.3), we have

$$
\begin{aligned}
\langle\uparrow\uparrow|\uparrow\uparrow\rangle &= \langle A\uparrow|A\uparrow\rangle\langle B\uparrow|B\uparrow\rangle = |A\uparrow|^2\,|B\uparrow|^2 = 1\times 1 = 1, \\
\langle\uparrow\uparrow|\uparrow\downarrow\rangle &= \langle A\uparrow|A\uparrow\rangle\,\langle B\uparrow|B\downarrow\rangle = 1\times 0 = 0, \\
\langle\uparrow\uparrow|\downarrow\uparrow\rangle &= \langle A\uparrow|A\downarrow\rangle\,\langle B\uparrow|B\uparrow\rangle = 0\times 1 = 0,\ \text{etc.}
\end{aligned}
\tag{5.5}
$$

The Hilbert space H of the composite system is also called the *tensor product* of H_A and H_B, denoted by $H_A \otimes H_B$. An element $|\Psi\rangle \in H$ is a linear combination with complex (although possibly real) coefficients of the basis vectors (5.4):

$$
|\Psi\rangle = c_{11}|\uparrow\uparrow\rangle + c_{12}|\uparrow\downarrow\rangle + c_{21}|\downarrow\uparrow\rangle + c_{22}|\downarrow\downarrow\rangle.
\tag{5.6}
$$

We presume that the coefficients c_{ij} fulfill the normalization condition:

$$
\sum_{i,j=1}^{2} |c_{i,j}|^2 = |c_{11}|^2 + |c_{12}|^2 + |c_{21}|^2 + |c_{22}|^2 = 1.
\tag{5.7}
$$

We designate by $|\Psi\rangle$ the state of the composite system. The dimension of H is 4. If we had chosen spaces H_A, H_B of dimension n, m, respectively, the dimension of $H_A \otimes H_B$ would have been $n \times m$. As an example, suppose $H_A = \mathbb{C}^2$ and $H_B = \mathbb{C}^3$; a basis in the product space consists of the following set of 6 vectors:

$$
|v_{11}\rangle = \left\|\begin{matrix}1\\0\end{matrix}\right\| \otimes \left\|\begin{matrix}1\\0\\0\end{matrix}\right\|,\quad
|v_{12}\rangle = \left\|\begin{matrix}1\\0\end{matrix}\right\| \otimes \left\|\begin{matrix}0\\1\\0\end{matrix}\right\|,\quad
|v_{13}\rangle = \left\|\begin{matrix}1\\0\end{matrix}\right\| \otimes \left\|\begin{matrix}0\\0\\1\end{matrix}\right\|,
$$

$$
|v_{21}\rangle = \left\|\begin{matrix}0\\1\end{matrix}\right\| \otimes \left\|\begin{matrix}1\\0\\0\end{matrix}\right\|,\quad
|v_{22}\rangle = \left\|\begin{matrix}0\\1\end{matrix}\right\| \otimes \left\|\begin{matrix}0\\1\\0\end{matrix}\right\|,\quad
|v_{23}\rangle = \left\|\begin{matrix}0\\1\end{matrix}\right\| \otimes \left\|\begin{matrix}0\\0\\1\end{matrix}\right\|.
\tag{5.8}
$$

To return to the case of the spins A, B of two electrons, we assume that the two spins are in a superposition state, such as

$$|A\rangle = a_1|A\uparrow\rangle + a_2|A\downarrow\rangle, \qquad |B\rangle = b_1|B\uparrow\rangle + b_2|B\downarrow\rangle. \tag{5.9}$$

By the distributive property, the product state $|A\rangle \otimes |B\rangle$ becomes

$$|A\rangle \otimes |B\rangle = \Big(a_1|A\uparrow\rangle + a_2|A\downarrow\rangle\Big) \otimes \Big(b_1|B\uparrow\rangle + b_2|B\downarrow\rangle\Big) =$$
$$a_1b_1|A\uparrow\rangle \otimes |B\uparrow\rangle + a_1b_2|A\uparrow\rangle \otimes |B\downarrow\rangle + a_2b_1|A\downarrow\rangle \otimes |B\uparrow\rangle + a_2b_2|A\downarrow\rangle \otimes |B\downarrow\rangle. \tag{5.10}$$

Comparing (5.6), (5.9), we have

$$c_{11} = a_1b_1, \qquad c_{12} = a_1b_2, \qquad c_{21} = a_2b_1, \qquad c_{22} = a_2b_2. \tag{5.11}$$

The question then arises: are all vectors $\Psi \in H$ of the form $|A\rangle \otimes |B\rangle$? The answer is no. Before giving a formal proof, we note that the simplest intuitive picture of $|A\rangle \otimes |B\rangle$ is that of two *independent* particles, since $|A\rangle \in H_A$, $|B\rangle \in H_B$. However, there exists other states $|\Psi\rangle \in H$, representing particle pairs which are *not* independent. To see this, note that (5.11) implies that

$$c_{11}c_{22} = c_{12}c_{21}, \tag{5.12}$$

which is therefore a necessary condition[1] for the vector $|\Psi\rangle$ to be a product state. States that do *not* fulfill the condition (5.12) include the following (see Eq. 5.4):

$$|\Psi_{0,0}\rangle = \frac{1}{\sqrt{2}}\Big(|\uparrow\downarrow\rangle - |\downarrow\uparrow\rangle\Big), \qquad |\Psi_{1,0}\rangle = \frac{1}{\sqrt{2}}\Big(|\uparrow\uparrow\rangle + |\downarrow\downarrow\rangle\Big), \tag{5.13}$$

where $\frac{1}{\sqrt{2}}$ is a normalization factor. For the state $|\Psi_{0,0}\rangle$: $c_{11}c_{22} = 0$, $c_{12}c_{21} = -\frac{1}{2}$, while for the state $|\Psi_{1,0}\rangle$, $c_{11}c_{22} = \frac{1}{2}$, $c_{12}c_{21} = 0$.

Nonfactorizing states, states that cannot be written in a product form, are said to be *entangled* (a term introduced by Schrödinger: *Verschränkung*). For example, $|\Psi_{0,0}\rangle$ and $|\Psi_{1,0}\rangle$ in Eq. (5.4) are entangled, in fact *maximally entangled*. These are called *Bell states*, for reasons which will become apparent later. Entanglement is a central concept in our understanding of quantum mechanics and is also the fundamental principle underlying quantum computing.

A metaphoric representation of quantum entanglement might be Indra's net, as shown in Fig. 5.1. In Buddhist and Hindu tradition, Indra's net is an infinite array of strands of jewels. The surface of each jewel reflects the infinity of all the other jewels, to symbolize a cosmos in which every part is mutually interconnected to every other part.

---

[1] The condition $c_{11}c_{22} = c_{12}c_{21}$ is also sufficient for the vector $|\Psi\rangle$ to be a product state, assuming that $c_{11} + c_{22} \neq 0$. The product state

$$\left(|A\uparrow\rangle + \frac{c_{21}}{c_{11}}|A\downarrow\rangle\right) \otimes \left(c_{11}|B\uparrow\rangle + c_{12}|B\downarrow\rangle\right) = c_{11}|\uparrow\uparrow\rangle + c_{12}|\uparrow\downarrow\rangle + c_{21}|\downarrow\uparrow\rangle + \frac{c_{21}c_{12}}{c_{11}}|\downarrow\downarrow\rangle$$

coincides with the state (5.6).

Fig. 5.1 Indra's net. (https://www.scienceandnonduality.com/the-indras-net/)

A product of Hilbert spaces can be extended to three or more component spaces: given three Hilbert spaces H_A, H_B, H_C, we define

$$H_A \otimes H_B \otimes H_C = (H_A \otimes H_B) \otimes H_C = H_A \otimes (H_B \otimes H_C). \tag{5.14}$$

For example, for a composite system of three electron spins, since each spin state is an element in a copy of $\mathbb{C}^2$, an orthonormal basis in $H_A \otimes H_B \otimes H_C$ is provided by $|A \uparrow\rangle \otimes |B \uparrow\rangle \otimes |C \uparrow\rangle$, $|A \uparrow\rangle \otimes |B \uparrow\rangle \otimes |C \downarrow\rangle$, etc. With an obvious extension of the notation (5.4), a basis is given by

$$|\uparrow\uparrow\uparrow\rangle, \ |\uparrow\uparrow\downarrow\rangle, \ |\uparrow\downarrow\uparrow\rangle, \ |\uparrow\downarrow\downarrow\rangle, \ |\downarrow\uparrow\uparrow\rangle, \ |\downarrow\uparrow\downarrow\rangle, \ |\downarrow\downarrow\uparrow\rangle, \ |\downarrow\downarrow\downarrow\rangle. \tag{5.15}$$

If O_A denotes a linear Hermitian operator acting on H_A, and O_B an operator acting on H_B, we define an operator $O_A \otimes O_B$, called a *tensor product*, acting on the product states according to the following rule:

$$(O_A \otimes O_B)(|A\rangle \otimes |B\rangle) = O_A|A\rangle \otimes O_B|B\rangle. \tag{5.16}$$

One can imagine that $O_A \otimes O_B$ acts separately on the spaces H_A, H_B, and the results are combined to produce a product state. Given two basis $\{|A_i\rangle \in H_A, \ i = 1, 2, \ldots, m\}$, $\{|B_j\rangle \in H_B, \ j = 1, 2, \ldots, n)\}$, the action of $O_A \otimes O_B$ on a general linear combination $|\Psi\rangle = \sum_{ij} c_{ij} |A_i\rangle \otimes |B_j\rangle$ is given by

$$(O_A \otimes O_B) |\Psi\rangle = \sum_{ij} c_{ij} \, O_A |A_i\rangle \otimes O_B |B_j\rangle. \tag{5.17}$$

The product of two tensor products, $O_{A_1} \otimes O_{B_1}$, $O_{A_2} \otimes O_{B_2}$, is given by

$$(O_{A_1} \otimes O_{B_1})(O_{A_2} \otimes O_{B_2}) = (O_{A_1} O_{A_2}) \otimes (O_{B_1} O_{B_2}). \qquad (5.18)$$

The next step is to find the matrix representing $(O_A \otimes O_B)$: given a 2×2 matrix M operating on the $\mathbb{C}^2$ space:

$$M = \left\| \begin{matrix} M_{11} & M_{12} \\ M_{21} & M_{22} \end{matrix} \right\|, \qquad (5.19)$$

we denote by $|e_1\rangle = \left\| \begin{matrix} 1 \\ 0 \end{matrix} \right\|$, $|e_2\rangle = \left\| \begin{matrix} 0 \\ 1 \end{matrix} \right\|$, the standard orthonormal basis vectors; then

$$M|e_1\rangle = \left\| \begin{matrix} M_{11} \\ M_{21} \end{matrix} \right\|, \qquad M|e_2\rangle = \left\| \begin{matrix} M_{12} \\ M_{22} \end{matrix} \right\|. \qquad (5.20)$$

Therefore,

$$M_{11} = \langle e_1|M|e_1\rangle, \ M_{21} = \langle e_2|M|e_1\rangle, \ M_{12} = \langle e_1|M|e_2\rangle, \ M_{22} = \langle e_2|M|e_2\rangle, \qquad (5.21)$$

which can be written as

$$M_{ij} = \langle e_i|M|e_j\rangle, \qquad i, j = 1, 2. \qquad (5.22)$$

Equation (5.22) can be generalized to arbitrary dimension n, such that $|e_1\rangle = (1, 0, 0, ...0)$, $|e_2\rangle = (0, 1, 0, ...0)$, ..., with $i, j = 1, 2, \ldots, n$.

Suppose now that O_A is represented by a $2 \otimes 2$ matrix $a_{ij} = \langle A_i|O_A|A_j\rangle$, $i, j = 1, 2$, where $|A_1\rangle = \left\| \begin{matrix} 1 \\ 0 \end{matrix} \right\|$, $|A_2\rangle = \left\| \begin{matrix} 0 \\ 1 \end{matrix} \right\|$ denote, for example, spin-up and spin-down states. Analogously, the operator O_B is represented by a $2 \otimes 2$ matrix $b_{ij} = \langle B_i|O_B|B_j\rangle$, $i, j = 1, 2$. Let us compute the matrix elements of $O_A \otimes O_B$ in the orthonormal basis $|A_i\rangle \otimes |B_j\rangle$. By (5.16) we can write

$$(O_A \otimes O_B)(|A_i\rangle \otimes |B_j\rangle) = O_A|A_i\rangle \otimes O_B|B_j\rangle, \qquad i, j = 1, 2. \qquad (5.23)$$

Let us now take the scalar product of this vector with $|A_m\rangle \otimes |B_n\rangle$, $m, n = 1, 2$. Using (5.1), (5.2) we find the 4×4 matrix that represents $O_A \otimes O_B$:

$$C_{mn,ij} = \langle A_m|O_A|A_i\rangle \langle B_n|O_B|B_j\rangle = a_{mi}b_{nj} \qquad \text{for} \qquad m, n, i, j = 1, 2. \qquad (5.24)$$

The matrix C is also called the *tensor product or Kronecker product* of O_A, O_B. Ordering the basis $|A_i\rangle \otimes |B_j\rangle$ as follows:

$$|A_1\rangle \otimes |B_1\rangle, \quad |A_1\rangle \otimes |B_2\rangle, \quad |A_2\rangle \otimes |B_1\rangle, \quad |A_2\rangle \otimes |B_2\rangle, \qquad (5.25)$$

the matrix C is given by

$$\left\| \begin{array}{cc} a_{11} \left\| \begin{array}{cc} b_{11} & b_{12} \\ b_{21} & b_{22} \end{array} \right\| & a_{12} \left\| \begin{array}{cc} b_{11} & b_{12} \\ b_{21} & b_{22} \end{array} \right\| \\ a_{21} \left\| \begin{array}{cc} b_{11} & b_{12} \\ b_{21} & b_{22} \end{array} \right\| & a_{22} \left\| \begin{array}{cc} b_{11} & b_{12} \\ b_{21} & b_{22} \end{array} \right\| \end{array} \right\| = \left\| \begin{array}{cccc} a_{11}b_{11} & a_{11}b_{12} & a_{12}b_{11} & a_{12}b_{12} \\ a_{11}b_{21} & a_{11}b_{22} & a_{12}b_{21} & a_{12}b_{22} \\ a_{21}b_{11} & a_{21}b_{12} & a_{22}b_{11} & a_{22}b_{12} \\ a_{21}b_{21} & a_{21}b_{22} & a_{22}b_{21} & a_{22}b_{22} \end{array} \right\|. \tag{5.26}$$

An operator O_A acting on a state $|A\rangle \otimes |B\rangle$ of a composite system can be denoted $O_A \otimes I$, with the operator O_B written as $I \otimes O_B$. Physicists prefer often to skip the direct product notation, since it is generally obvious on which part of the composite system an operator is acting.

We next consider the very fundamental example of an entangled state of two electrons a, b:

$$|\Psi_{0,0}\rangle = \frac{1}{\sqrt{2}}\Big(|\uparrow\downarrow\rangle - |\downarrow\uparrow\rangle\Big). \tag{5.27}$$

The following operator represents the sum of the z-components of the spins of the two electrons:

$$\Sigma_z = \sigma_{zA} \otimes I + I \otimes \sigma_{zB} = \left\| \begin{array}{cc} 1 & 0 \\ 0 & -1 \end{array} \right\| \otimes I + I \otimes \left\| \begin{array}{cc} 1 & 0 \\ 0 & -1 \end{array} \right\|. \tag{5.28}$$

We have

$$\sigma_{zA} \left\| \begin{array}{c} 1 \\ 0 \end{array} \right\|_A = \left\| \begin{array}{c} 1 \\ 0 \end{array} \right\|_A, \quad \sigma_{zA} \left\| \begin{array}{c} 0 \\ 1 \end{array} \right\|_A = - \left\| \begin{array}{c} 0 \\ 1 \end{array} \right\|_A,$$

$$\sigma_{zB} \left\| \begin{array}{c} 1 \\ 0 \end{array} \right\|_B = \left\| \begin{array}{c} 1 \\ 0 \end{array} \right\|_B, \quad \sigma_{zB} \left\| \begin{array}{c} 0 \\ 1 \end{array} \right\|_B = - \left\| \begin{array}{c} 0 \\ 1 \end{array} \right\|_B. \tag{5.29}$$

Therefore, recalling that $|\uparrow\downarrow\rangle = \left\| \begin{array}{c} 1 \\ 0 \end{array} \right\|_A \otimes \left\| \begin{array}{c} 0 \\ 1 \end{array} \right\|_B$, $|\downarrow\uparrow\rangle = \left\| \begin{array}{c} 0 \\ 1 \end{array} \right\|_A \otimes \left\| \begin{array}{c} 1 \\ 0 \end{array} \right\|_B$, we find

$$\Sigma_z|\Psi_{00}\rangle = \frac{1}{\sqrt{2}}\Big(|\uparrow\downarrow\rangle - |\uparrow\downarrow\rangle + |\downarrow\uparrow\rangle - |\downarrow\uparrow\rangle\Big) = 0. \tag{5.30}$$

This is quite reasonable, since the z spin components of $|\uparrow\downarrow\rangle$ and $|\downarrow\uparrow\rangle$ sum to zero.

We find a similar result after changing the direction of the spin projection from z to $\mathbf{r} = (x, y, z)$. Define the following observable (see Eq. 4.18):

$$\Sigma_{\mathbf{r}} = \left\| \begin{array}{cc} z & x - iy \\ x + iy & -z \end{array} \right\|_A \otimes I_B + I_A \otimes \left\| \begin{array}{cc} z & x - iy \\ x + iy & -z \end{array} \right\|_B, \tag{5.31}$$

representing the projection of the total spin in the $\mathbf{r}$ direction. We have

$$\Sigma_{\mathbf{r}}|\uparrow\downarrow\rangle = \left\|\begin{matrix} z & x-iy \\ x+iy & -z \end{matrix}\right\|_A \left\|\begin{matrix} 1 \\ 0 \end{matrix}\right\|_A \otimes \left\|\begin{matrix} 0 \\ 1 \end{matrix}\right\|_B + \left\|\begin{matrix} 1 \\ 0 \end{matrix}\right\|_A \otimes \left\|\begin{matrix} z & x-iy \\ x+iy & -z \end{matrix}\right\|_B \left\|\begin{matrix} 0 \\ 1 \end{matrix}\right\|_B =$$

$$\left\|\begin{matrix} z \\ x+iy \end{matrix}\right\|_A \otimes \left\|\begin{matrix} 0 \\ 1 \end{matrix}\right\|_B + \left\|\begin{matrix} 1 \\ 0 \end{matrix}\right\|_A \otimes \left\|\begin{matrix} x-iy \\ -z \end{matrix}\right\|_B =$$

$$(z)|\uparrow\downarrow\rangle + (x+iy)|\downarrow\downarrow\rangle + (x-iy)|\uparrow\uparrow\rangle + (-z)|\uparrow\downarrow\rangle.$$
(5.32)

Analogously,

$$\Sigma_{\mathbf{r}}|\downarrow\uparrow\rangle = \left\|\begin{matrix} z & x-iy \\ x+i & -z \end{matrix}\right\|_A \left\|\begin{matrix} 0 \\ 1 \end{matrix}\right\|_A \otimes \left\|\begin{matrix} 1 \\ 0 \end{matrix}\right\|_B + \left\|\begin{matrix} 0 \\ 1 \end{matrix}\right\|_A \otimes \left\|\begin{matrix} z & x-iy \\ x+iy & -z \end{matrix}\right\|_B \left\|\begin{matrix} 1 \\ 0 \end{matrix}\right\|_B =$$

$$\left\|\begin{matrix} x-iy \\ -z \end{matrix}\right\|_A \otimes \left\|\begin{matrix} 1 \\ 0 \end{matrix}\right\|_B + \left\|\begin{matrix} 0 \\ 1 \end{matrix}\right\|_A \otimes \left\|\begin{matrix} z \\ x+iy \end{matrix}\right\|_B =$$

$$(x-iy)|\uparrow\uparrow\rangle + (-z)|\downarrow\uparrow\rangle + (z)|\downarrow\uparrow\rangle + (x+iy)|\downarrow\downarrow\rangle.$$
(5.33)

Therefore, $\Sigma_{\mathbf{r}}|\Psi_{00}\rangle = \frac{1}{\sqrt{2}}\Sigma_{\mathbf{r}}(|\uparrow\downarrow\rangle - |\downarrow\uparrow\rangle) = 0$; thus $|\Psi_{00}\rangle$ is an eigenstate of $\Sigma_{\mathbf{r}}$ corresponding to the eigenvalue 0 for *all* values of x, y, z. Physically, this means that if we measure the total spin for the two-electron state $|\Psi_{00}\rangle$, we obtain the value 0, independent of the direction of $\mathbf{r}$. The state $|\Psi_{00}\rangle$ is called the *singlet state*. The second entangled state of (5.13), $|\Psi_{1,0}\rangle = \frac{1}{\sqrt{2}}(|\uparrow\downarrow\rangle + |\downarrow\uparrow\rangle)$ is called the *triplet state*. It is a composite of the three states labeled by two indexes S, M with $S = 1$ and $M = 1, 0, -1$:

$$|\Psi_{1,1}\rangle = |\uparrow\uparrow\rangle, \qquad |\Psi_{1,0}\rangle = \frac{1}{\sqrt{2}}\left(|\uparrow\downarrow\rangle + |\uparrow\downarrow\rangle\right), \qquad |\Psi_{1,-1}\rangle = |\downarrow\downarrow\rangle. \quad (5.34)$$

In general, for n particles with spin $\frac{1}{2}$, S denotes the maximum total spin of the system. The maximum is reached when all spins are aligned, with $S = 2n$. In the case of two electrons, $S = \frac{1}{2} + \frac{1}{2} = 1$. The index M is a quantum number denoting the projection of the total spin on the z-axis. From a semiclassical point of view, we can imagine that the two aligned spins have only three possible orientations with respect to the z-axis (see Fig. 5.2). The analogous picture for the singlet state is shown in Fig. 5.3. It is easy to verify that $|\Psi_{1,M}\rangle$ is eigenvector of the operator Σ_z (see 5.28) corresponding to the eigenvalue M:

$$\Sigma_z|\Psi_{1,M}\rangle = (\sigma_{zA} \otimes I + I \otimes \sigma_{zB})|\Psi_{1,M}\rangle = M|\Psi_{1,M}\rangle \quad \text{for} \quad M = 1, 0, -1. \quad (5.35)$$

When S is an integer, the possible values of M are $S, S-1, \ldots, 0, -1, \ldots, -S; 2S+1$ values in all. For example, for the case of 4 aligned electron spins, $S = 4 \times \frac{1}{2} = 2$ and the 5 eigenvalues of the operator Σ_z are $M = 2, 1, 0, -1, -2$. In (Fig. 5.4) the possible orientations of the aligned spins with respect to the z-axis are shown. A similar picture can be imagined for the possible z-components of the angular momentum of one electron in a D state, with a total angular momentum $L = 2$. We have considered here the specific case of particles with spin-$\frac{1}{2}$, since this is the simplest mathematically, yet the most relevant for the following discussion of the foundations of quantum mechanics.

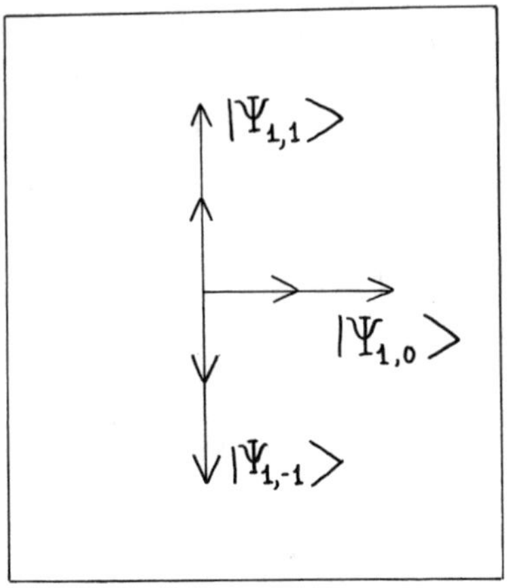

Fig. 5.2 Semiclassical
picture of the three possible
orientations, with respect to
the z-axis, of two aligned
spins, $S = 1$

Fig. 5.3 Semiclassical
picture of a singlet state of
two spins, $S = \frac{1}{2}$

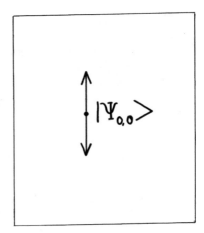

5.2   Bohm's Version of the Einstein–Podolsky–Rosen Experiment

A more practical version of the EPR experiment was suggested by Bohm (1951), based on correlations between electron spin states, rather than positions and momenta. It is useful then to consider the entangled singlet state $|\Psi_{00}\rangle$ of two electrons, a and b, adding a more complete description of the system by taking account, as well, of the spatial part of the wavefunction $\varXi(\mathbf{r}_A, \mathbf{r}_B)$. We have in mind a situation in which the two electrons are initially at a single point, say $\mathbf{r} = 0$, and then move apart in the direction of two detectors. Here, $\mathbf{R}_A$, $\mathbf{R}_B$ designate two points in the physical space $\mathbb{R}^3$. The wave function $\varXi$ gives the probability amplitude of finding the electron a near the point $\mathbf{R}_A$ and the electron b near the point $\mathbf{R}_B$. For two small volumes ΔV_A and ΔV_B around the points $\mathbf{R}_A$, $\mathbf{R}_B$, $|\varXi(\mathbf{r}_A, \mathbf{r}_B)|^2 \, \Delta V_A \, \Delta V_B$ gives the probability of finding electron a inside ΔV_A and electron b inside ΔV_B. We assume that the distance

$|\mathbf{R}_A - \mathbf{R}_B|$ is macroscopic (recall that one meter is about 2×10^{10} atomic units). We can combine the spatial description provided by the wave function $\Xi(\mathbf{r}_A, \mathbf{r}_B)$ with the spin state description by a simple product of the space and spin wavefunctions. Technically, the Hilbert space of the system is now the *tensor product* of the Hilbert space of spatial wave functions times the Hilbert space of the spin states. Assuming that the spin state is a singlet, the total wave function of the system is then given by

$$\Xi(\mathbf{r}_A, \mathbf{r}_B) \times \frac{1}{\sqrt{2}}\Big(|A\uparrow\rangle \otimes |B\downarrow\rangle - |A\downarrow\rangle \otimes |B\uparrow\rangle\Big), \qquad (5.36)$$

where $|A\uparrow\rangle = \left\|\begin{matrix}1\\0\end{matrix}\right\|_A$, $|B\downarrow\rangle = \left\|\begin{matrix}0\\1\end{matrix}\right\|_B$, etc. Suppose tentatively that the space part of the two-electron wavefunction $\Xi(\mathbf{r}_A, \mathbf{r}_B)$ can be factorized onto the product of separate one-electron functions $f(\mathbf{r}_A)\,g(\mathbf{r}_B)$, in such a way that electron a is located near $\mathbf{R}_A$, while electron b is near $\mathbf{R}_B$, as shown in Fig. 5.4. This, however, violates one of the fundamental tenets of QM, that electrons are *indistinguishable particles*. Electrons a and b must each be equally associated with $\mathbf{r}_A$ and $\mathbf{r}_B$. This can be realized by writing the spatial wavefunction as the sum:

$$\Xi(\mathbf{r}_A, \mathbf{r}_B) = \frac{1}{\sqrt{2}}\big[f(\mathbf{r}_A)g(\mathbf{r}_B) + g(\mathbf{r}_A)f(\mathbf{r}_b)\big]. \qquad (5.37)$$

The complete wavefunction for the singlet state, including both space and spin contributions, is then, as sketched in Fig. 5.5,

Fig. 5.4 Semiclassical picture of four aligned spins, $S = 2$

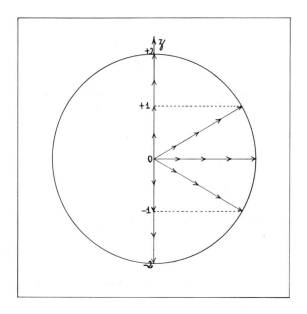

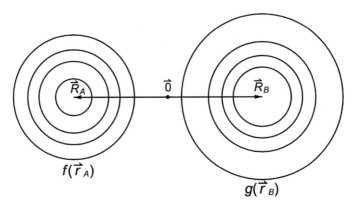

Fig. 5.5 The function $\Xi(\mathbf{r}_A, \mathbf{r}_B)$ is of significant magnitude only when $\mathbf{r}_A \approx \mathbf{R}_A$ and $\mathbf{r}_B \approx \mathbf{R}_B$

$$|\Psi\rangle = \frac{1}{\sqrt{2}}\Big(f(\mathbf{r}_A)g(\mathbf{r}_B) + g(\mathbf{r}_A)f(\mathbf{r}_B)\Big) \times \frac{1}{\sqrt{2}}\Big(|A \uparrow\rangle \otimes |B \downarrow\rangle - |A \downarrow\rangle \otimes |B \uparrow\rangle\Big).$$
$$(5.38)$$

Note that the total wavefunction is *antisymmetric* (i.e., it changes sign) with exchange of electron labels A and B. This is true for all fermions (in particular, particles of spin $\frac{1}{2}$), and leads to the Pauli exclusion principle for many-electron systems.

For a triplet spin state, the three possible wavefunctions are instead

$$|\Psi\rangle = \frac{1}{\sqrt{2}}\Big(f(\mathbf{r}_A)g(\mathbf{r}_B) - g(\mathbf{r}_A)f(\mathbf{r}_B)\Big) \times \begin{cases} |A \uparrow\rangle \otimes |B \uparrow\rangle \\ \frac{1}{\sqrt{2}}|A \uparrow\rangle \otimes |B \downarrow\rangle + |A \downarrow\rangle \otimes |B \uparrow\rangle \\ |A \downarrow\rangle \otimes |B \downarrow\rangle \end{cases}$$
$$(5.39)$$

Here, the antisymmetry (minus sign) is contained in the space factor.

In both the original EPR and Bohm's version of the *Gedankenexperiment*, the two particles enter two detectors, say at $\mathbf{r}_A = \mathbf{D}$ and $\mathbf{r}_B = -\mathbf{D}$. Thus the spatial part of the wavefunction $\Xi(\mathbf{D}, -\mathbf{D})$ reduces to a constant factor and can be neglected. From now on, the wavefunction $|\Psi\rangle$ will refer just to the spin part.

Suppose now that we measure the spin component of one of the electrons along the z-axis. Since the wave function $|\Psi\rangle$ contains only the terms $|A \uparrow\rangle \otimes |B \downarrow\rangle$ and $|A \downarrow\rangle \otimes |B \uparrow\rangle$, if the measurement of the A spin shows "up," we can infer that the two-particle spin state has collapsed to the state $|A \uparrow\rangle \otimes |B \downarrow\rangle$. If the A spin shows "down," the collapsed spin state must be $|A \downarrow\rangle \otimes |B \uparrow\rangle$. Thus a measurement of the spin state of one particle gives us precise information on the spin state of the other particle: B is "down" if A is "up" and "up" if A is "down." There is a perfect correlation between the measurements of the spin A and the measurement of the spin B. However, as J. S. Bell pointed out, this is not a simple correlation like those we encounter in everyday life. For example, if you find that you have with you only one glove, say the left one, you immediately know that you have forgotten the right glove at home. But in the case of the electrons, before the measurement, the spin state is a superposition of $(|A \uparrow\rangle \otimes |B \downarrow\rangle$ and $|A \downarrow\rangle \otimes |B \uparrow\rangle$ (see Eq. 5.38), and only after the measurement is there the instantaneous collapse into $|A \uparrow\rangle \otimes |B \downarrow\rangle$ or $|A \downarrow\rangle \otimes |B \uparrow\rangle$. Since the two electrons are macroscopically separated, it seems as some "spooky action at a distance" has occurred. The "message" of the A particle to the B particle, something like "my spin is down, your spin must be up" is evidently propagated faster than the speed of light, in contradiction to the central tenet of the theory of relativity.

Remarkably the situation is even more paradoxical. Actually the singlet state Ψ_{00} is rotationally invariant, and nothing prevents us from measuring the spin projections in an arbitrary direction, say (x, y, z). From Eqs. (4.18), (4.19), (4.21), we recall that the up and down spinors corresponding to the (x, y, z) direction are given by

$$|\Psi_1\rangle = \left\| \begin{matrix} \psi \\ \phi \end{matrix} \right\| = \alpha \left\| \begin{matrix} 1+z \\ x+iy \end{matrix} \right\|, \qquad |\Psi_2\rangle = \left\| \begin{matrix} \phi^* \\ -\psi^* \end{matrix} \right\| = \alpha^* \left\| \begin{matrix} x-iy \\ -(1+z) \end{matrix} \right\|. \quad (5.40)$$

Let us compute the singlet state corresponding to the new direction. Since the system is rotationally invariant we do not expect anything different. We have

$$|\Psi_{1A}\rangle \otimes |\Psi_{2B}\rangle - |\Psi_{2A}\rangle \otimes |\Psi_{1B}\rangle =$$

$$|\alpha|^2 \left\| \begin{matrix} 1+z \\ x+iy \end{matrix} \right\|_A \otimes \left\| \begin{matrix} x-iy \\ -(1+z) \end{matrix} \right\|_B - |\alpha|^2 \left\| \begin{matrix} x-iy \\ -(1+z) \end{matrix} \right\|_A \otimes \left\| \begin{matrix} 1+z \\ x+iy \end{matrix} \right\|_B .$$
$$(5.41)$$

Since $\mathbb{C}^2 \otimes \mathbb{C}^2 = \mathbb{C}^4$, (5.41) becomes the four-component vector:

$$|\alpha|^2 \left\| \begin{matrix} (1+z)(x-iy) - (x-iy)(1+z) \\ -(1+z)^2 - (x^2+y^2) \\ x^2+y^2+(1+z)^2 \\ -(x+iy)(1+z) + (1+z)(x+iy) \end{matrix} \right\| = |\alpha|^2(x^2+y^2+z^2+2z+1) \left\| \begin{matrix} 0 \\ -1 \\ 1 \\ 0 \end{matrix} \right\|.$$
$$(5.42)$$

Since the state vectors $|\Psi_1\rangle$ and $|\Psi_2\rangle$ are normalized, $|\alpha|^2(x^2+y^2+z^2+2z+1) = 1$, and we obtain

$$\left\| \begin{matrix} 1 \\ 0 \end{matrix} \right\|_A \otimes \left\| \begin{matrix} 0 \\ 1 \end{matrix} \right\|_B - \left\| \begin{matrix} 0 \\ 1 \end{matrix} \right\|_A \otimes \left\| \begin{matrix} 1 \\ 0 \end{matrix} \right\|_B = |A \uparrow\rangle \otimes |B \downarrow\rangle - |A \downarrow\rangle \otimes |B \uparrow\rangle = |\Psi_{00}\rangle, \quad (5.43)$$

which is the same as the singlet state with respect to the z-axis. Evidently, the singlet state does not depend on the choice of axis (x, y, z). However, measuring spin components of the particles A and B in two different directions, say $(0, 0, z)$ and $(x, 0, 0)$, there is no longer a simple correlation of the results. This is so since the operators

$$\sigma_z = \left\| \begin{matrix} 1 & 0 \\ 0 & -1 \end{matrix} \right\|, \qquad \sigma_x = \left\| \begin{matrix} 0 & 1 \\ 1 & 0 \end{matrix} \right\|, \quad (5.44)$$

representing the spin projections in the z- and x-directions, do not commute. Indeed the eigenstates $\left\|\begin{matrix} 1 \\ 0 \end{matrix}\right\|$, $\left\|\begin{matrix} 0 \\ 1 \end{matrix}\right\|$ of σ_z do not coincide with those of σ_x, which are, in fact, $\frac{1}{\sqrt{2}}\left\|\begin{matrix} 1 \\ 1 \end{matrix}\right\|$, $\frac{1}{\sqrt{2}}\left\|\begin{matrix} 1 \\ -1 \end{matrix}\right\|$. In conclusion, excluding the possibility of superluminal, instantaneous communications, we remain unable to answer the question: How does particle B "know" which component of the spin of A we measured?

5.3 Hidden Variables and Bell's Inequality

Before proceeding, let us say a few words about the historical development and philosophical implications of concepts such as the wave–particle duality, entangled states, etc. Einstein believed that distant objects cannot instantaneously influence one another; rather, every object is acted upon only by its contiguous surroundings. This principle is called *locality*. In his words: "The following idea characterizes the relative independence of objects far apart in space, A and B: external influence on A has no direct influence on B." This is known as the *principle of local action*, which forms the basis of classical and quantum field theories; for example, electromagnetic waves do not propagate instantaneously, but only at a velocity less than or equal to the vacuum speed of light c. Were it not for localization, formulation of empirical laws for finite systems would not be possible.

We recall from Sect. 3.7 that Einstein's epistemological orientation can be classified as *realistic*: he considered that physical entities, including possibly the wavefunction, possessed inherent properties regardless of whether they are being observed or not. Pais (1982) relates "I recall that during one walk Einstein suddenly stopped, turned to me and asked whether I really believed that the Moon exists only when I look at it." In their famous paper, Einstein, Podolsky and Rosen (EPR) (Einstein et al. 1935) wrote "If, without in any way disturbing a system, we can predict with certainty the value of a physical quantity, then there exists an element of physical reality corresponding to this physical quantity." Therefore, for the case of the entangled states just considered, it appears that Einstein's belief was that, prior to the measurement of the spin of particle B, the spin was already aligned in the observed direction. The combination of the locality principle and realism is called *local realism*. However, without entering too deeply into a discussion of epistemology or ontology, one might ask "What about believing in the existence of systems, spins before their measurement, or moons, or even entire universes, no one will ever see?"

We saw in Sect. 3.6 that the "collapse of the wavefunction" gives rise to a multitude of problems. Does the collapse happen when a detector records a result? Or, it is perhaps just something in the observer's mind when he becomes aware of the result of a measurement? Since animals have sense organs as we do, would the observation of a cat, or a bacterium, be sufficient to produce the collapse? And why does the wavefunction follows two different types of temporal evolution, one linear and continuous, governed by the time-dependent Schrödinger equation, and the other

Table 5.1 Hypothetical particles with instruction sets

Instruction set	Result of the measurement of σ_z	Result of the measurement of σ_x
(z^+, x^+)	+1	+1
(z^+, x^-)	+1	−1
(z^-, x^+)	−1	+1
(z^-, x^-)	−1	−1

nonlinear and discontinuous? In order to find a solution to these problems, Einstein and others thought that there had to be some unknown hidden variables, not yet accessible to experiment; these variables were supposed to predetermine, in some way, the results of our measurements. We have seen in Chap. 1 that classical statistical mechanics, taking the average of the motion of a multitude of molecules, explains the behavior of thermodynamic quantities, such as pressure or temperature; in an analogous way a hypothetical new theory of hidden variables should explain both the corpuscular and wave-like behavior of particles, and the paradoxical properties of entangled states.

Let us see how the existence of hidden variables could avoid faster-than-light propagation of signals between two spin-$\frac{1}{2}$ particles A, B considered above, making it possible, for example, for the particle B to have a spin aligned parallel, opposite to the measured direction of the spin of A. (Alternatively, for photons, the two possible states represent vertical and horizontal polarization.) Imagine that each particle carries a hidden "set of instructions" which determines the outcome of any measurement. For example, considering only the observables σ_z, σ_x, we have four types of "instruction sets," producing four variants of the particles (Table 5.1).

In this way, there is no need for superluminal actions at a distance, and the realistic picture of the world would be saved. In an analogous way, the assumption of hidden variables could, in principle, provide an resolution of the wave–particle duality. All right then; can we expect that sooner or later the hidden variables will be discovered? After all, our understanding of molecular motion came many years after the interrelationships among thermodynamic quantities were known. However, this expectation fails. An ingenious inequality proposed by Bell (1964) enabled experimental tests to verify whether or not a local deterministic interpretation of quantum mechanics is possible. There are several variants of Bell's inequality. We will present here a version due to Wigner,[2] which is perhaps the simplest.

Consider three unit vectors **a**, **b**, **c** in three-dimensional space, and the possible results of measurement of the spin projections $\sigma \cdot \mathbf{a}$, $\sigma \cdot \mathbf{b}$, $\sigma \cdot \mathbf{c}$, where $\sigma \cdot \mathbf{a} = \sigma_x x_a + \sigma_y y_a + \sigma_z z_a$, etc., together with the corresponding "instruction sets." In Table 5.2, we show a population of N_A pairs of particles, all pairs coupled in a singlet state, and partitioned into eight groups of $N_1, N_2, N_3, \ldots, N_8$ pairs (capital letters A, B denote the two particles of a pair, lower case letters **a**, **b**, **c** denote unit vectors

---

[2]EP Wigner (1970), Am J Phys 38:1005–1009.

Table 5.2 Results of spin measurements

Population	Instr. set of particle A	Instr. set of particle B	$\sigma_A \cdot \mathbf{a}$	$\sigma_A.\mathbf{b}$	$\sigma_A \cdot \mathbf{c}$	$\sigma_B \cdot \mathbf{a}$	$\sigma_B.\mathbf{b}$	$\sigma_B \cdot \mathbf{c}$
N_1	$(\mathbf{a}^+, \mathbf{b}^+, \mathbf{c}^+)$	$(\mathbf{a}^-, \mathbf{b}^-, \mathbf{c}^-)$	$+1$	$+1$	$+1$	-1	-1	-1
N_2	$(\mathbf{a}^+, \mathbf{b}^+, \mathbf{c}^-)$	$(\mathbf{a}^-, \mathbf{b}^-, \mathbf{c}^+)$	$+1$	$+1$	-1	-1	-1	$+1$
N_3	$(\mathbf{a}^+, \mathbf{b}^-, \mathbf{c}^+)$	$(\mathbf{a}^-, \mathbf{b}^+, \mathbf{c}^-)$	$+1$	-1	$+1$	-1	$+1$	-1
N_4	$(\mathbf{a}^+, \mathbf{b}^-, \mathbf{c}^-)$	$(\mathbf{a}^-, \mathbf{b}^+, \mathbf{c}^+)$	$+1$	-1	-1	-1	$+1$	$+1$
N_5	$(\mathbf{a}^-, \mathbf{b}^+, \mathbf{c}^+)$	$(\mathbf{a}^+, \mathbf{b}^-, \mathbf{c}^-)$	-1	$+1$	$+1$	$+1$	-1	-1
N_6	$(\mathbf{a}^-, \mathbf{b}^+, \mathbf{c}^-)$	$(\mathbf{a}^+, \mathbf{b}^-, \mathbf{c}^+)$	-1	$+1$	-1	$+1$	-1	$+1$
N_7	$(\mathbf{a}^-, \mathbf{b}^-, \mathbf{c}^+)$	$(\mathbf{a}^+, \mathbf{b}^+, \mathbf{c}^-)$	-1	-1	$+1$	$+1$	$+1$	-1
N_8	$(\mathbf{a}^-, \mathbf{b}^-, \mathbf{c}^-)$	$(\mathbf{a}^+, \mathbf{b}^+, \mathbf{c}^+)$	-1	-1	-1	$+1$	$+1$	$+1$

in physical space). Of course, in all cases $\sigma_A \cdot \mathbf{a} = -\sigma_B \cdot \mathbf{a}$, $\sigma_A \cdot \mathbf{b} = -\sigma_B \cdot \mathbf{b}$, etc. Let us now suppose that

$$\sigma_A \cdot \mathbf{a} = \sigma_B \cdot \mathbf{b} = +1. \tag{5.45}$$

Reading from Table 5.2, we see that only the groups of N_3 and N_4 pairs fulfill this condition. Choosing a pair at random, the probability that (5.45) is verified is given by

$$P(\mathbf{a}^+, \mathbf{b}^+) = \frac{N_3 + N_4}{N}. \tag{5.46}$$

The probability that $\sigma_A \cdot \mathbf{b} = \sigma_B \cdot \mathbf{c} = +1$ is given by

$$P(\mathbf{b}^+, \mathbf{c}^+) = \frac{N_2 + N_6}{N}, \tag{5.47}$$

and the probability that $\sigma_A \cdot \mathbf{a} = \sigma_B \cdot \mathbf{c} = +1$ is

$$P(\mathbf{a}^+, \mathbf{c}^+) = \frac{N_2 + N_4}{N}. \tag{5.48}$$

From (5.46), (5.47), (5.48) it follows that

$$P(\mathbf{a}^+, \mathbf{b}^+) + P(\mathbf{b}^+, \mathbf{c}^+) - P(\mathbf{a}^+, \mathbf{c}^+) = \frac{N_3 + N_6}{N} \geq 0. \tag{5.49}$$

Therefore, the following "triangle inequality" holds:

$$P(\mathbf{a}^+, \mathbf{b}^+) + P(\mathbf{b}^+, \mathbf{c}^+) \geq P(\mathbf{a}^+, \mathbf{c}^+). \tag{5.50}$$

Let us now compute the corresponding quantum mechanical results. Consider a singlet state of two-electron spins A and B, denoted $|\Psi_{0,0}\rangle$. By virtue of its rotational symmetry, this can be expressed in terms of quantization about any axis. Let us consider the two directions **a** and **b**. The two alternative forms are given by

$$|\Psi_{0,0}\rangle = \frac{1}{\sqrt{2}}\Big(|A_a \uparrow\rangle|B_a \downarrow\rangle - |A_a \downarrow\rangle|B_a \uparrow\rangle\Big) = \frac{1}{\sqrt{2}}\Big(|A_b \uparrow\rangle|B_b \downarrow\rangle - |A_b \downarrow\rangle|B_b \uparrow\rangle\Big).$$
(5.51)

Recalling Eqs. (4.52) and (4.53), the **b** basis functions can be expressed in terms of the **a** basis using analogous relations. We require just the transformations for spin B:

$$|B_b \uparrow\rangle = \cos\frac{\theta_{ab}}{2}|B_a \uparrow\rangle + \sin\frac{\theta_{ab}}{2}|B_b \downarrow\rangle,$$
$$|B_b \downarrow\rangle = -\sin\frac{\theta_{ab}}{2}|B_a \uparrow\rangle + \cos\frac{\theta_{ab}}{2}|B_b \downarrow\rangle,$$
(5.52)

where θ_{ab} is the angle between the directions **a** and **b**. Now, substituting (5.52) into (5.51), we obtain

$$|\Psi_{0,0}\rangle = \frac{1}{\sqrt{2}}\Big(-\sin\frac{\theta_{ab}}{2}|A_b \uparrow\rangle|B_a \uparrow\rangle + \cos\frac{\theta_{ab}}{2}|A_b \uparrow\rangle|B_b \downarrow\rangle$$
$$-\cos\frac{\theta_{ab}}{2}|A_b \downarrow\rangle|B_a \uparrow\rangle - \sin\frac{\theta_{ab}}{2}|A_b \downarrow\rangle|B_b \downarrow\rangle\Big).$$
(5.53)

The wave function can be interpreted as a superposition of four possible outcomes. The amplitude for the joint event of particle A registering spin-up in direction **a** and particle B registering spin-up in direction **b** is equal to $-\frac{1}{\sqrt{2}}\sin\frac{\theta_{ab}}{2}$. The probability of this event, which we designate $P^{QM}(\mathbf{a}^+, \mathbf{b}^+)$, is then given by the square of the amplitude $\frac{1}{2}\sin^2\frac{\theta_{ab}}{2}$. The four probabilities, which are listed in the last column of Table 5.2, can then be found:

$$P^{QM}(\mathbf{a}^+, \mathbf{b}^+) = P^{QM}(\mathbf{a}^-, \mathbf{b}^-) = \frac{1}{2}\sin^2\frac{\theta_{ab}}{2},$$
$$P^{QM}(\mathbf{a}^+, \mathbf{b}^-) = P^{QM}(\mathbf{a}^-, \mathbf{b}^+) = \frac{1}{2}\cos^2\frac{\theta_{ab}}{2}.$$
(5.54)

These four probabilities add up to 1, as they should. Analogous results will be found for the combinations involving **b** and **c**, **a** and **c**.

Assuming that the three vectors **a**, **b**, **c** are in anticlockwise order around the z-axis, we find, for the angles, $\theta_{ac} = \theta_{ab} + \theta_{bc}$. We have now found the quantum analogs of the classical probabilities in Eq. (5.50), namely,

$$P^{QM}(\mathbf{a}^+, \mathbf{b}^+) = \frac{1}{2}\sin^2\frac{\theta_{ab}}{2}, \qquad P^{QM}(\mathbf{b}^+, \mathbf{c}^+) = \frac{1}{2}\sin^2\frac{\theta_{bc}}{2},$$
$$P^{QM}(\mathbf{a}^+, \mathbf{c}^+) = \frac{1}{2}\sin^2\frac{\theta_{ab} + \theta_{bc}}{2}.$$
(5.55)

You can show that, for $0 \leq \theta_1, \theta_2 \leq \frac{\pi}{2}$,

$$\sin^2 \frac{\theta_1 + \theta_2}{2} \geq \sin^2 \frac{\theta_1}{2} + \sin^2 \frac{\theta_2}{2}. \tag{5.56}$$

Therefore

$$P^{QM}(\mathbf{a}^+, \mathbf{c}^+) \geq P^{QM}(\mathbf{a}^+, \mathbf{b}^+) + P^{QM}(\mathbf{b}^+, \mathbf{c}^+), \tag{5.57}$$

which is the exact *reverse* of Bell's inequality (5.50). Although there are certain combinations of angles for which the inequality (5.57) fails and Bell's inequality is satisfied, the fact that Bell's inequality can be violated at all is very compelling evidence that quantum mechanics is inconsistent with any physical theory based on local realism. We arrive thereby at *Bell's theorem*.

Theorem 5.1 *Bell's inequality (5.50) is violated by quantum mechanics.*

We have thus proved that no discrete hidden variable of the type considered by Wigner, no simple "instruction sets" for particles A and B, can account for the quantum mechanical results. This means that quantum mechanics predicts greater than expected correlations between events that are out of range by classical causality. Let us illustrate with a particular example. Let $\theta_{ab} = \theta_{bc} = \pi/6$, so that $\theta_{ac} = \pi/3$, as shown in Fig. 5.6. Then $P^{QM}(\mathbf{a}^+, \mathbf{b}^+) = P^{QM}(\mathbf{b}^+, \mathbf{c}^+) \approx 0.0335$ and $P^{QM}(\mathbf{a}^+, \mathbf{c}^+) = 0.125$. Thus $0.125 > 0.0335 + 0.0335$, and the inequality (5.57) is verified.

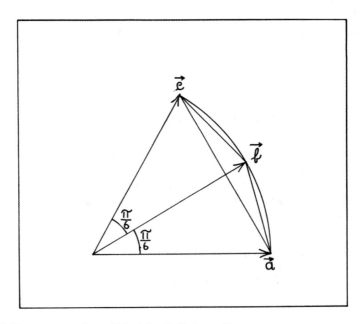

Fig. 5.6 Three vectors **a, b, c** which violate Bell's inequality

5.4 Generalized Bell's Inequality

Wigner's proof can easily be generalized, even when the set of "hidden variables" is very large, including the case when its description requires one or more continuous variables. Bell's inequality still remains valid. For the proof we must continue to conform to the locality principle: no "message" can be instantly transmitted from one particle of the pair to the other. For simplicity, consider the case of a single continuous hidden variable λ associated with the pair. The probability $\rho(\lambda)$ of the value λ for the hidden variable must satisfy the condition:

$$\int \rho(\lambda)d\lambda = 1, \tag{5.58}$$

since the total probability must equal 1. With the integral running over the set of all possible values of λ, the *average* $\langle F \rangle$ of a quantity $F(\lambda)$ is given by

$$\langle F \rangle = \int F(\lambda)\rho(\lambda)d\lambda. \tag{5.59}$$

Denoting by m, M, the minimum and maximum values of F, we have

$$m = m \int \rho(\lambda)d\lambda \leq \int F(\lambda)\rho(\lambda)d\lambda \leq M \int \rho(\lambda)d\lambda = M. \tag{5.60}$$

Of course, $m \leq \langle F \rangle \leq M$. The quantum mechanical analog of the average is the expectation value $\langle \Psi | F | \Psi \rangle$, where $|\Psi\rangle$ represents the state of the system.

Let us now define a quantum mechanical *correlation function*. Given a state $|\Psi\rangle$ of a pair of particles A, B, with spin directions $\mathbf{a}$, $\mathbf{b}$, the quantum correlation function is written $E^{QM}(\mathbf{a}, \mathbf{b})$, as the expectation value of the product of the two commuting operators $\sigma_A \cdot \mathbf{a}$, $\sigma_B \cdot \mathbf{b}$, given by

$$E^{QM}(\mathbf{a}, \mathbf{b}) = \langle \Psi | (\sigma_A \cdot \mathbf{a})(\sigma_B \cdot \mathbf{b}) | \Psi \rangle. \tag{5.61}$$

The corresponding *classical* correlation function, assuming that the quantum mechanical results can be obtained by taking the average over the hidden variable λ:

$$E(\mathbf{a}, \mathbf{b}) = \int \alpha(\mathbf{a}, \lambda)\,\beta(\mathbf{b}, \lambda)\rho(\lambda)d\lambda, \tag{5.62}$$

where $\alpha(\mathbf{a}, \lambda) = \pm 1$, $\beta(\mathbf{b}, \lambda) = \pm 1$ are the results of the measurements of the spin projections of the a and b particles in the $\mathbf{a}$, $\mathbf{b}$ directions, respectively. The analogous quantum mechanical values $\alpha(\mathbf{a}, \lambda)$, $\beta(\mathbf{b}, \lambda)$ are the eigenvalues of the operators $\sigma_A \cdot \mathbf{a}$, $\sigma_B \cdot \mathbf{b}$.

Before proving the generalized Bell's inequality, we make use of the result:

Lemma 5.1 *Let $|\Psi\rangle$ be a singlet state of two particles; then $E^{QM}(\mathbf{a}, \mathbf{b})$ is given by*

Table 5.3 Quantum mechanical results

Eigenvectors	Eigenvalues Q	Probabilities
$\lvert \uparrow\uparrow \rangle = \lvert A_a \uparrow \rangle \otimes \lvert B_b \uparrow \rangle$	$(+1)(+1) = +1$	$P^{QM}(\mathbf{a}^+, \mathbf{b}^+) =$ $\lvert \langle \uparrow\uparrow \lvert \Psi \rangle \rvert^2 = \frac{1}{2} \sin^2 \frac{\theta_{ab}}{2}$
$\lvert \uparrow\downarrow \rangle = \lvert A_a \uparrow \rangle \otimes \lvert B_b \downarrow \rangle$	$(+1)(-1) = -1$	$P^{QM}(\mathbf{a}^+, \mathbf{b}^-) =$ $\lvert \langle \uparrow\downarrow \lvert \Psi \rangle \rvert^2 = \frac{1}{2} \cos^2 \frac{\theta_{ab}}{2}$
$\lvert \downarrow\uparrow \rangle = \lvert A_a \downarrow \rangle \otimes \lvert B_b \uparrow \rangle$	$(-1)(+1) = -1$	$P^{QM}(\mathbf{a}^-, \mathbf{b}^+) =$ $\lvert \langle \downarrow\uparrow \lvert \Psi \rangle \rvert^2 = \frac{1}{2} \cos^2 \frac{\theta_{ab}}{2}$
$\lvert \downarrow\downarrow \rangle = \lvert A_a \downarrow \rangle \otimes \lvert B_b \downarrow \rangle$	$(-1)(-1) = +1$	$P^{QM}(\mathbf{a}^-, \mathbf{b}^-) =$ $\lvert \langle \downarrow\downarrow \lvert \Psi \rangle \rvert^2 = \frac{1}{2} \sin^2 \frac{\theta_{ab}}{2}$

$$E^{QM}(\mathbf{a}, \mathbf{b}) = -\mathbf{a} \cdot \mathbf{b}. \tag{5.63}$$

Proof To compute the expectation value of the operator $Q = (\boldsymbol{\sigma}_A \cdot \mathbf{a})(\boldsymbol{\sigma}_B \cdot \mathbf{b})$ in the singlet state $\lvert \Psi \rangle$, we use the quantum mechanical formula to find the expectation value of an observable F; we multiply the eigenvalues of F by their probabilities and sum the results. The eigenvalues of the operators $\boldsymbol{\sigma}_A \cdot \mathbf{a}$, $\boldsymbol{\sigma}_B \cdot \mathbf{b}$ are ± 1; the eigenvectors and eigenvalues of the product $Q = (\boldsymbol{\sigma}_A \cdot \mathbf{a})(\boldsymbol{\sigma}_B \cdot \mathbf{b})$ are shown in Table 5.3. In Eq. (5.54), we gave the probabilities as functions of the angle θ_{ab}: $P^{QM}(\mathbf{a}^+, \mathbf{b}^+) = \frac{1}{2} \sin^2 \frac{\theta_{ab}}{2}$ and $P^{QM}(\mathbf{a}^+, \mathbf{b}^-) = \frac{1}{2} \cos^2 \frac{\theta_{ab}}{2}$. Multiplying the eigenvalues by the probabilities, we obtain finally

$$\begin{aligned} E^{QM}(\mathbf{a}, \mathbf{b}) &= P^{QM}(\mathbf{a}^+, \mathbf{b}^+) - P^{QM}(\mathbf{a}^+, \mathbf{b}^-) \\ &- P^{QM}(\mathbf{a}^-, \mathbf{b}^+) + P^{QM}(\mathbf{a}^-, \mathbf{b}^-) = \sin^2 \frac{\theta}{2} - \cos^2 \frac{\theta}{2} = -\cos\theta, \end{aligned} \tag{5.64}$$

where θ is the angle between $\mathbf{a}$ and $\mathbf{b}$. Since $\mathbf{a} \cdot \mathbf{b} = \lvert a \rvert \lvert b \rvert \cos\theta$, and $\lvert a \rvert = \lvert b \rvert = 1$; the lemma is proved.

We can now prove the generalization of Bell's theorem.

Theorem 5.2 *No local hidden variables theory can reproduce the quantum mechanical values of the correlation function.*

Proof Let $E(\mathbf{a}, \mathbf{b})$ be the "classical" correlation function (see Eq. 5.62), which should reproduce $E^{QM}(\mathbf{a}, \mathbf{b})$. Consider the hidden variable expression

$$\begin{aligned} S &= E(\mathbf{a}, \mathbf{b}) + E(\mathbf{a}, \mathbf{b}') + E(\mathbf{a}', \mathbf{b}) - E(\mathbf{a}', \mathbf{b}') = \\ &\int \rho(\lambda) \Big[\alpha(\mathbf{a})\beta(\mathbf{b}) + \alpha(\mathbf{a})\beta(\mathbf{b}') + \alpha(\mathbf{a}')\beta(\mathbf{b}) - \alpha(\mathbf{a}')\beta(\mathbf{b}') \Big] d\lambda, \end{aligned} \tag{5.65}$$

where $\alpha(\mathbf{a}) = \alpha(\mathbf{a}, \lambda)$, $\beta(\mathbf{b}) = \beta(\mathbf{b}, \lambda)$. We have

$$S = \int \rho(\lambda) \Big[\big(\alpha(\mathbf{a}) + \alpha(\mathbf{a}') \big)\beta(\mathbf{b}) + \big(\alpha(\mathbf{a}) - \alpha(\mathbf{a}') \big)\beta(\mathbf{b}') \Big] d\lambda. \tag{5.66}$$

Four cases are possible:

(1) $\alpha(\mathbf{a}) = +1$ and $\beta(\mathbf{b}) = +1$, then $E = +1$
(2) $\alpha(\mathbf{a}) = +1$ and $\beta(\mathbf{b}) = -1$, then $E = -1$
(3) $\alpha(\mathbf{a}) = -1$ and $\beta(\mathbf{b}) = +1$, then $E = -1$
(4) $\alpha(\mathbf{a}) = -1$ and $\beta(\mathbf{b}) = -1$, then $E = +1$

Comparing with (5.60) and choosing $F(\lambda)$ equal to $\alpha(\mathbf{a})\beta(\mathbf{b})$, we see that the minimum and maximum of S are $m = -2$ and $M = +2$, so that

$$-2 \leq S \leq +2. \tag{5.67}$$

This was the form of Bell's formula tested by Clauser, Horne, Shimony and Holt (CHSH) (Clauser et al. 1969). This inequality has been proved assuming that the correlation functions are averages over some unknown hidden variables. But it is easy to prove that inequality (5.67) is violated in quantum mechanics. Let E^{QM} be the quantum analog of E (see Eq. 5.65):

$$S^{QM} = E^{QM}(\mathbf{a}, \mathbf{b}) + E^{QM}(\mathbf{a}, \mathbf{b}') + E^{QM}(\mathbf{a}', \mathbf{b}) - E^{QM}(\mathbf{a}', \mathbf{b}') =$$
$$- \mathbf{a} \cdot \mathbf{b} - \mathbf{a} \cdot \mathbf{b}' - \mathbf{a}' \cdot \mathbf{b} + \mathbf{a}' \cdot \mathbf{b}', \tag{5.68}$$

and assume that the pair is in a *singlet* state. Since (5.67) holds for *any* choice of $\mathbf{a}$, $\mathbf{b}$, $\mathbf{a}'$, $\mathbf{b}'$, we can adopt the particular choice suggested by Mermin (1985), which is particularly suitable to dramatize the counterintuitive aspects of quantum mechanics. Consider three coplanar unit vectors $\mathbf{x}$, $\mathbf{u}$, $\mathbf{v}$ (see Fig. 5.7) with coordinates:

Fig. 5.7 Mermin's choice of $\mathbf{a}$, $\mathbf{b}$, $\mathbf{a}'$, $\mathbf{b}'$

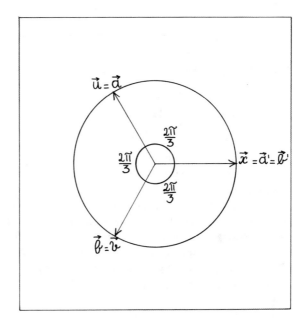

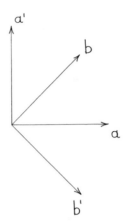

Fig. 5.8 The choice by Clauser and Shimony for **a**, **b**, **a'**, **b'**

$$\mathbf{x} = (1, 0), \quad \mathbf{u} = \left(-\frac{1}{2}, \frac{\sqrt{3}}{2}\right), \quad \mathbf{v} = \left(-\frac{1}{2}, -\frac{\sqrt{3}}{2}\right). \tag{5.69}$$

Using the elementary rule $\mathbf{x} \cdot \mathbf{u} = x_1 u_1 + x_2 u_2$, etc., we have

$$\mathbf{x} \cdot \mathbf{u} = \mathbf{x} \cdot \mathbf{v} = \mathbf{u} \cdot \mathbf{v} = -\frac{1}{2}. \tag{5.70}$$

Mermin suggests the choice $\mathbf{a} = \mathbf{u}$, $\mathbf{b} = \mathbf{v}$, $\mathbf{a'} = \mathbf{b'} = \mathbf{x}$. Then, using the Lemma and Eq. (5.68), we find

$$S^{QM} = -\mathbf{u} \cdot \mathbf{v} - \mathbf{u} \cdot \mathbf{x} - \mathbf{x} \cdot \mathbf{v} + \mathbf{x} \cdot \mathbf{x} = \frac{1}{2} + \frac{1}{2} + \frac{1}{2} + 1 = 2.5 > 2. \tag{5.71}$$

Therefore, the inequality (5.67) is violated, and the theorem is proved.

Clearly, many different choices for **a**, **b**. **a'**, **b'** are possible. The optimal choice has been suggested by Clauser and Shimony, namely $\mathbf{a} = (1, 0)$, $\mathbf{b} = \left(\frac{1}{\sqrt{2}}, \frac{1}{\sqrt{2}}\right)$, $\mathbf{a'} = (0, 1)$, $\mathbf{b'} = \left(\frac{1}{\sqrt{2}}, -\frac{1}{\sqrt{2}}\right)$, as shown in Fig. 5.8. With this choice, $S^{QM} = 2\sqrt{2} > 2$, so that Bell's inequality is maximally violated.

The beauty and significance of Bell's inequality lies not only in its elegant mathematical formulation, but also in the fact that it permits experimental verification. One could argue that it is not that extraordinary for a mathematical formula concerning physical quantities can be tested experimentally. But the remarkable fact here is that we are dealing with an answer to an almost "metaphysical" question: we are asking: "Do hidden variables in the sense of Bell's inequality exist?" or, equivalently, "Does God play dice?" In simpler terms, do the fundamental laws of Nature contain a primary, irreducible, and unavoidable probabilistic aspect?

Before describing the very elegant experiments that conclusively provide an answer to this question, let us review one of the most extraordinary accomplishments of the human intellect, namely, Maxwell's discovery of the equations of the electromagnetic field. These equations, which revealed the fundamental nature of light, will help us to understand in greater depth the crucial experiments associated with Bell's inequality.

5.5 Maxwell's Equations, the Nature of Light, and All That

Let us begin by describing two experiments which were instrumental in the development of electromagnetic theory. (1) Hans Christian Ørsted discovered that an electric current flowing through a wire produces a magnetic field circulating around the wire, which he mapped using a compass. This was the first observed connection between electricity and magnetism. The phenomenon is illustrated in Fig. 5.9. If the wire is wound into a helix, the superposition of the magnetic field ringlets produces a *solenoidal* magnetic field (similar to that of a bar magnet). If the coils of wire are wound around a soft iron core and connected to an AC power supply, the result is an electromagnet, with its field oscillating sinusoidally at the same frequency as the current, as shown in Fig. 5.10. (2) As shown in Fig. 5.11, we insert a permanent magnet into a solenoid connected to a sensitive galvanometer. When the magnet is moved inside the solenoid, the galvanometer indicates a current flow. We conclude that a magnetic field **B** which varies in time can generate an electric field **E** (which moves the electrons in the wires of the solenoid). This phenomenon, *magnetic induction*, was first discovered by Michael Faraday. The three-dimensional vector fields **E** and **B** exist at all the points in space. These two experiments described above can be represented qualitatively by the differential relations[3]:

$$(1) \ \nabla \times \mathbf{B} = \mu_0 \varepsilon_0 \frac{\partial \mathbf{E}}{\partial t} \qquad (2) \ \nabla \times \mathbf{E} = -\frac{\partial \mathbf{B}}{\partial t}. \qquad (5.72)$$

We omit the full details, but it can be surmised that the curl operator $\nabla \times$ is associate with the *circulation* of a vector field, while $\frac{\partial}{\partial t}$ represents a variation with time. In both of the above equations, the electric field produces a current in the same direction. Here ε_0 represents the electric permittivity of free space, and μ_0, the magnetic permeability of free space.

These and many other experiments show that the spatial and temporal variations of **E** and **B** are closely interconnected; this is true not only inside wires and solenoids, but also in empty space. Maxwell produced a synthesis of the work of his several brilliant predecessors, including, Faraday, Coulomb, Ampere, Gauss, Ørsted. He proposed a set of equations which describe the spacial and temporal interrelationships between the electric and magnetic fields. By manipulating these equations, Maxwell derived an equation describing the propagation of waves, a phenomenon which was already familiar, but in different contexts. This had the form of the *wave equation*, which can

---

[3] These are, in fact, the 3rd and 4th of Maxwell's equation in free space. The 1st and 2nd are $\nabla \cdot \mathbf{E} = 0$ and $\nabla \cdot \mathbf{B} = 0$.

Fig. 5.9 An electric current
I in a straight wire produces
a circulating magnetic field
B. The *right-hand rule*
determines the relative
orientations of **I** and **B**

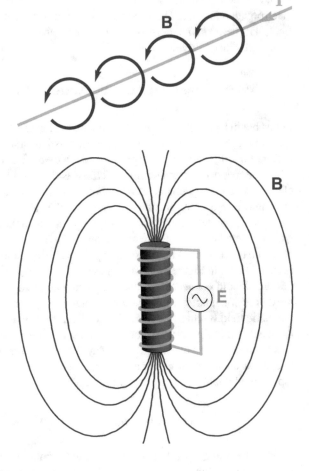

Fig. 5.10 A solenoidal
electromagnet

represent, for example, waves on the surface of a lake produced by a falling stone
(see Fig. 1.1) or pressure variations in the air produced by sound.

The wave equation for **E** can be derived from Eq. (5.72). Applying $\frac{\partial}{\partial t}$ to (1) and
$\nabla \times$ to (2) and adding, we obtain

$$\nabla^2 \mathbf{E} - \varepsilon_0 \mu_0 \frac{\partial^2 \mathbf{E}}{\partial t^2} = 0. \tag{5.73}$$

Reversing the operations on (1) and (2), we obtain an analogous equation for **B**. From
analogies with other wave equations, (5.73) can be recognized as a description of
waves propagating with velocity $c = 1/\sqrt{\varepsilon_0 \mu_0}$. Since the values of ε_0 and μ_0 were
known, Maxwell found $c \simeq 300,000$ km/s. At that time it was possible to measure
the speed of light, with results approximating this value. Maxwell concluded that
light consists of propagating electromagnetic waves, one of the most extraordinary

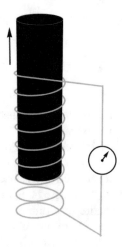

Fig. 5.11 A magnet moving inside a solenoid induces a current flow, detected by a galvanometer

achievements in the history of science. Maxwell's epoch-making discovery had thus succeeded in unifying the diverse phenomena of electricity, magnetism, and optics.

Let us consider first the electromagnetic wave equation for the simple case of a wave propagating in vacuum, in the positive z-direction, with the electric field **E** directed along the x-axis, and the magnetic field **B** directed along the y-axis, as shown in Fig. 5.12. The wave equation for the electric-field component E_x is given by

$$\frac{\partial^2 E_x}{\partial z^2} = \frac{1}{c^2} \frac{\partial^2 E_x}{\partial t^2}.$$
(5.74)

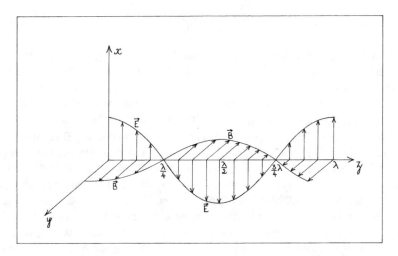

Fig. 5.12 A transverse electromagnetic wave with **E** parallel to the x-axis

where t represents the time. An identical wave equation holds for the magnetic-field component B_y:

$$\frac{\partial^2 B_y}{\partial z^2} = \frac{1}{c^2} \frac{\partial^2 B_y}{\partial t^2}. \tag{5.75}$$

Two comments are in order. Man has been able to discover a law so remote from everyday experience, in view of the enormously high speed of light and the microscopically small dimension of its wavelength. Yet, the nature of light is the same throughout all the known universe (which extends outward at least 14 billion light years). Furthermore, light is so important for us, that we would be blind if it did not exist. Even for those with a secular perspective, it is fascinating that in Genesis, among the first words spoken by God are "Let there be light." After Maxwell's discovery, further exploration revealed the entire spectrum of electromagnetic waves. Typical values of the wavelengths (in meters) are the following: radio 10^3 m, microwave 10^{-2} m, infrared 10^{-5} m, visible 5×10^{-7} m, ultraviolet 10^{-8} m, X-rays 10^{-10} m, gamma rays 10^{-12} m. It is, of course, superfluous to mention the vast number of technical applications of electromagnetic waves in modern life.

Returning to the wave equation (5.74), it is easy to verify that it can be satisfied by a simple periodic solution of the form

$$E_y = E_y^0 \cos\left[\frac{2\pi}{\lambda}(z - ct)\right], \tag{5.76}$$

where E_y^0 is a constant. Indeed we have

$$\frac{\partial^2 E_y}{\partial z^2} = -\left(\frac{2\pi}{\lambda}\right)^2 E_y, \qquad \frac{\partial^2 E_y}{\partial t^2} = -\left(\frac{2\pi c}{\lambda}\right)^2 E_y, \tag{5.77}$$

so that

$$\frac{\partial^2 E_y}{\partial z^2} - \frac{1}{c^2} \frac{\partial^2 E_y}{\partial t^2} = 0, \tag{5.78}$$

which is the one-dimensional wave equation. Since replacing z in (5.76) by $z + \lambda$ gives the same value of E_y, the solution is periodic in space with the wavelength λ representing the distance between two successive maxima (or minima or zeros) of the wave. Since the propagation velocity is c, the period of oscillation is $T = \frac{\lambda}{c}$. Therefore, the *frequency* $v = 2\pi\omega = \frac{1}{T}$. The two particular solutions of the wave equations (5.74), (5.75) represent (1) $\mathbf{E}$ parallel to the x-axis and $\mathbf{B}$ parallel to the y-axis (see Fig. 5.12) and (2) $\mathbf{E}$ parallel to the y-axis and $\mathbf{B}$ parallel to the x-axis (see Fig. 5.13). If $\mathbf{k}$ is the vector along the propagation direction, then in both cases, $\mathbf{E}$, $\mathbf{B}$, $\mathbf{k}$ form a right-handed system. In fact, $\mathbf{k} = \text{const } \mathbf{E} \times \mathbf{B}$. The plane containing the vectors $\mathbf{E}$ and $\mathbf{k}$ is called the *plane of polarization*. Since wave equations are *linear*, a more general solution can be obtained taking a linear combination of (1) and (2).

From now on, we can focus just on the electric field in these waves, since the accompanying perpendicular magnetic field is then determined by Maxwell's equations. We write the particular solutions (1), (2), respectively, in the following form:

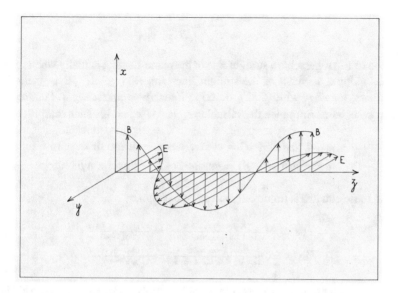

Fig. 5.13 A transverse electromagnetic wave with **E** parallel to the y-axis

$$E_x = A \cos\left[\frac{2\pi}{\lambda}(ct - z) + \alpha)\right], \qquad E_y = B \cos\left[\frac{2\pi}{\lambda}(ct - z) + \beta)\right], \qquad (5.79)$$

where A, B are constants determining the intensity of the field. Since there is no restriction for the two components to be in phase, α and β are not, in general, equal. Taking the sum of the particular solutions $(E_x, 0, 0)$ and $(0, E_y, 0)$ we obtain the vector $\mathbf{E}(t) = (E_x, E_y, 0)$ which represents the electric field of a light wave. It is convenient to define the radian frequency

$$\omega = \frac{2\pi}{\lambda} c. \qquad (5.80)$$

The variation of **E** for a fixed value of z, say $z = 0$, is then given by

$$E_x = A \cos(\omega t + \alpha), \qquad E_y = B \cos(\omega t + \beta). \qquad (5.81)$$

Clearly, **E** lies in a plane orthogonal to the direction of propagation.

5.6 Light Polarization and the Spin of the Photon

Before discussing the general case, let us see what happens if $\alpha = \beta$. Then from (5.81) we have, for $E_y \neq 0$:

$$\frac{E_x}{E_y} = \frac{A}{B}.$$ (5.82)

Equation (5.82) represents a straight line in the plane E_x, E_y; actually when t varies, **E** varies along a segment of the straight line with $|E_x| \leq A$, $|E_y| \leq B$. These inequalities, together with (5.82), describe *linearly polarized light*. For the more general case, let us introduce the variables $e_x = \frac{E_x}{A}$, $e_y = \frac{E_y}{B}$, such that

$$
\begin{aligned}
e_x &= \cos(\omega t + \alpha) = \cos \omega t \; \cos \alpha - \sin \omega t \; \sin \alpha, \\
e_y &= \cos(\omega t + \beta) = \cos \omega t \; \cos \beta - \sin \omega t \; \sin \beta.
\end{aligned}
$$ (5.83)

Solving these equations for $\cos \omega t$, $\sin \omega t$, we obtain

$$
\begin{aligned}
\cos \omega t &= \frac{e_x \sin \beta - e_y \sin \alpha}{\cos \alpha \sin \beta - \cos \beta \sin \alpha} = \frac{e_x \sin \beta - e_y \sin \alpha}{\sin(\beta - \alpha)} \\
\sin \omega t &= \frac{e_x \cos \beta - e_y \cos \alpha}{\cos \alpha \sin \beta - \cos \beta \sin \alpha} = \frac{e_x \cos \beta - e_y \cos \alpha}{\sin(\beta - \alpha)}.
\end{aligned}
$$ (5.84)

Squaring, then adding the two equations, and multiplying by $\sin^2(\beta - \alpha)$, we obtain

$$\sin^2(\beta - \alpha) = e_x^2 + e_y^2 - 2e_x e_y (\sin \beta \sin \alpha + \cos \beta \cos \alpha) = e_x^2 + e_y^2 - 2e_x e_y \cos(\beta - \alpha).$$ (5.85)

Therefore, in the plane $z = 0$, the point $\mathbf{E} = (E_x, E_y)$ traces out a conic:

$$\left(\frac{E_x}{A}\right)^2 + \left(\frac{E_y}{B}\right)^2 - 2\frac{E_x E_y}{A B} \cos(\beta - \alpha) = \sin^2(\beta - \alpha).$$ (5.86)

This is the equation of an ellipse; the light is then described as *elliptically polarized*. Figure 5.14 shows different ellipses and degenerate line segments obtained by varying $\beta - \alpha$. An important particular case occurs when $A = B$, and $\beta - \alpha = (2k + 1)\frac{\pi}{2}$, with integer k; then $\cos(\beta - \alpha) = 0$ and $\sin^2(\beta - \alpha) = 1$. Equation (5.85) becomes the equation of a circle of radius A:

$$E_x^2 + E_y^2 = A^2.$$ (5.87)

In this case, the light is *circularly polarized*: *left-circularly polarized* if the vector **E**, seen from the propagation direction **k** (the light ray coming toward you), rotates anticlockwise, and *right-circularly polarized* if it rotates clockwise.

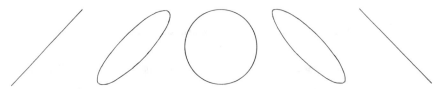

Fig. 5.14 Ellipses and degenerate line segments obtained by varying the angle $\beta - \alpha$

We now introduce the concept of the *spin of the photon*, implying, in essence, that a light quantum carries angular momentum. Let us see how the polarization plane changes under a rotation R_z around the **k** axis; R_z can be written using Eq. (4.39):

$$R_z(\theta) = \begin{Vmatrix} \cos\theta & -\sin\theta & 0 \\ \sin\theta & \cos\theta & 0 \\ 0 & 0 & 1 \end{Vmatrix}, \tag{5.88}$$

where θ is the rotation angle about the z-axis ($0 \le \theta < 2\pi$). As expected, $R_z(2\pi) = R_z(0) = I_3$. We know from Eqs. (4.62), (4.63) that for small $\delta\theta$, $R_z(\delta\theta) \simeq I_3 - i\delta\theta J_3$ where $J_3 = i\Lambda_3$ (using units with $\hbar = 1$). J_3 is the z-component of angular momentum and the generator for rotation about the z-axis and I_3 is the identity matrix in three dimensions. Note that the behavior of R_z is analogous to that of a wavefunction $\Psi(\theta)$ under infinitesimal rotations; we can write

$$\Psi(\theta + \delta\theta) \simeq \Psi(\theta) + \frac{d\Psi}{d\theta}\delta\theta = \Psi(\theta) - i\delta\theta\, J_3\Psi(\theta). \tag{5.89}$$

As in Chap. 4, we are applying the group of rotations about the z-axis. Only now, the vector space is different: J_3 acts in the physical 3D space $\mathbb{R}^3$, while earlier, J_3 acted in the Hilbert space of spinor wavefunctions Ψ. However, the operations of the rotation group SO(3) in the two cases are isomorphic and we can use the same matrix representations. A 3×3 matrix representation for angular momentum was given in Sect. 3.4.4. Replacing L by J, this corresponds to an angular momentum with $J = 1$, in a basis with J_3 diagonalized. The three eigenvalues $M = -1, 0, +1$ then appear in the diagonal of the matrix:

$$J_3 = \begin{Vmatrix} 1 & 0 & 0 \\ 0 & 0 & 0 \\ 0 & 0 & -1 \end{Vmatrix}. \tag{5.90}$$

From this, we can conclude that the photon has an intrinsic angular momentum $J = 1$, and it is a *spin-1 particle*. By contrast, we had found that the spin wave-functions for particles including electrons, protons, and neutrons transform under rotations according to the SU(2) group. We found the relations $J_i = \frac{1}{2}\sigma_i$, consistent with our classification of these as spin-$\frac{1}{2}$ particles.

5.7   The Hilbert Space of One Photon and Aspect's Experiment

5.7.1   *Photon Polarization*

The eigenvalue 0 of J_3 corresponds to the eigenvector $\mathbf{e}_3 = (0, 0, 1)$. Physically, it would represent a *longitudinal* component of the electric field, one directed along the propagation direction. Since in free space $\mathbf{E}$ is always *transverse*, thus orthogonal to $\mathbf{k}$, longitudinal electric fields occur only in constrained geometries, such as waveguides. Our concern is only with the transverse fields of electromagnetic radiation in free space. By specifying only the wavelength λ and the propagation direction $\mathbf{k}$ of the field, a "small" Hilbert space $\mathbb{C}^2$ is sufficient to describe the possible states of one photon. For a photon in free space, corresponding to a transverse field, $\mathbf{E} = (E_x, E_y)$ is given by

$$E_x = A\cos(kz - \omega t + \alpha) = A \ Re\left[e^{i(kz - \omega t + \alpha)}\right],$$
$$E_y = B\cos(kz - \omega t + \beta) = B \ Re\left[e^{i(kz - \omega t + \beta)}\right]. \tag{5.91}$$

We can define, using $E = \sqrt{E_x^2 + E_y^2}$:

$$a = \frac{A}{E}e^{i\alpha}, \qquad b = \frac{B}{E}e^{i\beta}, \tag{5.92}$$

so that

$$|a|^2 + |b|^2 = \frac{A^2|e^{i\alpha}|^2 + B^2|e^{i\beta}|^2}{A^2 + B^2} = 1, \tag{5.93}$$

with $\left\|\begin{matrix} a \\ b \end{matrix}\right\|$ representing a unit vector in the Hilbert space $\mathbb{C}^2$. *This represents the quantum state of the photon.* Note that this a two-dimensional *vector*, and *not* a spinor. The vector $\left\|\begin{matrix} 1 \\ 0 \end{matrix}\right\|$ corresponds to vertical polarization ($\mathbf{E}$ directed along the x-axis), and $\left\|\begin{matrix} 0 \\ 1 \end{matrix}\right\|$ corresponds to horizontal polarization ($\mathbf{E}$ directed along the y-axis). In general, the electric field can be written as

$$\mathbf{E}(z, t) = E \ Re\left[\left\|\begin{matrix} a \\ b \end{matrix}\right\| e^{i(kz - \omega t)}\right]. \tag{5.94}$$

Given the state $\left\|\begin{matrix} a \\ b \end{matrix}\right\|$ of the photon, Eq. (5.94) gives the corresponding electric field. If a and b are *real*, we have $E_x = Ea\cos(kz - \omega t)$, $E_y = Eb\cos(kz - \omega t)$, with

$$\frac{E_x}{E_y} = \frac{a}{b}, \tag{5.95}$$

showing that the photon is linearly polarized. More generally, for complex $a = a_1 + ia_2$, $b = b_1 + ib_2$, we have, from (5.94):

$$E_x = E \, Re\left[(a_1 + ia_2)(\cos(kz - \omega t) + i \sin(kz - \omega t)\right] =$$
$$E\left[a_1 \cos(kz - \omega t) - a_2 \sin(kz - \omega t)\right],$$
$$E_y = E \, Re\left[(b_1 + ib_2)(\cos(kz - \omega t) + i \, \sin(kz - \omega t)\right] =$$
$$E\left[b_1 \cos(kz - \omega t) - b_2 \sin(kz - \omega t)\right]. \quad (5.96)$$

For the particular case, with $a = a_1 = \frac{1}{\sqrt{2}}$, $b = ib_2 = \frac{i}{\sqrt{2}}$, we have, from Eq. (5.96), that

$$E_x = \frac{1}{\sqrt{2}} E \cos(kz - \omega t) = \frac{1}{\sqrt{2}} E \cos(\omega t - kz),$$
$$E_y = -\frac{1}{\sqrt{2}} E \sin(kz - \omega t) = \frac{1}{\sqrt{2}} E \sin(\omega t - kz). \quad (5.97)$$

Here $E_x^2 + E_y^2 = E^2$, and the photon is *circularly* polarized; more precisely, left-circularly polarized (see Fig. 5.14). Alternatively, with $a = a_1 = \frac{1}{\sqrt{2}}$, $b = ib_2 = -\frac{i}{\sqrt{2}}$, the photon turns out to be right-circularly polarized. In the general case with a, b, complex and $|a|^2 + |b|^2 = 1$, the photon is elliptically polarized.

The two-dimensional vectors $\left\| \begin{matrix} a \\ b \end{matrix} \right\|$ representing photon polarization (known as *Jones vectors*) include the following special cases of vertical and horizontal polarization:

$$|V\rangle = \left\| \begin{matrix} 1 \\ 0 \end{matrix} \right\|, \qquad |H\rangle = \left\| \begin{matrix} 0 \\ 1 \end{matrix} \right\|. \quad (5.98)$$

More relevant to the representation of photon angular momentum are the left- and right-circularly polarized states:

$$|L\rangle = \frac{1}{\sqrt{2}} \left\| \begin{matrix} 1 \\ i \end{matrix} \right\|, \qquad |R\rangle = \frac{1}{\sqrt{2}} \left\| \begin{matrix} 1 \\ -i \end{matrix} \right\|. \quad (5.99)$$

The operator for the z-component of angular momentum can be taken from the first $2 \otimes 2$ block of Λ_3 in Eq. (4.63). Note that this is the direction of the propagation vector $\mathbf{k}$. We can write

$$J_3 = i \left\| \begin{matrix} 0 & -1 \\ 1 & 0 \end{matrix} \right\| = \left\| \begin{matrix} 0 & -i \\ i & 0 \end{matrix} \right\|. \quad (5.100)$$

It is then simple to verify that

$$\left\| \begin{matrix} 0 & -i \\ i & 0 \end{matrix} \right\| \left\| \begin{matrix} 1 \\ \pm i \end{matrix} \right\| = \pm \left\| \begin{matrix} 1 \\ \pm i \end{matrix} \right\|, \quad (5.101)$$

showing that $|L\rangle$ and $|R\rangle$ are eigenvectors of J_3 with eigenvalues ± 1, respectively. The eigenvector for $J_3 = 0$ is missing, since this would correspond to the nonexistent longitudinal polarization.

The two-photon state is described by the wavefunction

$$|\Psi_0\rangle = \frac{1}{\sqrt{2}} \Big(|R_A\rangle \otimes |L_B\rangle + |R_B\rangle \otimes |L_A\rangle \Big). \tag{5.102}$$

Since $J_3|L\rangle = |L\rangle$ and $J_3|R\rangle = -|R\rangle$, the single photon states $|L\rangle, |R\rangle$ correspond to angular momentum eigenvectors with $J_3 = +1$, $J_3 = -1$, respectively, representing left- and right-circularly polarized photons.

5.7.2 Aspect's Experiment

We now describe a famous experiment which actually attempts to answer the "metaphysical" question of quantum realism. In 1981–1982 Alain Aspect and collaborators performed definitive tests of Bell's inequality (Aspect et al. 1981). They used a cascade emission of two photons A, B, from electronic transitions from an excited level of calcium atoms, as shown in Fig. 5.15. The lifetime of the intermediate state is only 5 ns; this interval of time is so small that we can assume that the two photons are emitted simultaneously (light travels 30 cm in one nanosecond).

A diagram of the experiment is shown in Fig. 5.16. The two photons A, B are emitted by the source S in the state $|\Psi_0\rangle$. After traveling a distance L they are incident on a pair of two-channel polarizers P_A, P_B, each of which can convert a circularly polarized photon to either a vertical or a horizontal polarization. The directions of two polarizers are randomly changed approximately every 20 ns. Photons exiting the polarizers can be detected by four photomultipliers $D^\pm$, located as shown in Fig. 5.16. (A photomultiplier is an extremely sensitive phototube that multiplies the very weak current produced by light about 100 million times.)

When the photons strike one of the polarizers P (with polarization directions $\mathbf{a}^\pm$, $\mathbf{b}^\pm$), they can either pass straight through or be deflected in the orthogonal direction; in the latter case they are detected by the photomultipliers D^- (with polarization directions $\mathbf{a}^-$, $\mathbf{b}^-$). The two paths (transmission or reflection) occur with equal probabilities $\frac{1}{2}$. Finally, an electronic counter module CM monitors the coincidences $(\mathbf{a}^+, \mathbf{b}^+)$, $(\mathbf{a}^+, \mathbf{b}^-)$, $(\mathbf{a}^-, \mathbf{b}^+)$, and $(\mathbf{a}^-, \mathbf{b}^-)$. Thus the correlation coefficient $E(\mathbf{a}, \mathbf{b})$ (see 5.65, 5.68) for the hidden variables and the corresponding quantum mechanical

Fig. 5.15 Cascade emission of two photons (*green* and *blue*) by calcium atom after laser excitation. The photons, entangled with opposite circular polarizations, are ejected in opposite directions

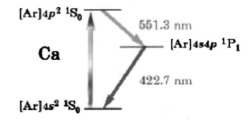

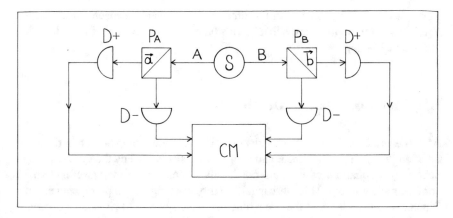

Fig. 5.16 Experimental apparatus for testing Bell's inequality

expressions can be measured, and Bell's inequality (5.67) thereby tested. As we anticipated earlier, the quantum mechanical prediction is in agreement with the measurement, with the CHSH parameter $S = E(\mathbf{a}, \mathbf{b}) + E(\mathbf{a}, \mathbf{b}') + E(\mathbf{a}', \mathbf{b}) - E(\mathbf{a}', \mathbf{b}')$ for many choices of $\mathbf{a}$, $\mathbf{a}'$ for the polarizer P_A, and $\mathbf{b}$, $\mathbf{b}'$ for the polarizer P_B. Note that the distance L between the source S and the polarizers was 6 m or longer.[4] Due to the rapid switching of P_A, P_B, no signal with speed less than or equal to that of light can "inform" the polarizer P_B (or the detector near P_B) of the direction of the vector $\mathbf{a}$ chosen by the polarizer P_A. The same goes for the path (D^+ or D^-) chosen by the photon.

The experimental result gave a value of $S_{\exp} = 2.70 \pm 0.05$, in remarkable agreement with the prediction of quantum mechanics. We conclude this brief description with a quote by Alain Aspect[5]:

> The experimental violation of Bell's inequality confirms that a pair of entangled photons separated by hundreds of meters must be considered a single non-separable object, it is impossible to assign local physical reality to the state of each photon.

The conclusion that can be drawn from these experiments can be formalized as *Bell's theorem*: *No local realistic physical theory can reproduce the predictions of quantum mechanics*. Local realism is the worldview in which physical properties of objects exist independently of their measurement and where interactions among objects cannot travel faster than the speed of light. Bell's theorem states that this

---

[4]The distance $2L$ was increased up to 400 m by Weihs et al. '(1998) Phys Rev Lett 81:5039–5043. Entangled photons were transmitted over 144 km by Zeilinger et al. (2007) Proc of SPIE 6780:67800B.

[5]Aspect (1999) Bell's inequality test: more ideal than ever. Nature 398:189–190.

worldview is incompatible with the predictions of quantum mechanics.[6] Particle physicist Henry Stapp believes that "Bell's theorem is the most profound discovery of science." (Stapp 1975).

5.8 Measurement and Decoherence

As we described in Chap. 3, there exist two different modes of evolution of the wavefunction $|\Psi(t)\rangle$. The first, linear and continuous, is governed by the time-dependent Schrödinger equation, while the second, called "collapse of the wavefunction," is nonlinear and random. This inescapable duality has long been a perplexing enigma. When does the collapse of the wave function actually occur? Is it when a conscious observer "takes note" of the new state? Or, perhaps more reasonably, when an inanimate physical mechanism triggers the "quantum jump"? To try to answer these and similar questions, consider a microscopic system S, a measurement apparatus (or detector) M and the environment E, consisting of the enormous number of atoms and photons that surround the measuring apparatus. According to the definition of Zurek (2002): "Environments can be external (such as particles of air or photons that scatter off, say, the apparatus pointer) or internal (such as collections of phonons or other excitations in the materials from which an apparatus is constructed)." It follows from this definition that the boundary between M and E is somewhat nebulous and largely dependent on our arbitrary choice. The part of the environment that does not contribute significantly to the evolution of the state of the apparatus can be ignored. M is a macroscopic object, but often the interaction with S involves one of its microscopic components (Doplicher 2012). The enormous number of degrees of freedom necessary to describe both M and E causes the interaction with S to give rise to an irreversible process, somewhat similar to the situation in statistical thermodynamics.

In principle, we can regard M and E as intrinsic parts of a composite quantum mechanical system, even if it is not possible to exactly solve the time-dependent Schrödinger equation for a realistic system with so many degrees of freedom. For simplicity, assume that S can be in only one of two states, for example, the states $J_3 = \pm 1$ of a photon. To these states there correspond the vectors $|m \uparrow\rangle$, $|m \downarrow\rangle$ of the detector M; we will call these *pointer states* since they allow us to read the result of a measurement on S. Attempting to treat the apparatus M as a quantum mechanical system might be tricky, since M is usually pictured as a device governed by *classical* mechanics. It is easy to prove that $\langle m \uparrow | m \downarrow \rangle = 0$. If the normalized vectors $|m \uparrow\rangle$, $|m \downarrow\rangle$ correspond to different positions of a pointer, then displacement of each of the

---

[6]Scrupulous analysis of these tests of Bell's theorem identified two possible "loopholes" to their universal validity. The *locality loophole* arises from the possibility that, if the measurements are too slow, the photon detectors might be communicating with one another in some way (by a yet unknown mechanism). The *detection loophole* is the possibility that the observed events might represent a skewed sample of all the emitted photon pairs, since detectors are less than 100% efficient. However, in late 2015, three research groups (at Delft University of Technology, University of Vienna and NIST) have succeeded in carrying out "loophole-free" Bell tests.

molecules in the pointer by just a few millimeters certainly produces orthogonality, since the product of the two wavefunctions vanishes everywhere. Assume now that the system S is in a superposition state $|\phi_s\rangle$:

$$|\phi_s\rangle = a|\uparrow\rangle + b|\downarrow\rangle \tag{5.103}$$

and the state of the detector is initially $|m \downarrow\rangle$. After the measurement, the state of the composite system S, M, becomes

$$|\phi_c\rangle = a|\uparrow\rangle \otimes |m \uparrow\rangle + b|\downarrow\rangle \otimes |m \downarrow\rangle. \tag{5.104}$$

The density matrix corresponding to this state is

$$\rho_c = |\phi_c\rangle\langle\phi_c| = (a|\uparrow\rangle \otimes |m \uparrow\rangle + b|\downarrow\rangle \otimes |m \downarrow\rangle)(a^*\langle\uparrow| \otimes \langle m \uparrow| + b^*\langle\downarrow| \otimes \langle m \downarrow|) =$$
$$|a|^2 |\uparrow\rangle\langle\uparrow| \otimes |m \uparrow\rangle\langle m \uparrow| + |b|^2 |\downarrow\rangle\langle\downarrow| \otimes |m \downarrow\rangle\langle m \downarrow| +$$
$$ab^*|\uparrow\rangle\langle\downarrow| \otimes |m \uparrow\rangle\langle m \downarrow| + ba^*|\downarrow\rangle\langle\uparrow| \otimes |m \downarrow\rangle\langle m \uparrow|. \tag{5.105}$$

Let us denote by $|up\rangle$ the state $|\uparrow\rangle \otimes |m \uparrow\rangle$ and by $|down\rangle$ the state $|\downarrow\rangle \otimes |m \downarrow\rangle$. Since $|\Psi_c\rangle$ is a superposition of $|up\rangle$ and $|down\rangle$, the following block of the matrix representing $\rho_c = |\Psi_c\rangle\langle\Psi_c|$ contains the non-diagonal elements ab^* and ba^*:

$$\left\| \begin{matrix} \langle up|\rho_c|up\rangle & \langle up|\rho_c|down\rangle \\ \langle down|\rho_c|up\rangle & \langle down |\rho_c|down\rangle \end{matrix} \right\| = \left\| \begin{matrix} |a|^2 & ab^* \\ ba^* & |b|^2 \end{matrix} \right\|. \tag{5.106}$$

Suppose now that we want to discard the quantum description of the apparatus M, leaving unaltered the probabilities of the $|\uparrow\rangle$ and $|\downarrow\rangle$ states. A simple way to do it is just to take the expectation value of ρ_c over the two states $|m \uparrow\rangle$, $|m \downarrow\rangle$, and sum the results. This is equivalent to taking a partial trace of ρ_c with respect to a basis of M. This corresponds to performing an average over the effects of M. The result, taking into account the orthonormality of $|m \uparrow\rangle$ and $|m \downarrow\rangle$, is a new density matrix ρ_r acting on the Hilbert space of the system S:

$$\rho_r = \langle m \uparrow |\rho_c|m \uparrow\rangle + \langle m \downarrow |\rho_c|m \downarrow\rangle = |a|^2 |\uparrow\rangle\langle\uparrow| + |b|^2 |\downarrow\rangle\langle\downarrow|. \tag{5.107}$$

The matrix ρ_r is represented by the following block:

$$\left\| \begin{matrix} \langle\uparrow |\rho_r| \uparrow\rangle & \langle\uparrow |\rho_r| \downarrow\rangle \\ \langle\downarrow |\rho_r| \uparrow\rangle & \langle\downarrow |\rho_r| \downarrow\rangle \end{matrix} \right\| = \left\| \begin{matrix} |a|^2 & 0 \\ 0 & |b|^2 \end{matrix} \right\|, \tag{5.108}$$

where the non-diagonal matrix elements of (5.107) have disappeared; ρ_r is called a *reduced density matrix* (Von Neumann 1932). It has the advantage of allowing a simple interpretation of the probabilities, $|a|^2$ for the $|\uparrow\rangle$ state, and $|b|^2$ for the $|\downarrow\rangle$ state, in agreement with Born's interpretation of probability in quantum mechanics. Furthermore, no superposition of $|m \uparrow\rangle$ and $|m \downarrow\rangle$ appears. This seems reasonable, since nobody has ever seen a half-alive and half-dead cat (we will come back to this later in connection with quantum superposition of macroscopic objects).

We recognize that a spontaneous evolution of ρ_c to ρ_r (or, more precisely, of ρ_c to $\rho_r \otimes I$)[7] is impossible. To prove this, we note that $\rho_c^2 = \rho_c$, since

$$\rho_c^2 = |\phi_c\rangle \langle \phi_c | | \phi_c \rangle \langle \phi_c | = |\phi_c\rangle \langle \phi_c | = \rho_c. \tag{5.109}$$

Denote by $U(t)$, the unitary operator for time evolution according to the time-dependent Schrödinger equation. Since $|\phi_c(t)\rangle = U(t)|\phi_c\rangle$, it is evident that $\rho_c(t) = U(t)\rho_c U(t)^{-1}$ (note that the time evolution of the density matrix differs from the time evolution of observables). Then, $\rho_c(t)^2 = \rho_c(t)$, since

$$\rho_c(t)^2 = U(t)\rho_c U(t)^{-1} U(t)\rho_c U(t)^{-1} = U(t)\rho_c^2 U(t)^{-1} = U(t)\rho_c U(t)^{-1} = \rho_c(t). \tag{5.110}$$

But Eq. (5.110) does not hold for ρ_r, since

$$\rho_r^2 = |a|^4| \uparrow \rangle \langle \uparrow | + |b|^4| \downarrow \rangle \langle \downarrow | \neq \rho_r, \tag{5.111}$$

which completes the proof.

Physicists have tentatively discovered a possible effect that gives rise to the transition $\rho_c \to \rho_r$. This effect is called *decoherence*; much work remains in order to formulate a theory of decoherence on a rigorous mathematical basis. The general idea is the following: in the description of the complete system, including S, M, and the environment E, assume that the state $|E \uparrow\rangle$ of the environment is associated to the pair $| \uparrow\rangle$, $|m \uparrow\rangle$, and the state $|E \downarrow\rangle$ of the environment is associated to the pair $| \downarrow\rangle$, $|m \downarrow\rangle$. Furthermore $\langle E \uparrow |E \downarrow\rangle = 0$ (we will come back in a moment to show that this is reasonable). Of course, we can always normalize the two states: $\langle E \uparrow |E \uparrow\rangle = \langle E \downarrow |E \downarrow\rangle = 1$. The wave function of the global system is now

$$|\Psi\rangle = a| \uparrow\rangle \otimes |m \uparrow\rangle \otimes |E \uparrow\rangle + b| \downarrow\rangle \otimes |m \downarrow\rangle \otimes |E \downarrow\rangle = a|\text{UP}\rangle + b|\text{DOWN}\rangle, \tag{5.112}$$

where $|\text{UP}\rangle = | \uparrow\rangle \otimes |m \uparrow\rangle \otimes |E \uparrow\rangle$ and $|\text{DOWN}\rangle = | \downarrow\rangle \otimes |m \downarrow\rangle \otimes |E \downarrow\rangle$. Recall that $|\text{up}\rangle = | \uparrow\rangle \otimes |m \uparrow\rangle$, $|\text{down}\rangle = | \downarrow\rangle \otimes |m \downarrow\rangle$; in $|\text{up}\rangle$ and $|\text{down}\rangle$ (lower case), the state of E does not appear. Since the states $|E \uparrow\rangle$, $|E \downarrow\rangle$ are orthogonal, we can repeat the same reasoning as before, tracing over[8] the degrees of freedom of the environment. The result is the following reduced density matrix:

$$\rho_R = |a|^2|\text{up}\rangle \langle \text{up}| + |b|^2|\text{down}\rangle \langle \text{down}|, \tag{5.113}$$

in which no superposition term appears.

---

[7] I denotes the identity in the Hilbert space of the measuring apparatus.
[8] Taking the expectation values on $|E \uparrow\rangle$ and $|E \downarrow\rangle$ and summing the results.

We are, however, faced with two crucial issues:

(1) Why are the two states of the environment $|E \uparrow\rangle$ and $|E \downarrow\rangle$ orthogonal to each other?

(2) What is the order of magnitude of the time needed for the superposition terms to disappear, to give a reduced density matrix (5.113)?

We can answer question (1) with an simplified argument in which we neglect the effect of Fermi–Dirac statistics (the antisymmetry of the electronic wavefunctions), which does not change the essentials of the argument. Consider a macroscopic system of N particles, for example, N molecules of a gas. Let us denote by $|E\rangle$, $|F\rangle$ two quantum states of the gas that are assumed to be *product states*:

$$|E\rangle = |e_1\rangle \otimes |e_2\rangle \dots \otimes |e_N\rangle, \qquad |F\rangle = |f_1\rangle \otimes |f_2\rangle \dots \otimes |f_N\rangle, \qquad (5.114)$$

where $|e_i\rangle$, $|f_i\rangle$ are normalized states of the ith molecule. The scalar product $\langle F|E\rangle = \langle f_1|e_1\rangle\langle f_2|e_2\rangle \dots \langle f_N|e_N\rangle$ contains an enormous number of factors whose absolute value is less or equal to 1, since by Schwarz's inequality:

$$|\langle f_i|e_i\rangle|^2 \leq \langle f_i|f_i\rangle\langle e_i|e_i\rangle = 1. \qquad (5.115)$$

Thus, choosing the factors $|\langle f_i|e_i\rangle|$ in the interval $[0, 1]$ at random, one obtains a vanishingly small result. Furthermore, the states $|e_i\rangle$, $|f_i\rangle$ correspond to displaced positions of the ith molecule, giving a physical reason for the scalar product $\langle f_i|e_i\rangle$ to be small. The weak point of this argument is that its validity is restricted to *product states*. A more general state can consist of a huge linear combination of product states. A deeper analysis of the interaction between the system S and the macroscopic systems M and E is needed, which we will not attempt here. In conclusion then, there is a general belief that we can safely use the reduced density matrix ρ_R, in which only diagonal matrix elements appear. It is also quite difficult to answer question (2) in a rigorous way. However, many particular solvable models have been examined, with the result that the *decoherence time* between macroscopically separated positions is shown to be extremely small.

Consider a very simple solvable model that describes the interaction of a two-state apparatus M (or a microscopic part of M) with a very large number N of spins in the environment E. We denote by $|m \uparrow\rangle$, $|m \downarrow\rangle$ the two states of M, and by $|k \uparrow\rangle$, $|k \downarrow\rangle$ the two states of the kth environmental spin. Let L_M be the space spanned by $|m \uparrow\rangle$, $|m \downarrow\rangle$, and L_k, the space spanned by $|k \uparrow\rangle$, $|k \downarrow\rangle$. We define the third spin component σ_M and the third spin component σ_k by the same matrix:

$$\sigma_M = \sigma_k = \left\| \begin{matrix} 1 & 0 \\ 0 & -1 \end{matrix} \right\|. \qquad (5.116)$$

This matrix represents an operator acting on L_M, while σ_k represents an operator acting on L_k. The matrix is the same, but they act in different spaces. The coupling Hamiltonian H_{ME} is given by

$$H_{ME} = \sum_k g_k \, \sigma_M \otimes \sigma_k, \tag{5.117}$$

where the g_k are coupling constants. An alternative way to write the same Hamiltonian defines the projection operators (Zurek 2003)

$$P_{M\uparrow} = |m \uparrow\rangle\langle m \uparrow|, \ \ P_{M\downarrow} = |m \downarrow\rangle\langle m \downarrow|, \ \ P_{k\uparrow} = |k \uparrow\rangle\langle k \uparrow|, \ \ P_{k\downarrow} = |k \downarrow\rangle\langle k \downarrow|, \tag{5.118}$$

so that the interaction Hamiltonian becomes

$$H_{ME} = (P_{M\uparrow} - P_{M\downarrow}) \otimes \sum_k g_k (P_{k\uparrow} - P_{k\downarrow}). \tag{5.119}$$

The initial state of M and E is

$$|\Psi(0)\rangle = (a|m \uparrow\rangle + b|m \downarrow\rangle) \otimes \prod_k \otimes (\alpha_k |k \uparrow\rangle + \beta_k |k \downarrow\rangle). \tag{5.120}$$

Since $|\Psi(0)\rangle$ is normalized, $|a|^2 + |b|^2 = |\alpha_k|^2 + |\beta_k|^2 = 1 \ \forall \ k$. To obtain $|\Psi(t)\rangle$, apply the evolution operator $e^{-iH_{ME}t/\hbar}$ to $|\Psi(0)\rangle$. In so doing, we neglect the evolution due to the free (non-interacting) Hamiltonians for M and E.

The operators σ_k commute, since they act on different factors of a tensor product; it follows that the operators $\sigma_M \otimes \sigma_k$ also commute. If A and B commute, the following simple exponential operator formula is valid:

$$e^{A+B} = e^A e^B. \tag{5.121}$$

The identity (5.121) can be demonstrated by series expansions of the exponentials. The algebraic steps are the same as if A, B were real or complex numbers. Using (5.121), we can write the evolution operator $e^{-iH_{ME}t/\hbar}$ in the form:

$$e^{-iH_{ME}t/\hbar} = e^{-i(\sum_k \sigma_M \otimes \sigma_k)t/\hbar} = \prod_k e^{-ig_k \sigma_M \otimes \sigma_k t/\hbar}. \tag{5.122}$$

Consider now the four-dimensional space L_k spanned by the vectors

$$|m \uparrow\rangle \otimes |k \uparrow\rangle, \ \ |m \uparrow\rangle \otimes |k \downarrow\rangle, \ \ |m \downarrow\rangle \otimes |k \uparrow\rangle, \ \ |m \downarrow\rangle \otimes |k \downarrow\rangle. \tag{5.123}$$

The matrix representing $\sigma_M \otimes \sigma_k$, an operator acting on L_k, is diagonal, with matrix elements $+1, -1, -1, +1$. Therefore, the operator $e^{-i\sigma_M \otimes \sigma_k g_k t/\hbar}$ is represented by the matrix:

$$\left\|\begin{matrix} e^{-ig_k t/\hbar} & 0 & 0 & 0 \\ 0 & e^{ig_k t/\hbar} & 0 & 0 \\ 0 & 0 & e^{ig_k t/\hbar} & 0 \\ 0 & 0 & 0 & e^{-ig_k t/\hbar} \end{matrix}\right\|. \tag{5.124}$$

The initial state $|\Psi(0)\rangle$ of M and E (see 5.120) can be written as the tensor product:

$$|\Psi(0)\rangle = \prod_k \otimes \Big(a\alpha_k |m \uparrow\rangle \otimes |k \uparrow\rangle + a\beta_k |m \uparrow\rangle \otimes |k \downarrow\rangle +$$
$$b\alpha_k |m \downarrow\rangle \otimes |k \uparrow\rangle + b\beta_k |m \downarrow\rangle \otimes |k \downarrow\rangle \Big) = \prod_k \otimes |v_k\rangle, \tag{5.125}$$

where the components of $|v_k\rangle \in L_k$ are $(a\alpha_k,\ a\beta_k,\ b\alpha_k,\ b\beta_k)$. Since time evolution does not mix the spaces L_k, we obtain $|\Psi(t)\rangle$ by simply applying the matrix (5.124) to $|v_k\rangle$ and writing the resulting tensor product. This gives

$$|\Psi(t)\rangle = \otimes_k \Big(e^{-ig_k t/\hbar} a\alpha_k |m \uparrow\rangle \otimes |k \uparrow\rangle + e^{ig_k t/\hbar} a\beta_k |m \uparrow\rangle \otimes |k \downarrow\rangle +$$
$$e^{ig_k t/\hbar} b\alpha_k |m \downarrow\rangle \otimes |k \uparrow\rangle + e^{-ig_k t/\hbar} b\beta_k |m \downarrow\rangle \otimes |k \downarrow\rangle \Big) = \tag{5.126}$$
$$a |m \uparrow\rangle \otimes |E \uparrow (t)\rangle + b |m \downarrow\rangle \otimes |E \downarrow (t)\rangle,$$

where

$$|E \uparrow (t)\rangle = \otimes_k^N \Big(\alpha_k e^{-ig_k t/\hbar} |k \uparrow\rangle + \beta_k e^{\frac{itg_k}{\hbar}} |k \downarrow\rangle \Big),$$
$$|E \downarrow (t)\rangle = \otimes_k^N \Big(\alpha_k e^{ig_k t/\hbar} |k \uparrow\rangle + \beta_k e^{-ig_k t/\hbar} |k \downarrow\rangle \Big). \tag{5.127}$$

Clearly then, $|E \downarrow (t)\rangle = |E \uparrow (-t)\rangle$. Denoting by $E\uparrow\uparrow = |E \uparrow (t)\rangle\langle E \uparrow (t)|$, $E\uparrow\downarrow = |E \uparrow (t)\rangle\langle E \downarrow (t)|$, $E\downarrow\uparrow = |E \downarrow (t)\rangle\langle E \uparrow (t)|$, and $E\downarrow\downarrow = |E \downarrow (t)\rangle\langle E \downarrow (t)|$, respectively, the four operators appearing in the expression for the density matrix $\rho = |\Psi(t)\rangle\langle\Psi(t)|$, we can write

$$\rho = \Big(a |m \uparrow\rangle \otimes |E \uparrow (t)\rangle + b |m \downarrow\rangle \otimes |E \downarrow (t)\rangle \Big) \times$$
$$\Big(a^* \langle m \uparrow | \otimes \langle E \uparrow (t)| + b^* \langle m \downarrow | \otimes \langle E \downarrow (t)| \Big) =$$
$$|a|^2 |m \uparrow\rangle\langle m \uparrow | E \uparrow\uparrow + ab^* |m \uparrow\rangle\langle m \downarrow | E \uparrow\downarrow +$$
$$ba^* |m \downarrow\rangle\langle m \uparrow | E \downarrow\uparrow + |b|^2 |m \downarrow\rangle\langle m \downarrow | E \downarrow\downarrow. \tag{5.128}$$

In order to compute the reduced density matrix, taking the trace over all E, we will make use of the following:

Lemma 5.2 *Suppose that l_k denotes the two-dimensional space spanned by $|k \uparrow\rangle$, $|k \downarrow\rangle$, and L_E denotes the product space $\prod_k \otimes l_k$. Given two vectors in L_E, $|v\rangle = |v_1\rangle \otimes |v_2\rangle \ldots \otimes |v_N\rangle$, $|w\rangle = |w_1\rangle \otimes |w_2\rangle \ldots \otimes |w_N\rangle$, where $|v_k\rangle$, $|w_k\rangle$ belong to l_k, we consider the operator $A = |v\rangle\langle w|$. It follows then that*

$$\text{Tr}_E\, A = \prod_k \Big(\langle k \uparrow | v \rangle \langle w | k \uparrow \rangle + \langle k \downarrow | v \rangle \langle w | k \downarrow \rangle \Big) = \prod_k \langle w_k | v_k \rangle. \qquad (5.129)$$

We will just prove the Lemma for $N = 3$, the generalization to arbitrary N being straightforward. It is convenient to denote by $|k, 1\rangle$, $|k, 2\rangle$ the vectors $|k \uparrow\rangle$, $|k \downarrow\rangle$, for $k = 1, 2, 3$. A generic vector $|v_k\rangle \in l_k$ can be written as $|v_k\rangle = v_{k,1}|k, 1\rangle + v_{k,2}|k, 2\rangle$. The same notation holds for $|w_k\rangle = w_{k,1}|k, 1\rangle + w_{k,2}|k, 2\rangle$. Denote by A the operator $|v\rangle\langle w|$; since for any $|x\rangle$, $A|x\rangle = |v\rangle\langle w|x\rangle$, the definition of Hermitian scalar product implies that numerical coefficients in the "functional" $\langle w|$ can be extracted, remembering to take the complex conjugate as appropriate. Thus the operator $|v\rangle\langle w|$ can be written as

$$A = |v\rangle\langle w| = \sum_{i,j,p=1}^{2} \sum_{l,m,n=1}^{2} v_{1,i}\, v_{2,j}\, v_{3,p}\, (w_{1,l}\, w_{2,m}\, w_{3,n})^* \,|1, i\rangle|2, j\rangle|3, p\rangle \,\langle 1, l|\langle 2, m|\langle 3, n|.$$
$$(5.130)$$

Now substituting Eq. (5.130) into the following expression for the trace:

$$\text{Tr}_E\, A = \sum_{r=1}^{2}\sum_{s=1}^{2}\sum_{t=1}^{2} \langle 1, r| \otimes \langle 2, s| \otimes \langle 3, t|A|1, r\rangle \otimes |2, s\rangle \otimes |3, t\rangle, \qquad (5.131)$$

results in the following scalar products:

$$\langle 1, r|1, i\rangle, \ \langle 2, s|2, j\rangle, \ \langle 3, t|3, p\rangle, \ \langle 1, l|1, r\rangle, \ \langle 2, m|2, s\rangle, \ \langle 3, n|3, t\rangle. \quad (5.132)$$

Since the vectors $|k1\rangle$, $|k2\rangle$ are orthogonal for any value of k, we have the equalities $i = r = l$, $j = s = m$, $p = t = n$. Thus (5.131) becomes

$$\text{Tr}_E\, A = \sum_{i=1}^{2} v_{1i} w_{1i}^* \sum_{j=1}^{2} v_{2j} w_{2j}^* \sum_{p=1}^{2} v_{3p} w_{3p}^* = \langle w_1|v_1\rangle \langle w_2|v_2\rangle \langle w_3|v_3\rangle, \quad (5.133)$$

and the lemma is proven.

In order to compute $\text{Tr}_E\, E{\uparrow}{\downarrow}$, let us specialize $|w\rangle = |v\rangle = |E \uparrow (t)\rangle$, and apply the lemma. Since $\langle w_k|v_k\rangle = \langle v_k|v_k\rangle = |v_k|^2 = |\alpha_k|^2 + |\beta_k|^2 = 1$, we obtain $\text{Tr}_E\, E{\uparrow}{\uparrow} = \prod_{k=1}^{N} 1 = 1$. In the same way it can be shown that $\text{Tr}_E\, E{\downarrow}{\downarrow} = 1$. The analogous computation for the operator $E{\uparrow}{\downarrow} = |E \uparrow (t)\rangle\langle E \downarrow (t)|$ is more interesting: the trace Tr_k over the two states $|k \uparrow\rangle$, $|k \downarrow\rangle$ gives

$$\text{Tr}_k\left[\left(\alpha_k e^{-\frac{itg_k}{\hbar}}|k \uparrow\rangle + \beta_k e^{\frac{itg_k}{\hbar}}|k \downarrow\rangle \right) \left(\alpha_k^* e^{-\frac{itg_k}{\hbar}}\langle k \uparrow | + \beta_k^* e^{\frac{itg_k}{\hbar}}\langle k \downarrow | \right) \right] =$$
$$e^{-ig_k t/\hbar}\alpha_k\, e^{-ig_k t/\hbar}\alpha_k^* + e^{ig_k t/\hbar}\beta_k\, e^{ig_k t/\hbar}\beta_k^* =$$
$$|\alpha_k|^2 \left(\cos\frac{2g_k t}{\hbar} - i\sin\frac{2g_k t}{\hbar} \right) + |\beta_k|^2 \left(\cos\frac{2g_k t}{\hbar} + i\sin\frac{2g_k t}{\hbar} \right) =$$
$$\cos\frac{2g_k t}{\hbar} + i\left(|\beta_k|^2 - |\alpha_k|^2 \right)\sin\frac{2g_k t}{\hbar}. \qquad (5.134)$$

Denoting by $z(t)$ the trace of $E_{\uparrow\downarrow}$ over the whole environment, we have

$$z(t) = \mathrm{Tr}_E \, |E \uparrow (t)\rangle\langle E \downarrow (t)| = \langle E \downarrow (t)|E \uparrow (t)\rangle =$$

$$\prod_k^N \left(\cos \frac{2g_k t}{\hbar} + i(|\beta_k|^2 - |\alpha_k|^2) \sin \frac{2g_k t}{\hbar} \right) = \prod_k^N f_k. \quad (5.135)$$

Therefore, tracing over the whole environment, we obtain from (5.128) the reduced density matrix:

$$\rho_R(t) = \mathrm{Tr}_E \, \rho = |a|^2 |m \uparrow\rangle\langle m \uparrow | + |b|^2 |m \downarrow\rangle\langle m \downarrow | +$$
$$z(t) \, a \, b^* \, |m \uparrow\rangle\langle m \downarrow | + z(t)^* \, a^* \, b \, |m \downarrow\rangle\langle m \uparrow |. \quad (5.136)$$

We see that $z(t)$ appears in the two non-diagonal matrix elements of ρ_R. The square of the modulus of f_k then works out to

$$|f_k|^2 = \left(\cos \frac{2g_k t}{\hbar} \right)^2 + (\beta_k^2 - \alpha_k^2)^2 \left(\sin \frac{2g_k t}{\hbar} \right)^2 =$$

$$\left(\cos \frac{2g_k t}{\hbar} \right)^2 + (\beta_k^4 + 2\beta_k^2\alpha_k^2 + \alpha_k^4) \left(\sin \frac{2g_k t}{\hbar} \right)^2 - 4\beta_k^2\alpha_k^2 \left(\sin \frac{2g_k t}{\hbar} \right)^2.$$
$$(5.137)$$

But $\beta_k^4 + 2\beta_k^2\alpha_k^2 + \alpha_k^4 = (\beta_k^2 + \alpha_k^2)^2 = 1$. Therefore,

$$|z(t)|^2 = \prod_{k=1}^N \left[1 - 4|\alpha_k|^2|\beta_k|^2 \left(\sin \frac{2g_k t}{\hbar} \right)^2 \right]. \quad (5.138)$$

Setting $|\alpha_k| = \cos\theta_k$, $|\beta_k| = \sin\theta_k$ we see that

$$|f_k|^2 = 1 - \sin^2(2\theta_k) \sin^2 \left(\frac{2g_k t}{\hbar} \right) < 1. \quad (5.139)$$

Thus it is not surprising that for large N, $|z(t)|^2$, as a product of many factors less than 1, becomes very small, and the reduced density matrix ρ_R becomes practically diagonal. At the same time, from (5.135), we see that the states $|E \uparrow (t)\rangle$ and $|E \downarrow (t)\rangle$ become practically orthogonal.

For a simple particular case, set $s = \sin\theta_k$, independent of k, and $g_k = g$. Then from (5.138), it is clear that the function $|z(t)|$ becomes *periodic*, with period $T = h/2g$:

$$|z(t)|^2 = \left[1 - s^2 \sin^2 \left(\frac{2gt}{\hbar} \right) \right]^N. \quad (5.140)$$

Supposing further that $gt/\hbar$ is very small, so that $\sin(2gt/\hbar) \simeq 2gt/\hbar$, we can write

$$|z(t)|^2 \simeq \left(1 - 4g^2t^2s^2/\hbar^2\right)^N. \tag{5.141}$$

For large N, we can approximate

$$\left(1 - \frac{x}{N}\right)^N \simeq e^{-x}, \tag{5.142}$$

where $x = (4g^2t^2s^2/\hbar^2)N$. Thus $\prod_{k=1}^N |f_k|^2 = e^{-(4g^2t^2s^2/\hbar^2)N}$ and $z(t) = e^{-(2g^2t^2s^2/\hbar^2)N}$. The width Δ of the Gaussian is given by $\Delta = \hbar/\sqrt{2}gs\sqrt{N}$. Neglecting $\sqrt{2}s$, which is of the order of the unity, $\Delta \simeq \hbar/g\sqrt{N}$. If we choose g of the order of the hyperfine transition energy of cesium-133, $g \simeq 6 \times 10^{-24}$ J, we have $\frac{\hbar}{2g} \simeq 10^{-11}$ s. Thus in order to be able to use the approximation $\sin(2g_k t/\hbar) \simeq 2\,gt/\hbar$, we must limit ourselves to extremely short times, say, $t \ll 10^{-11}$ s. Of course, if the model simulates a macroscopic system, N can be taken to be large, of the order of Avogadro's number 6×10^{23}. Then $\sqrt{N} = 8 \times 10^{11}$ and the width $\Delta \simeq 10^{-22}$ s is indeed many orders of magnitude smaller than $t \ll 10^{-11}$ s. Since $z(t)$ is a periodic function with period $T = h/2g \simeq 6 \times 10^{11}s$, $z(t)$ can be pictured as a series of very narrow Gaussians, with Δ very small, separated by time intervals of period T. Therefore, the model predicts *decoherence*, since the off-diagonal matrix elements of ρ_R are, in most cases, essentially vanishing. A slightly more realistic model is obtained assuming that the environment is "disordered," meaning that the coupling constants are randomly distributed. But this does not change the conclusions.

Other models have been proposed. In the beautiful book by Omnes (1994), a macroscopic oscillator (such as a pendulum) interacts with an environment of microscopic oscillators (phonons). Omitting the details, we describe only the main result. Let the macroscopic oscillator be initially in a superposition of two quantum states, corresponding to positions x_1, x_2, with $|x_1 - x_2| \simeq 1$ micron. We assume also that the pendulum's mass m is 1 g, and its period 1 s, the damping time T due to friction is very large, say, of the order of 1 h. Then the frequency ω is 2π and the characteristic decoherence time is of the order of

$$4\hbar T \left(\frac{m\omega^2}{|x_2 - x_1|^2}\right) \simeq 10^{-20} \text{ s}, \tag{5.143}$$

which shows the extreme instability of a quantum superposition of the states of a macroscopic object.

However, it is possible, albeit with some difficulty, to obtain superpositions of macroscopic states, which are interesting objects of study and have possible technical applications. There are two important examples of such superposition states, which resemble one another, as they describe macroscopic objects made up of a huge number of microscopic systems, *all in the same quantum state*. The first example is a Bose–Einstein condensate, in which the microscopic objects are atoms in their ground state. In a Bose–Einstein condensate, the atoms lose their individuality, and the whole system can be described by an eikonal wavefunction $\sqrt{\rho}\,e^{i\alpha}$, with density ρ and phase α. When two condensates have different phases, interference effects can be observed. The second example is a Josephson junction, consisting of supercurrents, separated by a thin barrier. The microscopic objects that constitute the supercurrents are *Cooper pairs* of electrons. A Cooper pair consists two electrons, with their spins coupled in a singlet state. If $\mathbf{r}_1$, σ_1, $\mathbf{r}_2$, σ_2 denote positions and spins of the two electrons, the wave function of one pair is given by

$$\Psi(\mathbf{r}_1, \sigma_1; \mathbf{r}_2, \sigma_2) = e^{i\theta}\frac{1}{\sqrt{2}}\Big(|1\uparrow\rangle|2\downarrow\rangle - |1\downarrow\rangle|2\uparrow\rangle\Big)\phi(|\mathbf{r}_1 - \mathbf{r}_2|), \qquad (5.144)$$

where θ is a phase that is usually unobservable. In order to add a constant momentum $\hbar\mathbf{k}$ to the pair, multiply Eq. (5.144) by $e^{i\hbar\mathbf{k}\cdot\mathbf{R}}$, where $\mathbf{R} = \frac{\mathbf{r}_1+\mathbf{r}_2}{2}$ is the center of mass of the pair.

In 1962, Josephson predicted that if the overall phase of the state is θ_1 on the right of the junction, and θ_2 on the left, a supercurrent can flow through the junction by tunneling of Cooper pairs. The supercurrent can be detected experimentally since it induces a magnetic field. An important technical application of the Josephson effect is a device called a SQUID (superconducting quantum interference device) which is sensitive enough to detect extremely weak magnetic fields, of magnitude 10^{-17} tesla or less; by comparison, the magnetic field of the Earth ranges from 25 to 65 microtesla, and the magnetic field produced by the human brain is of the order of 10^{-12} tesla.

Chapter 6
Digital and Quantum Computers

Abstract The fundamentals of digital and quantum computers are presented. Binary numbers, modular arithmetic, and Boolean algebra are reviewed. Logic gates for both classic and quantum computers are introduced and combined in some simple circuits for carrying out useful computations. The possibility for efficient factoring of large numbers using quantum computers is discussed in detail. This requires a lengthy detour into number theory. The most obvious application is to cryptography and secure communication.

Keywords Binary number · Boolean algebra · Logic gates · Quantum computer · Modular arithmetic · Number theory · Cryptography · RSA

6.1 Binary Number System

Before dealing with quantum computers, we will review some basic features of conventional digital computers. Computers are devices that can manipulate and transmit bits of information. A bit (binary digit) is a unit of information that can assume just two possible values, for example, an electrical signal that can have two different voltages. We will indicate the two values by 1 and 0. The algebra that admits only bits as variables is called a *Boolean algebra*, which we will discuss in the next section. The first and most simple example of application of binary digits consists of the *binary numbers*, which only use the digits 1 and 0 (as opposed to decimal numbers, that use the digits 0, 1, 2, 3, 4, 5, 6, 7, 8, 9). Let us show a few examples, limiting ourselves to numbers with 4 digits. In the decimal number system,

$$a_3 a_2 a_1 a_0 = a_3 10^3 + a_2 10^2 + a_1 10^1 + a_0 10^0. \tag{6.1}$$

For example, $2014 = 2 \times 1000 + 0 \times 100 + 1 \times 10 + 4 \times 1$. We write this, for brevity, as 2014_{10}. In the binary number system,

$$b_3 b_2 b_1 b_0 = b_3 2^3 + b_2 2^2 + b_1 2^1 + b_0 2^0. \tag{6.2}$$

© Springer International Publishing AG 2017
G. Fano and S.M. Blinder, *Twenty-First Century Quantum Mechanics:
Hilbert Space to Quantum Computers*, UNITEXT for Physics,
DOI 10.1007/978-3-319-58732-5_6

Table 6.1 Binary numbers

Binary	Decimal
000	0
001	1
010	2
011	3
100	4
101	5
110	6
111	7

The binary number $1101 = 1 \times 2^3 + 1 \times 2^2 + 0 \times 2 + 1 \times 2^0 = 13$. An n-digit binary number $b_n b_{n-1} \ldots b_2 b_1 b_0$ is equal to $\sum_{j=0}^{n-1} b_j 2^j$. For example, the binary numbers with three digits are listed in Table 6.1. If you want, you can omit the zeros to the left, and write simply 0, 1, 10, 11, 100, 101, 110, 111.

Bits are also called *logical variables*. The reason for this name is that simple propositions can be true or false, so the digits 1 and 0 are sometimes referred to as *true* or *false*. For example, the propositions "Rome is in Italy" or "5 is a prime number" are true, the propositions "Paris is in Italy" and "10 is a prime number" are false.

6.2 Boolean Algebra

As in the case of a real function of real variables, you can define a *binary function* or *operation* from n bits to one bit; thus the value of a binary function is 1 or 0. The simplest and most relevant cases are those with $n = 1$ or $n = 2$. If $n = 1$, the most important operation is *negation*, denoted by NOT. The result of the negation of the bit a is denoted by $\bar{a}$. The truth table for $NOT(a) = \bar{a}$ is shown in Table 6.2. For $n = 2$, with a and b representing two logical variables, the basic operations are as follows:

(1) The *logical product* AND of a and b, also called *conjunction*; it is denoted by a AND b or $a \wedge b$; when there are many logical operations to perform, we will write, in simplified notation, ab. Its value is 1 only if both a and b are 1. The truth table is Table 6.3.

Table 6.2 NOT

a	$\bar{a}$
1	0
0	1

Table 6.3 AND

a	b	$a \wedge b$
1	1	1
1	0	0
0	1	0
0	0	0

Table 6.4 OR

a	b	$a \vee b$
1	1	1
1	0	1
0	1	1
0	0	0

Table 6.5 XOR

a	b	$a \oplus b$
1	1	0
1	0	1
0	1	1
0	0	0

(2) The *logical sum* OR of a and b, also called *disjunction*; it is denoted by aORb or $a \vee b$. Its value is 1 if at least one of the two bits a, b is equal to 1. The truth table of OR is Table 6.4.

The algebra of logical variables 1, 0 with the three operations AND, OR, NOT and their associated truth tables, is called *Boolean algebra*. Besides the three basic operations there are other operations that can be obtained by combinations. For example, the operation XOR, called *exclusive or*, and denoted by $a \oplus b$, is shown in Table 6.5. From the truth table of XOR we see that in the last three cases $a \oplus b$ coincides with the usual arithmetic sum, while $1 \oplus 1 = 0$ in the first case; now recalling that the number 2 is written 10 in binary notation, we see that $a \oplus b$ coincides with the last digit obtained by adding two bits. Other relevant operations are obtained by applying the NOT operation (negation) to the operations AND, OR, XOR. In this way, we obtain the operations NAND, NOR, XNOR, whose truth tables are shown in Table 6.6.

The logical product $\wedge$ and the logical sum $\vee$ are similar, but not identical, to the usual numerical sums and products; the basic laws that are the same for both logical and numerical variables are as follows:

(1) $a \vee b = b \vee a, \quad a \wedge b = b \wedge a$ (commutativity)
(2) $a \vee (b \vee c) = (a \vee b) \vee c, \quad a \wedge (b \wedge c) = (a \wedge b) \wedge c$ (associativity)
(3) $a \wedge (b \vee c) = (a \wedge b) \vee (a \wedge c)$ (distributivity)
(4) $a \vee 0 = a$ (identity for the sum)
(5) $a \wedge 1 = a$ (identity for the product)
(6) $a \wedge 0 = 0$ (annihilator for the product)

Table 6.6 Boolean truth tables

NAND			NOR			XNOR		
a	b	$\overline{a \wedge b}$	a	b	$\overline{a \vee b}$	a	b	$\overline{a \oplus b}$
1	1	0	1	1	0	1	1	1
1	0	1	1	0	0	1	0	0
0	1	1	1	1	0	0	1	0
0	0	1	0	0	1	0	0	1

Laws that hold only for Boolean algebra are:

(7) $a \vee a = a$, $a \wedge a = a$ (idempotence)

(8) $a \wedge (a \vee b) = a$, $a \vee (a \wedge b) = a$ (absorption)

(9) $a \vee (b \wedge c) = (a \vee b) \wedge (a \vee c)$ (distributivity of $\vee$ over $\wedge$)

(10) $a \vee 1 = 1$. Note, for example, that if $a = 1$, laws (7) and (10) are incorrect in ordinary arithmetic, since $1 + 1 = 2$. One must be careful in using the simplified forms, since, for example, the seventh law becomes $a + a = a$ and $aa = a$. Laws (1) through (10) can be easily checked using the truth tables for $\vee$ and $\wedge$. Properties of the negation that sends a to $\overline{a}$ are as follows:

(1) $a \wedge \overline{a} = 0$, $a \vee \overline{a} = 1$ (complementation)

(2) $\overline{\overline{a}} = a$ (double negation)

(3) $\overline{a \wedge b} = \overline{a} \vee \overline{b}$, $\overline{a \vee b} = \overline{a} \wedge \overline{b}$ (De Morgan's laws).

6.2.1 Venn Diagrams

There is another way to check the validity of the laws on logical variables, the method of Venn diagrams; this approach appeals to our visual intuition, as well as to logic. Given two subsets A, B of a set S, it is possible to define the *union* $A \cup B$ and the *intersection* $A \cap B$. Furthermore, the set of points of S not belonging to A is called the *complement* of A and denoted by $\complement A$. Let us show some examples: A and B indicate two circles, included in the rectangle S; ϕ is the void (or empty) set (see Fig. 6.1). It is simple to verify that the algebra of sets is a Boolean algebra. In fact there is a close correspondence between subsets of S (denoted by capital letters A, B, C, X, Y, Z, ...) and logical variables (denoted by $a, b, c, x, y, z, ...$). The interior of a set A corresponds to the value $a = 1$, the exterior to the value $a = 0$. The three last rows of Table 6.7 justify the correspondence of the sets S and ϕ, respectively, with 1 and 0 (see Fig. 6.1). In Fig. 6.2, Venn diagrams corresponding to XOR, NAND, etc., are shown. Venn diagrams can be drawn for any number of regions. For example, drawing three circles A, B, C, we can visualize the distributive law $A \cap (B \cup C) = (A \cap B) \cup (A \cap C)$ (see Fig. 6.3).

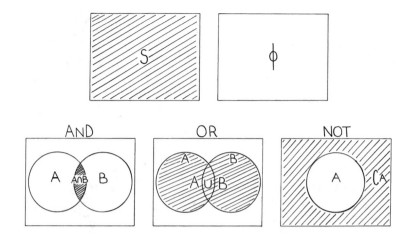

Fig. 6.1 The *shaded area* of the *rectangle* represents the whole space S, the *unshaded rectangle* represents the empty set ϕ

Table 6.7 Logical variables

Sets	Logical variables
$A, B, C..$	$a, b, c..$
$A \cap B$	$a \wedge b$
$A \cup B$	$a \vee b$
$\complement A$	$\bar{a}$
S	1
ϕ	0
$A \cap \complement A = \phi$	$a \wedge \bar{a} = 0$

Consider again three circles (see Fig. 6.4) included in a rectangle. Given the one-to-one correspondence between regions and logical variables, we can associate three variables a, b, c with the three circles; then the rectangle is partitioned into $2^3 = 8$ regions, and we can associate with each region a logical product in which the variables a, b, c appear once either in the complemented or uncomplemented form.

The 8 products of all 3 variables, in direct or complemented form, are called *minterms*. For example, $a \wedge b \wedge c$, $a \wedge b \wedge \bar{c}$, ... (or in simplified notation $a\,b\,c$, $a\,b\,\bar{c}$, ...). Any logical function f of a, b, c can be written as a sum of subsets of minterms. For example,

$$b \wedge c = (a \wedge b \wedge c) \vee (\bar{a} \wedge b \wedge c). \tag{6.3}$$

In the general case of n variables $a_1, a_2, \ldots, a_n$, the number of minterms is 2^n. The corresponding 2^3 logical sums $a \vee b \vee c, a \vee b \vee \bar{c}, \ldots$, are called *maxterms*. A

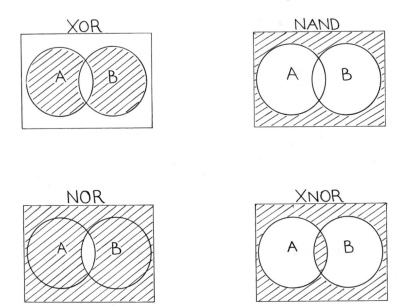

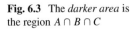

Fig. 6.2   Venn diagrams corresponding to XOR, NAND, NOR, and XNOR

Fig. 6.3   The *darker area* is
the region $A \cap B \cap C$

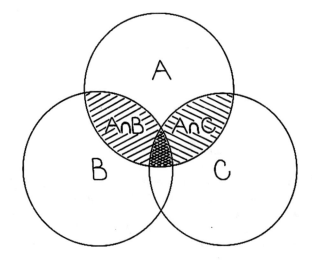

maxterm is obtained from a minterm by replacing the symbol $\wedge$ with $\vee$. Some logical functions are simpler using minterms, while others are simpler using maxterms. A minterm is equal to 1 on exactly one row of the truth table, while a maxterm is equal to 0 on exactly one row.

Fig. 6.4 Sets (and bits) are
denoted by simplified
notation. For example
$a\,\bar{b}\,\bar{c} = a \wedge \bar{b} \wedge \bar{c}$, etc.

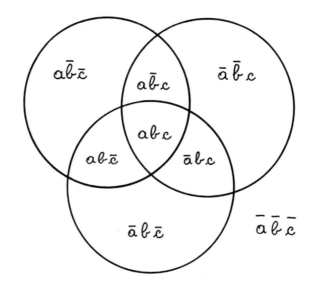

6.3 Classical Computers

6.3.1 Logic Gates

A *logic gate* is a physical device implementing a Boolean logical operation. On
a circuit diagram the elementary gates corresponding to the operations AND, OR,
NAND, etc., have two incoming wires and one outgoing wire; for example, the higher
voltage of a wire corresponds to 1, while the lower voltage corresponds to 0. Denoting
by x, y the two input variables, commonly used symbols are shown in Fig. 6.5. The
symbol corresponding to the negation has one input and one output wire, as shown
in Fig. 6.6. We note that the function $x \oplus y$ can be obtained by means of the circuit
shown in Fig. 6.7. Of course, the same logical function can be realized with different
circuits. The skill of the computer engineer consists in creating the optimal designs.

We have already noted that the logical operation $x \oplus y$ appears in the arithmetic
sum of two bits; however, in executing such a sum, there is a second operation that
we have to perform, namely the *carry*. For example, suppose you want to add the
binary numbers 101 and 111 (5_{10} and 7_{10}):

$$
\begin{array}{r}
1\,0\,1 \\
1\,1\,1 \\
\hline
1\,1\,0\,0
\end{array}
$$

Let us denote by S the arithmetic sum $x \oplus y$ of two bits x, y, and by C the carry.
Since the value of the carry is 1 if $x = y = 1$, and it is 0 otherwise, it coincides with
the logical product $x \wedge y$. Therefore,

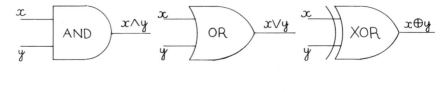

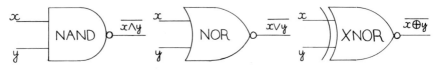

Fig. 6.5 Logic gates with two inputs and one output

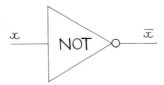

Fig. 6.6 Symbol of the negation gate

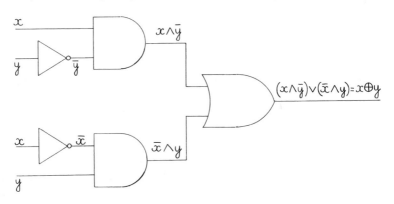

Fig. 6.7 A circuit implementing the $x \oplus y$ operation

$$S = x \oplus y, \qquad C = x \wedge y. \tag{6.4}$$

Let us examine the above sum of 101 and 111: adding the first digits $1+1$ from the right, we get $S = 0$ and carryover $C = 1$; for the second digits we must add three numbers 0, 1, and the carry $C = 1$. Finally, we do the sum $1+1+1$ and get 11.

How can we devise a digital circuit capable of performing these operations? Two simple circuits are employed: the *half-adder* and the *full-adder*. A half-adder (denoted by the symbol HA) is a digital circuit with two input bits x, y and two outputs S, C. It can be implemented as shown in Fig. 6.8. We have seen in the previous example, $5 + 7$ in binary digits, that it is necessary to sum three bits x, y, z. The circuit that

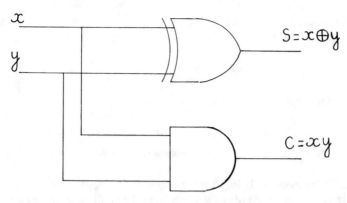

$$S = x \oplus y$$

$$C = xy$$

Fig. 6.8 Half-adder

Table 6.8 Full-adder

x	y	z	S_{10}	S_2	S	C
0	0	0	0	0	0	0
0	0	1	1	1	1	0
0	1	0	1	1	1	0
0	1	1	2	10	0	1
1	0	0	1	1	1	0
1	0	1	2	10	0	1
1	1	0	2	10	0	1
1	1	1	3	11	1	1

performs this operation is called a full-adder (denoted by the symbol FA). Usually x and y denote two bits to be added, and z is the carry from a previous sum. In the truth Table 6.8, S_{10} and S_2 are the arithmetic sums written in decimal and binary form, respectively; S is the last significant bit of $x + y + z$ and C denotes the new carry.

Unlike the half-adder, the full-adder has three inputs x, y, z and two outputs S, C. Let us use the simplified notation ($x \wedge y$ becomes simply xy, and $x \vee y$ becomes $x + y$). The function $x \oplus y$ is $x\bar{y} + \bar{x}y$. S is the arithmetic sum ($x \oplus y) \oplus z$. However, it would be incorrect to write $S = x \vee y \vee z = x + y + z$. Rather, we have

$$S = z \oplus (x \oplus y) = z\overline{(x\bar{y} + \bar{x}y)} + \bar{z}(x\bar{y} + \bar{x}y) = z(\bar{x} + y)(x + \bar{y}) + \bar{z}x\bar{y} + \bar{z}\bar{x}y, \tag{6.5}$$

where we have used the De Morgan law $\overline{ab} = \bar{a} + \bar{b}$. Since $x\bar{x} = y\bar{y} = 0$, we have

$$S = z\bar{x}\bar{y} + zyx + \bar{z}x\bar{y} + \bar{z}\bar{x}y = xyz + x\bar{y}\bar{z} + \bar{x}y\bar{z} + \bar{x}\bar{y}z. \tag{6.6}$$

We see that S is the sum of four minterms, corresponding to the values 111, 100, 010, and 001 of the full-adder table. Clearly, the four minterms vanish for several triples. This suggests writing the minterms directly corresponding to the carry C.

From Table 6.8, we see that the triples corresponding to $C = 1$ are 011, 101, 110, and 111. Thus we have

$$C = \bar{x}yz + x\bar{y}z + xy\bar{z} + xyz. \tag{6.7}$$

Using the identity $a + a = a$ with $a = xyz$:

$$C = \bar{x}yz + x\bar{y}z + xy\bar{z} + xyz + xyz + xyz =$$
$$(\bar{x} + x)yz + x(\bar{y} + y)z + xy(\bar{z} + z) = yz + xz + xy. \tag{6.8}$$

The Venn diagrams shown in Fig. 6.9 can help to understand the meaning of formulas (6.5) and (6.6). Next, we construct a digital circuit with three input digits x, y, z and two outputs S, C (see (6.4) and (6.6)). A possible realization is shown in Fig. 6.10. A circuit which can perform the sum of two binary numbers, each of four digits, such as

$$
\begin{array}{cccc}
a_3 & a_2 & a_1 & a_0 \\
b_3 & b_2 & b_1 & b_0 \\
\hline
C_4 \; S_3 & S_2 & S_1 & S_0
\end{array}
$$

can be easily constructed using FA gates. We can imagine a hypothetical carry $C_0 = 0$ entering in the circuit before adding the digits a_0, b_0. Then a digital circuit with four full-adders, shown in Fig. 6.12, carries out the sum.

Fig. 6.9 The *dashed area* corresponds to the carry $C = xy + xz + yz$

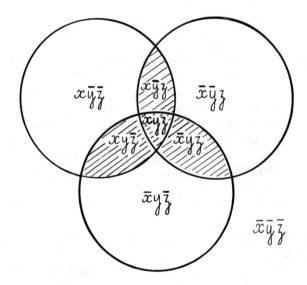

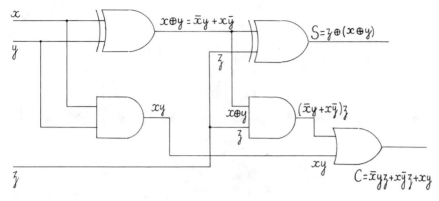

Fig. 6.10 Full-adder

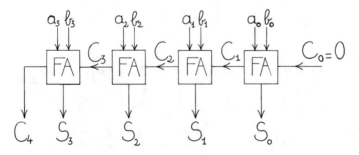

Fig. 6.11 The inputs x, y of a full-adder can enter from different sides

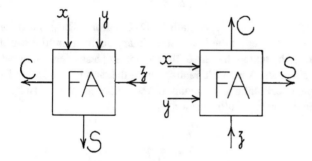

Fig. 6.12 A circuit implementing a 4-digit adder

6.3.2 Binary Multiplication

AND, half-adder, and full-adder logic gates can be easily connected to build a binary multiplier. As an example, consider the simple multiplication $6 \times 6 = 36$. Using binary numbers, the computation is the following:

$$
\begin{array}{r}
1\,1\,0 \\
1\,1\,0 \\
\hline
0\,0\,0 \\
1\,1\,0 \\
1\,1\,0 \\
\hline
1\,0\,0\,1\,0\,0
\end{array}
$$

In the general case of two 3-bit integers, we can write, using the simplified notation,

$$
\begin{array}{r}
a_2\,a_1\,a_0 \\
b_2\,b_1\,b_0 \\
\hline
a_2 b_0\ a_1 b_0\ a_0 b_0 \\
a_2 b_1\ a_1 b_1\ a_0 b_1 \\
a_2 b_2\ a_1 b_2\ a_0 b_2 \\
\hline
S_5\ S_4\ S_3\ S_2\ S_1\ S_0
\end{array}
$$

All the products $a_0 b_0, a_1 b_0, \dots$, contain AND operations. The products of the second and third row are shifted to the left as in ordinary decimal multiplication. The final result is obtained by taking the sum of the elements of a column, using half-adder and full-adder logic gates. A possible realization of the circuit is shown in Fig. 6.13. Let us do a simple check, taking the square of the binary number 111 (7_{10}); our circuit must perform the binary multiplication:

$$
\begin{array}{r}
1\,1\,1 \\
1\,1\,1 \\
\hline
1\,1\,1 \\
1\,1\,1 \\
1\,1\,1 \\
\hline
1\,1\,0\,0\,0\,1
\end{array}
$$

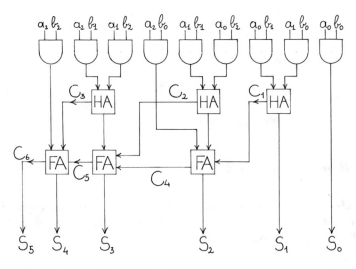

Fig. 6.13 A circuit implementing a 3-digit multiplier

All the products $a_i b_j$ $(i, j = 1, 2, 3)$ are equal to 1; also the carries $C_1, C_2, \ldots, C_6$ are equal to 1. Let us see how this happens: all gates HA of Fig. 6.13 have two outputs: a sum 0 and a carry 1. Therefore, $C_1 = C_2 = C_3 = 1$. Furthermore, $S_0 = a_0 b_0 = 1$ and $S_1 = 0$. Now consider the first FA from the right: it has three inputs, $C_1 = 1$, $a_2 b_0 = 1$, and a bit 0 coming from the second HA; the two outputs are the sum $S_2 = 0$ and the carry $C_4 = 1$. The second FA has three input, $C_2 = C_4 = 1$ and a bit 0 coming from the third HA; the two outputs are the sum $S_3 = 0$ and the carry $C_5 = 1$. Finally, the inputs of the last FA are $C_3 = C_5 = a_2 b_2 = 1$ and the outputs are $S_4 = 1$ and $C_6 = S_5 = 1$. Thus the result 110001 (49_{10}) is obtained.

6.3.3 A Living Computer

An interesting educational experiment can be performed in a school classroom to illustrate how a computer works. We build a "living computer" with children as logic gates: if a child raises his arm, this corresponds to the bit 1; if he keeps his arm lowered, this corresponds to the bit 0. For example, in order to simulate the circuit of Fig. 6.12, we need 4 children to indicate the number $a_3\, a_2\, a_1\, a_0$, and 4 other children for the number $b_3\, b_2\, b_1\, b_0$. Then we teach other children to build logic gates, as follows: to simulate an AND gate, a child is to look at two companions, say Dick and Jane, and raise one arm only if both Dick and Jane both raise one arm. Similarly to simulate a XOR gate, another child must watch two other companions, and raise his arm if only one of them raises his arm, and so on. Thereby a full-adder can be simulated with 5 children. Adding the first child to represent the bit $C_0 = 0$, we see that a school class of 29 children is sufficient to simulate the circuit of Fig. 6.12.[1]

In this section, we have given a simplified description of the basic operations of a classical digital computer; the next step would be a discussion of memory cells, CPUs, etc., but we stop here. We turn, instead, to consider the fundamental principles of quantum computation.

6.4 Quantum Computation

Digital computers, as we have seen, are based on the established principles of classical logic (although the contributions of quantum theory, such as transistors, have been indispensable). Since about 1980, a new paradigm of computation, based more directly on quantum physics, has been proposed (Feynman 1985), potentially with exponentially greater power than classical computers.

---

[1] GF has personally done this educational experiment; an interesting result was that after the children had understood what they had to do, the information could run very fast through the "computer" and the correct result was rapidly obtained.

The structure of quantum computation is based on a very simple concept: the replacement of the elementary unit of information, the *bit*, with the *qubit*, a much more complex object which, in principle, can contain an enormous amount of information. The qubit can, however, be fairly simply defined as follows:

Definition 6.1 A *qubit* is a normalized quantum state vector belonging to a complex two-dimensional Hilbert space.

Thus, from a mathematical point of view, a qubit is simply an element of $\mathbb{C}^2$ of norm 1. Choosing the $\mathbb{C}^2$ basis vectors,

$$|0\rangle = \left\| \begin{array}{c} 1 \\ 0 \end{array} \right\|, \qquad |1\rangle = \left\| \begin{array}{c} 0 \\ 1 \end{array} \right\| \tag{6.9}$$

a generic element of $\mathbb{C}^2$ can be written as a complex linear combination of two classical states:

$$|\Psi\rangle = \alpha|0\rangle + \beta|1\rangle, \tag{6.10}$$

where α and β are complex numbers. But a qubit should be regarded as an entirely different type of linear combination, which does not correspond to any well-defined classical entity. The norm $|\Psi|$ is assumed equal to 1; therefore, α, β must obey the normalization condition:

$$|\alpha|^2 + |\beta|^2 = 1. \tag{6.11}$$

Note, incidentally, that by setting $\alpha = a_1 + ia_2$, $\beta = b_1 + ib_2$ (where a_1, a_2, b_1, b_2 are real numbers), Eq. (6.11) becomes $a_1^2 + a_2^2 + b_1^2 + b_2^2 = 1$, which represents a 3-sphere in $\mathbb{R}^4$. In Eq. (4.55), we found a parametrization, in terms of two angles, for a two-state quantum system. This also applies nicely to an arbitrary qubit:

$$|\Psi\rangle = \cos\frac{\theta}{2}|0\rangle + \sin\frac{\theta}{2}\, e^{i\phi}|1\rangle. \tag{6.12}$$

The *Bloch sphere*, shown in Fig. 6.14, provides a useful geometrical representation of a qubit. In contrast to the Riemann sphere (used in our stereographic projection from the complex plane), antipodal points represent *orthogonal states*, rather than just negatives of the state vector. On the Bloch sphere, $|0\rangle$ is mapped onto the north pole ($\theta = 0$) and $|1\rangle$ onto the south pole ($\theta = \pi$). The equator, with $\theta = \pi/2$, contains four additional *cardinal points*, with $\phi = 0, \pi/2, \pi, 3\pi/2$, corresponding to $(|0\rangle + |1\rangle)/\sqrt{2}$, $(|0\rangle + i|1\rangle)/\sqrt{2}$, $(|0\rangle - |1\rangle)/\sqrt{2}$, $(|0\rangle - i|1\rangle)/\sqrt{2}$, respectively. There is a one-to-one correspondence between qubit states and points on the Bloch sphere.

Different physical systems can serve as realizations of a qubit: $|0\rangle$ and $|1\rangle$ can represent the two polarization states of a photon, the different alignments of an electron or nuclear spin, or even two atomic orbitals. Since θ and ϕ are real numbers, over a continuous range, a qubit can contain a vast amount of information. Quoting Nielsen and Chuang (Nielson and Chuang 2000, p. 15), "Paradoxically, there are an infinite number of points on the unit sphere, so that, in principle, one could store

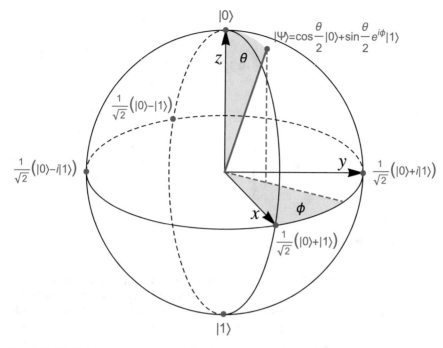

Fig. 6.14 Bloch sphere. The qubit $|\Psi\rangle$ and the six cardinal points are marked with *red dots*

an entire text of Shakespeare in the infinite binary expansion of θ." However, a *measurement* of the qubit Eq. (6.12) will give only one of two possible results. For example, if we make a measurement on the system represented by the matrix

$$Z = \begin{Vmatrix} 1 & 0 \\ 0 & -1 \end{Vmatrix}, \tag{6.13}$$

the possible results are ± 1, so that the measurement collapses the qubit to $|0\rangle$ or $|1\rangle$.

We next consider the case of multiple qubits. The Hilbert space $\mathbb{C}^{2n}$ of n qubits is the *tensor product* of n copies of $\mathbb{C}^2$, namely, $\mathbb{C}^2 \otimes \mathbb{C}^2, \ldots, \otimes \mathbb{C}^2$, n times. A useful basis in $\mathbb{C}^{2n}$ is the *computational basis*, with $|0\rangle = \begin{Vmatrix} 1 \\ 0 \end{Vmatrix}$, $|1\rangle = \begin{Vmatrix} 0 \\ 1 \end{Vmatrix}$, an orthonormal basis in *each* space $\mathbb{C}^2$. The computational basis is provided by the 2^n product states:

$$|i_1, i_2, \ldots, i_n\rangle = |i_1\rangle \otimes |i_2\rangle \otimes \cdots \otimes |i_n\rangle, \tag{6.14}$$

where $(i_1, i_2, \ldots, i_n)$ is a binary number with n bits. For example, in the space $\mathbb{C}^2 \otimes \mathbb{C}^2$, the four basis vectors are

$$|0\,0\rangle = |0\rangle \otimes |0\rangle, \quad |0\,1\rangle = |0\rangle \otimes |1\rangle, \quad |1\,0\rangle = |1\rangle \otimes |0\rangle, \quad |1\,1\rangle = |1\rangle \otimes |1\rangle. \tag{6.15}$$

Using the notation,

$$\begin{Vmatrix} \alpha_1 \\ \alpha_2 \end{Vmatrix} \otimes \begin{Vmatrix} \beta_1 \\ \beta_2 \end{Vmatrix} = \begin{Vmatrix} \alpha_1\beta_1 \\ \alpha_1\beta_2 \\ \alpha_2\beta_1 \\ \alpha_2\beta_2 \end{Vmatrix}, \tag{6.16}$$

we have

$$|0\,0\rangle = \begin{Vmatrix} 1 \\ 0 \\ 0 \\ 0 \end{Vmatrix}, \quad |0\,1\rangle = \begin{Vmatrix} 0 \\ 1 \\ 0 \\ 0 \end{Vmatrix}, \quad |1\,0\rangle = \begin{Vmatrix} 0 \\ 0 \\ 1 \\ 0 \end{Vmatrix}, \quad |1\,1\rangle = \begin{Vmatrix} 0 \\ 0 \\ 0 \\ 1 \end{Vmatrix}. \tag{6.17}$$

6.4.1 Quantum Gates

Once we have replaced bits with qubits, we can ask can we now construct quantum gates, analogs of logic gates for qubits? Let us consider the case of a single qubit; we are immediately aware of a major difference compared to digital computing: while there is only one single-bit logic gate, namely, the NOT gate, multiple *single-qubit* quantum gates are possible. A single-qubit quantum gate transforms the qubit $\begin{Vmatrix} \alpha \\ \beta \end{Vmatrix}$ into the qubit $\begin{Vmatrix} \alpha' \\ \beta' \end{Vmatrix}$. To satisfy Eq. (6.11), we must have

$$|\alpha'|^2 + |\beta'|^2 = |\alpha|^2 + |\beta|^2 = 1, \tag{6.18}$$

so that the *norm* of the state $\begin{Vmatrix} \alpha \\ \beta \end{Vmatrix}$ is invariant. Since we want to preserve linearity, there is only one class of operators that can serve as quantum gates, namely *unitary* operators (see Eq. 2.121). We recall that unitary operators U are a generalization of orthogonal operators for rotations, with complex matrix elements. They leave the norm invariant, so that $|U\Psi| = |\Psi| \; \forall \, \Psi$.

The path of each qubit in a quantum computer circuit is represented by a single horizontal line ——— which usually connects a sequence of gates. The qubit is understood to move along the "quantum wire" from left to right. A wire carrying a classical bit, 0 or 1, usually after a measurement, is indicated by a double line ===.

In Table 6.9, matrices representing some examples of single-qubit quantum gates are shown; these examples are relevant in the design of quantum circuits. Some comments are in order:

(1) The *Hadamard* gate H maps the qubits $|0\rangle = \begin{Vmatrix} 1 \\ 0 \end{Vmatrix}$, $|1\rangle = \begin{Vmatrix} 0 \\ 1 \end{Vmatrix}$, respectively, to the *superposition* states:

Table 6.9 Single-qubit quantum gates

Name	Symbol	Matrix
Hadamard		$\frac{1}{\sqrt{2}}\begin{Vmatrix} 1 & 1 \\ 1 & -1 \end{Vmatrix}$
Pauli X		$\begin{Vmatrix} 0 & 1 \\ 1 & 0 \end{Vmatrix}$
Pauli Z		$\begin{Vmatrix} 1 & 0 \\ 0 & -1 \end{Vmatrix}$
$\frac{\pi}{2}$-phase		$\begin{Vmatrix} 1 & 0 \\ 0 & i \end{Vmatrix}$
$\frac{\pi}{4}$-phase		$\begin{Vmatrix} 1 & 0 \\ 0 & e^{i\pi/4} \end{Vmatrix}$
Measurement		

$$H|0\rangle = \frac{1}{\sqrt{2}}(|0\rangle + |1\rangle) = \frac{1}{\sqrt{2}}\begin{Vmatrix} 1 \\ 1 \end{Vmatrix}, \quad H|1\rangle = \frac{1}{\sqrt{2}}(|0\rangle - |1\rangle) = \frac{1}{\sqrt{2}}\begin{Vmatrix} 1 \\ -1 \end{Vmatrix}.$$

(6.19)

Needless to say, this is something entirely different from the classical case, since the "bits" 0 and 1 are being mixed. The two superpositions in Eq. (6.19) can be defined as an alternative pair of basis functions (sometimes called the *Hadamard basis*):

$$|+\rangle = \frac{1}{\sqrt{2}}(|0\rangle + |1\rangle) \quad \text{and} \quad |-\rangle = \frac{1}{\sqrt{2}}(|0\rangle - |1\rangle).$$

(6.20)

(2) The quantum gate X maps the qubit $\begin{Vmatrix} \alpha \\ \beta \end{Vmatrix}$ to $\begin{Vmatrix} \beta \\ \alpha \end{Vmatrix}$. Since $X^2 = 1$, X is somewhat analogous, but not identical, to the classical NOT gate.

(3) The quantum gates T, S, Z change only the phase of β. The variation of the phase is $\pi/4$ for T, since $e^{i\pi/4} = \frac{1}{\sqrt{2}}(1 + i)$, $\pi/2$ for S, since $e^{i\pi/2} = i$ and π for Z, since $e^{i\pi} = -1$. Thus

$$T^2 = S, \quad S^2 = Z, \quad Z^2 = I.$$

(6.21)

Table 6.10 CNOT

Input	Output
$|00\rangle$	$|00\rangle$
$|01\rangle$	$|01\rangle$
$|10\rangle$	$|11\rangle$
$|11\rangle$	$|10\rangle$

Given their resemblance to Pauli spin matrices, X and Z are called *Pauli gates*.

Next we consider some two-qubit gates. Two qubits gates operate in the $\mathbb{C}^2 \otimes \mathbb{C}^2$ space. Thus they can be represented by 4×4 unitary matrices. Since the inverse of the unitary matrix U is the matrix $U^\dagger$, quantum gates are *reversible*, in contrast to classical logic gates. For a quantum gate, the input can be deduced from the output. But for a classical XOR gate $a \oplus b$, for the output 1, we do not know whether the input was $|10\rangle$ or $|01\rangle$. The output does not identify the input. Some essential two-qubit gates are the following:

(1) The CNOT (controlled-NOT) gate. We denote the two input qubits by a, b. The first qubit a is called the *control* qubit, the second b the *target* qubit. This target qubit flips if and only if $a = 1$. If $a = 0$ the second qubit remains unchanged. Table 6.10 is the truth table for CNOT: In a circuit, the CNOT gate is denoted by the symbol shown in Fig. 6.15. This symbol is easy to remember since for classical bits the sum $a \oplus b$ is the negation of b only if $a = 1$. The unitary matrix corresponding to the CNOT gate has the following representation in the computational basis (see Eq. (6.14)):

$$U = \begin{Vmatrix} 1 & 0 & 0 & 0 \\ 0 & 1 & 0 & 0 \\ 0 & 0 & 0 & 1 \\ 0 & 0 & 1 & 0 \end{Vmatrix}. \tag{6.22}$$

In fact, U leaves the states $|00\rangle$, $|01\rangle$ invariant, but swaps the states $|10\rangle$ and $|11\rangle$. If we exchange the control and target qubits, we obtain the circuit shown in Fig. 6.16. The corresponding matrix is

$$U' = \begin{Vmatrix} 0 & 1 & 0 & 0 \\ 1 & 0 & 0 & 0 \\ 0 & 0 & 1 & 0 \\ 0 & 0 & 0 & 1 \end{Vmatrix} \tag{6.23}$$

where now the states left invariant by U' are $|10\rangle$ and $|11\rangle$.

Quite generally, *controlled gates* act on 2 qubits, with the control qubit determining some operation on the target qubit. If the single-qubit gate V is represented by the 2×2 matrix:

$$V = \begin{Vmatrix} V_{11} & V_{12} \\ V_{21} & V_{22} \end{Vmatrix}, \tag{6.24}$$

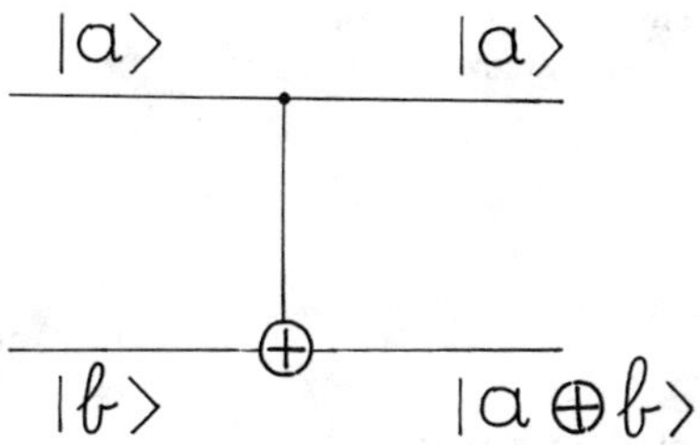

Fig. 6.15 CNOT gate. The *black dot* represents the control qubit

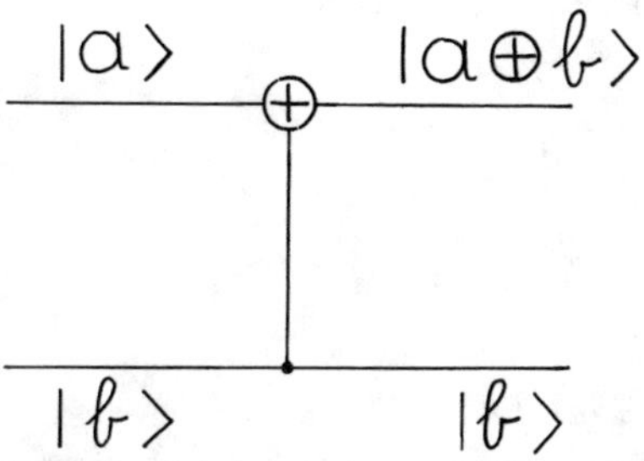

$$|a\rangle \qquad\qquad |a\rangle$$

$$|b\rangle \qquad\qquad |a \oplus b\rangle$$

Fig. 6.16 The CNOT gate with lower control qubit

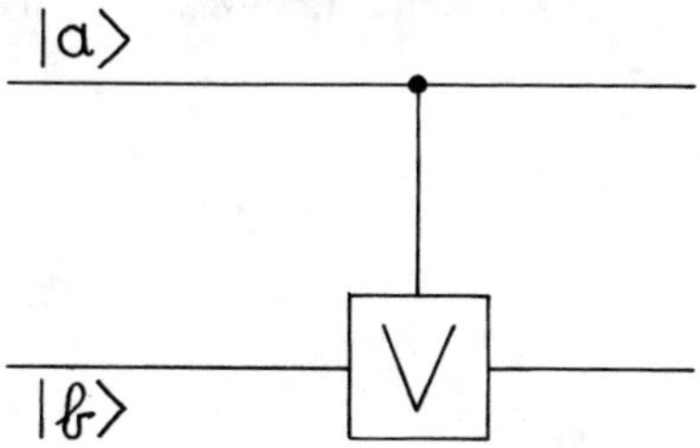

$$|a\rangle \qquad\qquad |a \oplus b\rangle$$

$$|b\rangle \qquad\qquad |b\rangle$$

Fig. 6.17 The gate V is applied only to $|b\rangle$, with $|a\rangle$ left unchanged

$$|a\rangle$$

$$|b\rangle \qquad\qquad \boxed{V}$$

the symbol of the controlled-V gate is shown in Fig. 6.17. The corresponding matrix is

$$U = \begin{Vmatrix} 1 & 0 & 0 & 0 \\ 0 & 1 & 0 & 0 \\ 0 & 0 & V_{11} & V_{12} \\ 0 & 0 & V_{21} & V_{22} \end{Vmatrix}. \tag{6.25}$$

Since

$$U|00\rangle = |00\rangle, \quad U|01\rangle = |01\rangle, \quad U|10\rangle = |1\rangle \otimes V|0\rangle, \quad U|11\rangle = |1\rangle \otimes V|1\rangle, \tag{6.26}$$

we see that V is applied to $|b\rangle$ *only if* $|a\rangle = |1\rangle$. The controlled-V gate should not be confused with the gate shown in Fig. 6.19, which is represented by the matrix $I \otimes V$.

Fig. 6.18 The controlled-V
gate with lower control qubit

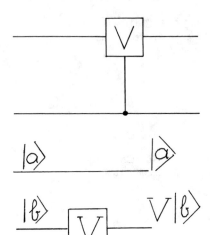

Fig. 6.19 Circuit
representing the matrix
$I \otimes V$

Exchanging the control and target qubits we obtain the *inverted controlled gate* in
Fig. 6.18, which is represented by the matrix:

$$U' = \begin{Vmatrix} 1 & 0 & 0 & 0 \\ 0 & V_{11} & 0 & V_{12} \\ 0 & 0 & 1 & 0 \\ 0 & V_{21} & 0 & V_{22} \end{Vmatrix}. \tag{6.27}$$

(2) Using the rules described above, one can write the matrices U_T, U_S, U_Z corresponding, respectively, to the gates controlled-T, controlled-S, and controlled-Z,
shown in Figs. 6.20, 6.21 and 6.22, as follows:

$$U_T = \begin{Vmatrix} 1 & 0 & 0 & 0 \\ 0 & 1 & 0 & 0 \\ 0 & 0 & 1 & 0 \\ 0 & 0 & 0 & e^{i\pi/4} \end{Vmatrix}, \quad U_S = \begin{Vmatrix} 1 & 0 & 0 & 0 \\ 0 & 1 & 0 & 0 \\ 0 & 0 & 1 & 0 \\ 0 & 0 & 0 & i \end{Vmatrix}, \quad U_Z = \begin{Vmatrix} 1 & 0 & 0 & 0 \\ 0 & 1 & 0 & 0 \\ 0 & 0 & 1 & 0 \\ 0 & 0 & 0 & -1 \end{Vmatrix}. \tag{6.28}$$

Fig. 6.20 Controlled-T gate

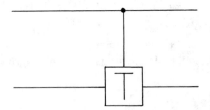

Fig. 6.21 Controlled-S gate

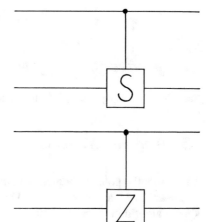

Fig. 6.22 Controlled-Z gate

Fig. 6.23 The swap gate

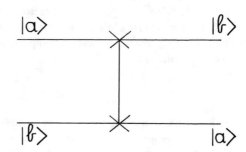

(3) The *swap* gate: this quantum gate swaps the states of the two input qubits, as shown in Fig. 6.23 and Eq. (6.29).

$$U = \begin{Vmatrix} 1 & 0 & 0 & 0 \\ 0 & 0 & 1 & 0 \\ 0 & 1 & 0 & 0 \\ 0 & 0 & 0 & 1 \end{Vmatrix}. \tag{6.29}$$

It is easy to verify the equivalence of the swap gate with the sequence of CNOT operations shown in Fig. 6.24; in fact, if a and b denote classical bits, $a \oplus (a \oplus b) = b$ and $b \oplus (a \oplus b) = a$. The same result follows from matrix multiplication:

$$\begin{Vmatrix} 1 & 0 & 0 & 0 \\ 0 & 1 & 0 & 0 \\ 0 & 0 & 0 & 1 \\ 0 & 0 & 1 & 0 \end{Vmatrix} \times \begin{Vmatrix} 1 & 0 & 0 & 0 \\ 0 & 0 & 0 & 1 \\ 0 & 0 & 1 & 0 \\ 0 & 1 & 0 & 0 \end{Vmatrix} \times \begin{Vmatrix} 1 & 0 & 0 & 0 \\ 0 & 1 & 0 & 0 \\ 0 & 0 & 0 & 1 \\ 0 & 0 & 1 & 0 \end{Vmatrix} = \begin{Vmatrix} 1 & 0 & 0 & 0 \\ 0 & 0 & 1 & 0 \\ 0 & 1 & 0 & 0 \\ 0 & 0 & 0 & 1 \end{Vmatrix}. \tag{6.30}$$

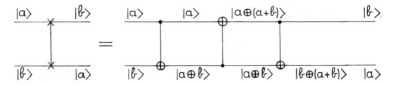

Fig. 6.24 The swap gate is equivalent to a sequence of CNOT operations

6.5   Deutsch's Algorithm

To explore the potentially greater computational power of a quantum computer, as compared to a classical computer, David Deutsch[2] proposed a highly contrived but simple algorithm. Consider four possible functions of a single-bit (or basis qubit), $x = 0$ or 1, which produce a single-bit result $y = 0$ or 1, as follows: $f_1(x) = 0$, $f_2(x) = 1$, $f_3(x) = x$, $f_4(x) = \overline{x}$. The first two functions are classified as "constant" (with $f(0) = f(1)$), while the last two are described as "balanced" (with $f(0) \neq f(1)$). Suppose now that a classical computer, idealized as a "black box", can perform the computation

$$x \rightarrow \boxed{\text{f}} \rightarrow y. \tag{6.31}$$

If it is desired to identify the function as one of f_1, f_2, f_3, f_4, it is necessary to run the classical program *twice*, with inputs $x = 0$ and $x = 1$.[3]

The key element of the circuit for running Deutsch's algorithm is the Cf gate, a modification of the CNOT gate (Fig. 6.15) in which the target qubit $|y\rangle$ is output as $|y \oplus f(x)\rangle$ (rather than $|y \oplus x\rangle$ for the standard CNOT). This is shown in Fig. 6.25.

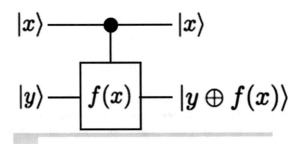

Fig. 6.25 Deutsch Cf gate. For inputs $|00\rangle, |01\rangle, |10\rangle, |11\rangle$, the outputs are $|0\ f(0)\rangle, |0\ \overline{f(0)}\rangle$, $|1\ f(1)\rangle, |1\ \overline{f(1)}\rangle$, respectively

---

[2]Deutsch D (1985) *Quantum Theory, the Church-Turing Principle and the Universal Quantum Computer*, Proceedings of the Royal Society of London A 400: 97–117.
[3]For example, with the input $x = 0$, suppose we find $f(x) = 1$. Then f can be either f_2 or f_4. We need a second run with $x = 1$ to determine which alternative is correct.

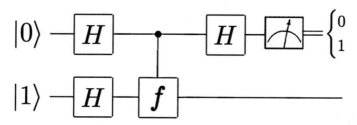

Fig. 6.26 Circuit for Deutsch's algorithm. The reading 0 or 1 identifies $f(x)$ as constant or balanced, respectively

The black box for evaluating the function $f(x)$ is assumed to be a component part of the circuit for this gate. Figure 6.26 shows the circuit for Deutsch's algorithm. The input 2-qubit state is $\psi_0 = |0\rangle \otimes |1\rangle = |01\rangle$. After the action of the two Hadamard gates, this becomes

$$\psi_1 = \frac{1}{2}\left(|00\rangle - |01\rangle + |10\rangle - |11\rangle\right). \tag{6.32}$$

After the Deutsch Cf gate,

$$\psi_2 = \frac{1}{2}\left(|\,0f(0)\rangle - |0\,\overline{f(0)}\rangle + |1\,f(1)\rangle - |1\,\overline{f(1)}\rangle\right) =$$
$$\frac{1}{2}\left(|0\rangle \otimes \left(|f(0)\rangle - |\overline{f(0)}\rangle\right) + |1\rangle \otimes \left(|f(1)\rangle - |\overline{f(1)}\rangle\right)\right). \tag{6.33}$$

We now consider the two possible cases. If $f(x)$ is constant, $f(0) = f(1)$. But if $f(x)$ is balanced, $f(0) \neq f(1)$; since f is a binary function, this implies that $\overline{f(0)} = f(1)$ and $\overline{f(1)} = f(0)$. The two cases for ψ_2 can be simplified to

$$\psi_2^{const} = \frac{1}{\sqrt{2}}\left(|0\rangle + |1\rangle\right) \otimes \frac{1}{\sqrt{2}}\left(|f(0)\rangle - |\overline{f(0)}\rangle\right)$$
$$\psi_2^{bal} = \frac{1}{\sqrt{2}}\left(|0\rangle - |1\rangle\right) \otimes \frac{1}{\sqrt{2}}\left(|f(0)\rangle - |f(1)\rangle\right). \tag{6.34}$$

Next, the Hadamard gate applied to the first qubit gives

$$H\frac{1}{\sqrt{2}}\left(|0\rangle + |1\rangle\right) = |0\rangle \quad \text{or} \quad H\frac{1}{\sqrt{2}}\left(|0\rangle - |1\rangle\right) = |1\rangle. \tag{6.35}$$

Finally, a measurement on the first qubit, giving 0 or 1, identifies $f(x)$ as constant or balanced, respectively. The second qubit is ignored. Thus, we have shown that using Deutsch's algorithm on a quantum computer, it is possible to identify f as constant or balanced in just a single run. This might not appear very impressive, but it represents a proof of principle, suggesting that a classical computation of complexity, say $O(N)$, might be accomplished by a quantum computation of complexity $O(\sqrt{N})$ or $O(\log N)$. This becomes highly significant for N of the order of thousands or millions.

6.6   Bell States

The superpositions $|\Phi^{\pm}\rangle = \frac{1}{\sqrt{2}}(|00\rangle \pm |11\rangle)$ and $|\Psi^{\pm}\rangle = \frac{1}{\sqrt{2}}(|01\rangle \pm |10\rangle)$ are called Bell (or EPR) states (see Eq. 5.13). These maximally entangled 2-qubit states form an orthonormal basis for $\mathbb{C}^4$. Bell states can be produced by the circuit shown in Fig. 6.27. The corresponding operator matrix can be obtained applying first $H \otimes I$, and then the matrix Eq. (6.22) representing the action of the CNOT gate:

$$\frac{1}{\sqrt{2}}\begin{Vmatrix} 1 & 0 & 0 & 0 \\ 0 & 1 & 0 & 0 \\ 0 & 0 & 0 & 1 \\ 0 & 0 & 1 & 0 \end{Vmatrix} \times \begin{Vmatrix} 1 & 0 & 1 & 0 \\ 0 & 1 & 0 & 1 \\ 1 & 0 & -1 & 0 \\ 0 & 1 & 0 & -1 \end{Vmatrix} = \frac{1}{\sqrt{2}}\begin{Vmatrix} 1 & 0 & 1 & 0 \\ 0 & 1 & 0 & 1 \\ 0 & 1 & 0 & -1 \\ 1 & 0 & -1 & 0 \end{Vmatrix}. \qquad (6.36)$$

Recall that products of matrices are evaluated in the order right to left. Thus, in the product XY of two matrices representing quantum gates, Y operates first. Applying the operator (6.36) to the computational basis produces one of the Bell states. Specifically, the inputs $|00\rangle, |10\rangle, |01\rangle, |11\rangle$ produce the Bell states $|\Phi^+\rangle$, $|\Phi^-\rangle, |\Psi^+\rangle, |\Psi^-\rangle$, respectively.

Fig. 6.27 Circuit producing Bell states

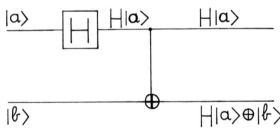

Fig. 6.28 Circuit to detangle a Bell state

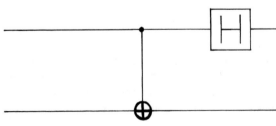

The 2-qubit circuit shown in Fig. 6.28 essentially reverses the creation of a Bell state. The corresponding operator can be obtained by first applying the matrix (6.22), representing the gate CNOT, and then $H \otimes I$. The result is as follows:

$$\frac{1}{\sqrt{2}} \begin{Vmatrix} 1 & 0 & 1 & 0 \\ 0 & 1 & 0 & 1 \\ 1 & 0 & -1 & 0 \\ 0 & 1 & 0 & -1 \end{Vmatrix} \times \begin{Vmatrix} 1 & 0 & 0 & 0 \\ 0 & 1 & 0 & 0 \\ 0 & 0 & 0 & 1 \\ 0 & 0 & 1 & 0 \end{Vmatrix} = \frac{1}{\sqrt{2}} \begin{Vmatrix} 1 & 0 & 0 & 1 \\ 0 & 1 & 1 & 0 \\ 1 & 0 & 0 & -1 \\ 0 & 1 & -1 & 0 \end{Vmatrix}. \tag{6.37}$$

You can thereby show that input of a Bell state $|\Phi^+\rangle$, $|\Psi^+\rangle$, $|\Phi^-\rangle$, $|\Psi^-\rangle$ produces the output $|00\rangle$, $|01\rangle$, $|10\rangle$, $|11\rangle$, respectively. The Bell state has thus become "detangled."

Suppose Alice and Bob share a known Bell state, each possessing one qubit. It is now possible for Alice to transmit *two* classical bits to Bob by sending him just a single qubit. This is known as *superdense coding*. As shown in Fig. 6.29, the Bell state $|\Phi^+\rangle$ is first created, beginning with the input $|00\rangle$, using the partial circuit of Fig. 6.27. One qubit is sent to Alice and the other to Bob. Alice then moves her qubit through one of four gates, I, X, Z, or XZ, enclosed by the rectangular box, and transmits the result to Bob. Including the qubit originally sent to him, Bob is now in possession of both qubits of one of the four Bell states $|\Phi^+\rangle$, $|\Psi^+\rangle$, $|\Phi^-\rangle$, or $|\Psi^-\rangle$. After using the partial circuit in Fig. 6.28, Bob's two measurements results in one of the disentangled states $|00\rangle$, $|01\rangle$, $|10\rangle$ or $|11\rangle$, respectively. Each of these is equivalent to two classical bits. The central steps in the superdense coding procedure have made use of the following transformations:

$$I|\Phi^+\rangle = |\Phi^+\rangle,$$
$$X|\Phi^+\rangle = X\frac{1}{\sqrt{2}}(|00\rangle + |11\rangle) = \frac{1}{\sqrt{2}}(|10\rangle + |01\rangle) = |\Psi^+\rangle,$$
$$Z|\Phi^+\rangle = Z\frac{1}{\sqrt{2}}(|00\rangle + |11\rangle) = \frac{1}{\sqrt{2}}(|00\rangle - |11\rangle) = |\Phi^-\rangle,$$
$$XZ|\Phi^+\rangle = XZ\frac{1}{\sqrt{2}}(|00\rangle + |11\rangle) = \frac{1}{\sqrt{2}}(|10\rangle - |01\rangle) = |\Psi^-\rangle. \tag{6.38}$$

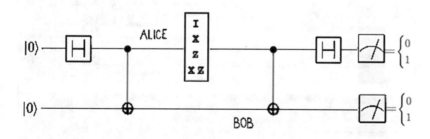

Fig. 6.29 Superdense coding

6.7   Quantum Teleportation

It is now possible to construct a circuit able to teleport the information contained in a quantum qubit, even over great distances.[4] However this must be accompanied by the conventional transmission of two classical bits. Suppose now that Alice wants to send Bob some wonderful special qubit, represented by the superposition $|\Psi\rangle = \left\| \begin{array}{c} \alpha \\ \beta \end{array} \right\|$. To accomplish this *teleportation*, Alice and Bob must again be simultaneously in possession of an entangled pair in one of the Bell states, with each holding (but not measuring) one spin. The procedure involves manipulation of three spins. Initially, spin 1 is Alice's $|\Psi\rangle$. From the Bell pair, Alice holds spin 2, while Bob holds spin 3: we denote these as $|\text{BELL}\rangle_A$ and $|\text{BELL}\rangle_B$, respectively. After the teleportation is complete, spin 3 will be in the state $|\Psi\rangle$. We still will not know the value of the coefficients α and β, but the state of spin 1, which was in initially in Alice's lab will have been transferred to spin 3, which is in Bob's lab. A circuit that can perform this teleportation is shown in Fig. 6.30.

The initial state of the composite quantum system can be represented by

$$|\Psi_0\rangle = |\Psi\rangle \otimes |\text{BELL}\rangle. \tag{6.39}$$

The two top lines in Fig. 6.30 pertain to Alice, while the bottom line pertains to Bob. The input qubit of the upper line is $|\Psi\rangle$, and we use the Bell state

$$|\text{BELL}\rangle = \frac{1}{\sqrt{2}}\big(|00\rangle + |11\rangle\big), \tag{6.40}$$

which is the input to the two bottom lines. Denoting by $|ijk\rangle$ $(i, j, k = 0, 1)$, the tensor product $|i\rangle \otimes |j\rangle \otimes |k\rangle$, where $|i\rangle$, $|j\rangle$, $|k\rangle$ are basis qubits corresponding to the top, middle and bottom lines, respectively, we have

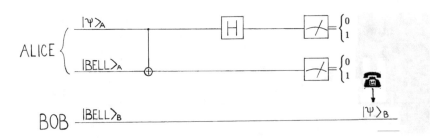

Fig. 6.30 Circuit implementing teleportation

[4]Bennett CH, Brassard G, Crépeau C, Jozsa R, Peres A, Wootters WK (1993) Teleporting an Unknown Quantum State via Dual Classical and Einstein-Podolsky-Rosen Channels. Phys Rev Lett 70:1895–1899.

$$|\Psi_0\rangle = (\alpha|0\rangle + \beta|1\rangle) \otimes \tfrac{1}{\sqrt{2}}(|00\rangle + |11\rangle) =$$

$$\tfrac{1}{\sqrt{2}}(\alpha|000\rangle + \alpha|011\rangle + \beta|100\rangle + \beta|111\rangle) =$$

$$\tfrac{1}{\sqrt{2}}(\alpha|00\rangle + \beta|10\rangle) \otimes |0\rangle + \tfrac{1}{\sqrt{2}}(\alpha|01\rangle + \beta|11\rangle) \otimes |1\rangle =$$

$$\tfrac{1}{\sqrt{2}}\left\|\begin{matrix}\alpha\\0\\\beta\\0\end{matrix}\right\| \otimes |0\rangle + \tfrac{1}{\sqrt{2}}\left\|\begin{matrix}0\\\alpha\\0\\\beta\end{matrix}\right\| \otimes |1\rangle, \tag{6.41}$$

where the four components vectors refer to Alice's qubits. Alice then successively applies the CNOT gate, followed by the transformation $H \otimes I$, on the first two qubits. This is represented by the matrix (6.37), which operates on $|\Psi_0\rangle$. Since the third qubit remains unaltered, we obtain the following state:

$$\tfrac{1}{2}\left\|\begin{matrix}\alpha\\\beta\\\alpha\\-\beta\end{matrix}\right\| \otimes |0\rangle + \tfrac{1}{2}\left\|\begin{matrix}\beta\\\alpha\\-\beta\\\alpha\end{matrix}\right\| \otimes |1\rangle =$$

$$\tfrac{1}{2}\left(\alpha|000\rangle + \beta|010\rangle + \alpha|100\rangle - \beta|110\rangle + \beta|001\rangle + \alpha|011\rangle - \beta|101\rangle + \alpha|111\rangle\right) =$$

$$\tfrac{1}{2}\left(|00\rangle \otimes \left\|\begin{matrix}\alpha\\\beta\end{matrix}\right\| + |01\rangle \otimes \left\|\begin{matrix}\beta\\\alpha\end{matrix}\right\| + |10\rangle \otimes \left\|\begin{matrix}\alpha\\-\beta\end{matrix}\right\| + |11\rangle \otimes \left\|\begin{matrix}-\beta\\\alpha\end{matrix}\right\|\right). \tag{6.42}$$

After Alice performs measurements on her two qubits, the global wave function collapses into one of four unentangled states, $|00\rangle, |01\rangle, |10\rangle$ or $|11\rangle$, each with a probability $\tfrac{1}{4}$. She then sends the result to Bob, by email, telephone, etc. If the message is $|00\rangle$, Bob knows that his qubit is the desired result $\left\|\begin{matrix}\alpha\\\beta\end{matrix}\right\|$. If the message is $|01\rangle$, his qubit is $\left\|\begin{matrix}\beta\\\alpha\end{matrix}\right\|$, but he can then apply the Pauli gate X, to recover $\left\|\begin{matrix}\alpha\\\beta\end{matrix}\right\|$. Analogously, if the message is $|10\rangle$, Bob applies the gate Z; if the message is $|11\rangle$, Bob applies $Z\,X = \left\|\begin{matrix}0 & 1\\-1 & 0\end{matrix}\right\|$. The end result is that the qubit $|\Psi\rangle$, which initially belonged to Alice, now belongs to Bob. The qubit has been teleported.

Since the collapse of the wavefunction occurs instantaneously, you may wonder if the quantum information contained in the qubit has, in fact, traveled at infinite speed; but the answer is no, because Alice also had to send Bob two bits by a classical channel, possibly electromagnetic signals, which cannot travel faster than light.

Note also that once Alice's qubit has been teleported, it has been destroyed. She has not been able to keep a "carbon copy." This is a consequence of the *no-cloning theorem*. Suppose there exists a magical linear unitary operator C which can copy Alice's special qubit $|\Psi\rangle = \alpha|0\rangle + \beta|1\rangle$ onto a "scratch" qubit, which we designate as $|\ \rangle$. This would involve the following hypothetical operation:

$$C(\alpha|0\rangle + \beta|1\rangle) \otimes |\ \rangle = (\alpha|0\rangle + \beta|1\rangle) \otimes (\alpha|0\rangle + \beta|1\rangle) =$$
$$\alpha^2|00\rangle + \alpha\beta|01\rangle + \beta\alpha|10\rangle + \beta^2|11\rangle. \qquad (6.43)$$

But C should also be able to clone a basis qubit: $C|0\rangle \otimes |\ \rangle = |0\rangle \otimes |0\rangle = |00\rangle$ and $C|1\rangle \otimes |\ \rangle = |1\rangle \otimes |1\rangle = |11\rangle$, so that we should also have

$$C(\alpha|0\rangle + \beta|1\rangle) \otimes |\ \rangle = \alpha|00\rangle + \beta|11\rangle. \qquad (6.44)$$

But the two alternative formulas for $C|\Psi\rangle$ are inconsistent, so that the attempted cloning must be impossible. This proves the no-cloning theorem.

6.8   The Toffoli Logic Gate

The circuit for the Toffoli gate is shown in Fig. 6.31. In the classical Toffoli gate, there are two control bits a and b, which remains unaltered. The third bit c flips if, and only if, $a = b = 1$, as shown in Table 6.11. This gate is also called a controlled–controlled-NOT gate (CCNOT).

Fig. 6.31   Toffoli gate

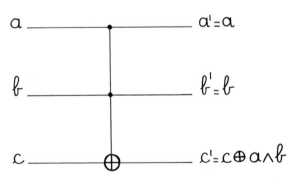

Table 6.11   Toffoli gate

a	b	c	a'	b'	c'
0	0	0	0	0	0
0	0	1	0	0	1
0	1	0	0	1	0
0	1	1	0	1	1
1	0	0	1	0	0
1	0	1	1	0	1
1	1	0	1	1	1
1	1	1	1	1	0

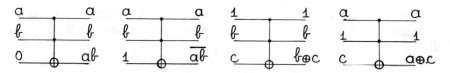

Fig. 6.32 Different gates constructed using Toffoli gate; *left* to *right* AND, NAND, XOR, XOR

Fig. 6.33 NOT gate obtained from Toffoli gate

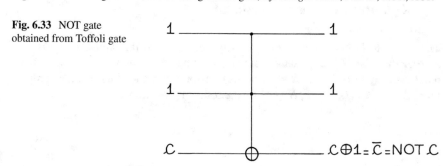

We turn now to the quantum Toffoli gate. The circuit is the same as that for the classical gate, but the bits $a, b, c \oplus ab$ are now replaced by the corresponding qubits $|a\rangle, |b\rangle, |c \oplus ab\rangle$ of the computational basis. From Table 6.11, we can find the unitary matrix representing the quantum Toffoli gate:

$$U_{\text{Tof}} = \begin{Vmatrix} 1 & 0 & 0 & 0 & 0 & 0 & 0 & 0 \\ 0 & 1 & 0 & 0 & 0 & 0 & 0 & 0 \\ 0 & 0 & 1 & 0 & 0 & 0 & 0 & 0 \\ 0 & 0 & 0 & 1 & 0 & 0 & 0 & 0 \\ 0 & 0 & 0 & 0 & 1 & 0 & 0 & 0 \\ 0 & 0 & 0 & 0 & 0 & 1 & 0 & 0 \\ 0 & 0 & 0 & 0 & 0 & 0 & 0 & 1 \\ 0 & 0 & 0 & 0 & 0 & 0 & 1 & 0 \end{Vmatrix}. \tag{6.45}$$

To build a circuit for the logic sum $a\mathrm{OR}b$, recall the fourth De Morgan law

$$a\mathrm{OR}b = a \vee b = \overline{(\overline{a} \wedge \overline{b})}. \tag{6.46}$$

Theorem 6.1 *The Toffoli gate is universal: NOR, AND, and OR gates can be constructed using only CCNOT gates.*

Proof The AND gate can be obtained by setting $c = 0$, since $0 \oplus a \wedge b = a \wedge b = a$ AND b. The NAND gate can be obtained by setting $c = 1$, since $1 \oplus a \wedge b = \overline{a \wedge b} = a$ NAND b. The XOR gate can be obtained by setting either $a = 1$ or $b = 1$ (see the third and fourth circuits of Fig. 6.32). The one-bit NOT gate is equivalent to the Toffoli gate with $a = b = 1$ (see Fig. 6.33). Thus the theorem is proved.

Since the Toffoli gate is universal, by Theorem 6.1, all classical computations can, in principle, be performed by a quantum computer (apart from inevitable practical difficulties).

6.9 Quantum Fourier Transform

The quantum Fourier transform is closely analogous to the well-known *discrete Fourier transform* (DFT), independent of quantum computation. A set of N complex numbers f_j ($j = 0, 1, \ldots, N - 1$) can be transformed into another set of N complex numbers F_k ($k = 0, 1, \ldots, N - 1$) by the following relations:

$$F_k = \frac{1}{\sqrt{N}} \sum_{j=0}^{N-1} f_j e^{2\pi i k j / N}. \tag{6.47}$$

The inverse DFT is then given by $f_j = \frac{1}{\sqrt{N}} \sum_{k=0}^{N-1} F_k e^{-2\pi i j k / N}$. Note that if the f_j are regarded as coordinates of a vector $\mathbf{f}$ and the F_k, coordinates of a vector $\mathbf{F}$, then we can write $\mathbf{F} = \text{DFT} \, \mathbf{f}$, where DFT acts as a unitary operator. A *quantum Fourier transform* QFT is equivalent to a DFT on the *amplitudes* of a quantum state, whereby

$$\text{QFT} \sum_{j=0}^{N-1} \alpha_j |j\rangle = \sum_{k=0}^{N-1} \beta_k |k\rangle, \quad \text{with} \quad \beta_k = \frac{1}{\sqrt{N}} \sum_{j=0}^{N-1} \alpha_j e^{2\pi i k j / N}. \tag{6.48}$$

This is reminiscent of the transformation of a Schrödinger wavefunction $\psi(x)$ to momentum representation $\phi(p)$.

A quantum Fourier transform can be carried out by a series of unitary operations on the computational basis $\{|j\rangle\}$. The result is a new basis set, composed of linear combinations of the original basis vectors. Let N be a positive integer; the complex number $\omega_N = e^{2\pi i / N}$ can be represented as a unit vector in the complex plane, making an angle $2\pi/N$ with the real axis. The quantum Fourier transform is then represented by the following matrix:

$$U_{jk} = U_{kj} = \frac{1}{\sqrt{N}} \omega_N^{jk} \quad \text{for} \quad j, k = 0, 1, 2, \ldots, N - 1. \tag{6.49}$$

The action of U on the vector $|j\rangle$ is then given by

$$U|j\rangle = \sum_{k=0}^{N-1} U_{kj} |k\rangle. \tag{6.50}$$

Let us prove that the matrix U is unitary:

$$(U^\dagger U)_{jl} = \sum_{k=0}^{N-1} U_{jk}^\dagger U_{kl} = \frac{1}{N}\sum_{k=0}^{N-1}\omega_N^{-jk+kl} = \frac{1}{N}\sum_{k=0}^{N-1}\omega_N^{k(l-j)},$$

$$j, l = 0, 1, 2, \ldots, N-1. \tag{6.51}$$

If $j = l$, $\omega_N^{k(l-j)} = \omega_N^0 = 1$ and the sum (6.51) is equal to $\frac{1}{N}N = 1$. If, instead, $j \neq l$, expression (6.51) becomes a sum of unit vectors from the origin to the vertices of a regular polygon of N sides, which sum to zero (see Fig. 6.34). Therefore $(U^\dagger U)_{jl} = \delta_{jl}$, so that $U^\dagger U = I$. It is interesting to note that the Fourier transform of every vector $|j\rangle$ is an equally weighted superposition of all the computational basis states, since, from Eq. (6.49), $|U_{jk}|^2 = 1/N$. A pure basis state in Hilbert space, say an energy eigenstate, exhibits a behavior somewhat similar to a classical system, in that the measurement of energy gives a reproducible result. By contrast, Fourier-transformed states are maximal superpositions, and results of measurement are highly probabilistic; this might be described as "extreme quantum" behavior.

Further manipulations take a more compact form if N equals a power of 2: $N = 2^n$, where n is integer. Let $|j\rangle$ $(j = 0, 1, \ldots, N-1)$ denote an element of the computational basis for a system of N qubits; then the following theorem holds:

Theorem 6.2 *The state $U|j\rangle$ can be written as the product state:*

$$U|j\rangle = \prod_{k=0}^{N-1} U_{kj}|k\rangle = \frac{1}{\sqrt{2^n}} \otimes_{r=0}^{n-1}\left(|0\rangle + \omega_N^{2^r j}|1\rangle\right). \tag{6.52}$$

Proof We will demonstrate the theorem for the case $n = 3$ ($N = 8$), but the argument can easily be extended to arbitrary n. Write k as the binary number $k_1 k_2 k_3$, and the state $|k\rangle$ as the product state $|k_1\rangle \otimes |k_2\rangle \otimes |k_3\rangle$. Then $k = 4k_1 + 2k_2 + k_3$, and we have

$$U|j\rangle = \frac{1}{\sqrt{8}}\sum_{k=0}^{7}\omega_8^{kj}|k\rangle =$$

$$\frac{1}{\sqrt{8}}\sum_{k_1=0}^{1}\sum_{k_2=0}^{1}\sum_{k_3=0}^{1}\omega_8^{(4k_1+2k_2+k_3)j}|k_1\rangle \otimes |k_2\rangle \otimes |k_3\rangle =$$

$$\frac{1}{\sqrt{8}}\left(\sum_{k_1=0}^{1}\omega_8^{4k_1 j}|k_1\rangle\right) \otimes \left(\sum_{k_2=0}^{1}\omega_8^{2k_2 j}|k_2\rangle\right) \otimes \left(\sum_{k_3=0}^{1}\omega_8^{k_3 j}|k_3\rangle\right) =$$

$$\frac{1}{\sqrt{8}}\left(|0\rangle + \omega_8^{4j}|1\rangle\right) \otimes \left(|0\rangle + \omega_8^{2j}|1\rangle\right) \otimes \left(|0\rangle + \omega_8^{j}|1\rangle\right). \tag{6.53}$$

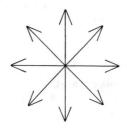

Fig. 6.34 The vectors representing the Nth roots of unity in the complex plane sum to zero: $\sum_{k=0}^{N-1} e^{2\pi i k/N} = 0$. These could be the vertices of an N-pointed star polygon

Thus the theorem is proved for the case $n = 3$.

Writing j as the binary number $j_1 j_2 j_3$, recall that the basis state $|j\rangle$ can be written as the product

$$|j\rangle = |j_1\rangle \otimes |j_2\rangle \otimes |j_3\rangle. \tag{6.54}$$

Noting that $\omega_8^8 = 1$, it is implied that

$$\begin{aligned}
\omega_8^{4j} &= \omega_8^{4(4j_1+2j_2+j_3)} = \omega_8^{4j_3}, \\
\omega_8^{2j} &= \omega_8^{2(4j_1+2j_2+j_3)} = \omega_8^{4j_2+2j_3}, \\
\omega_8^{j} &= \omega_8^{4j_1+2j_2+j_3}.
\end{aligned} \tag{6.55}$$

Thus Eq. (6.53) simplifies to

$$U|j\rangle = \frac{1}{\sqrt{8}}\left(|0\rangle + \omega_8^{4j_3}|1\rangle\right) \otimes \left(|0\rangle + \omega_8^{4j_2+2j_3}|1\rangle\right) \otimes \left(|0\rangle + \omega_8^{4j_1+2j_2+j_3}|1\rangle\right). \tag{6.56}$$

The last expression will help us to find a circuit implementing the quantum Fourier transform. Using the original form for the matrix U_{kj} (Eq. 6.49) and the identities

$$\omega_8^8 = 1, \quad \omega_8^9 = \omega_8, \quad \omega_8^{10} = \omega_8^2, \quad \omega_8^{11} = \omega_8^3, \quad \text{etc.,} \tag{6.57}$$

we find (for simplicity, writing $\omega_8 = \omega$)

$$U = \frac{1}{\sqrt{8}}\begin{Vmatrix}
1 & 1 & 1 & 1 & 1 & 1 & 1 & 1 \\
1 & \omega & \omega^2 & \omega^3 & \omega^4 & \omega^5 & \omega^6 & \omega^7 \\
1 & \omega^2 & \omega^4 & \omega^6 & 1 & \omega^2 & \omega^4 & \omega^6 \\
1 & \omega^3 & \omega^6 & \omega & \omega^4 & \omega^7 & \omega^2 & \omega^5 \\
1 & \omega^4 & 1 & \omega^4 & 1 & \omega^4 & 1 & \omega^4 \\
1 & \omega^5 & \omega^2 & \omega^7 & \omega^4 & \omega & \omega^6 & \omega^3 \\
1 & \omega^6 & \omega^4 & \omega^2 & 1 & \omega^6 & \omega^4 & \omega^2 \\
1 & \omega^7 & \omega^6 & \omega^5 & \omega^4 & \omega^3 & \omega^2 & \omega
\end{Vmatrix}. \tag{6.58}$$

An efficient realization of a circuit for implementing the quantum Fourier transform circuit is shown in Fig. 6.35. In order to prove that the circuit implements the operator U in (6.58), we begin with the following.

Lemma 6.1 *Partial circuit 1 of Fig. 6.36 sends the input state $|j_1\rangle \otimes |j_2\rangle$ into the state:*

$$\left(|0\rangle + \omega^{4j_1+2j_2}|1\rangle\right) \otimes |j_2\rangle. \tag{6.59}$$

Proof The first gate operates only on the first qubit leaving the second qubit unchanged; therefore the action on $|j_1\rangle \otimes |j_2\rangle$ is represented by the tensor product $H \otimes I$, which gives the following matrix:

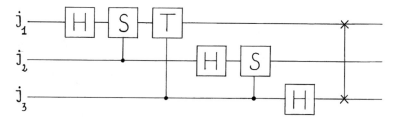

Fig. 6.35 Circuit implementing the quantum Fourier transform for $n = 3$, $N = 8$

Fig. 6.36 Partial circuit 1

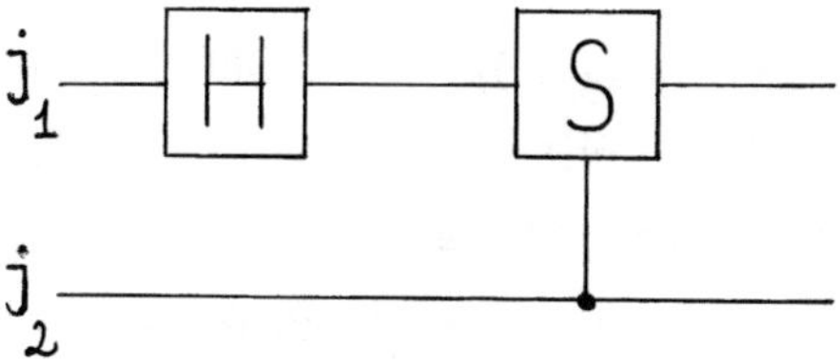

$$H \otimes I = \frac{1}{\sqrt{2}} \begin{Vmatrix} 1 & 1 \\ 1 & \omega^4 \end{Vmatrix} \otimes \begin{Vmatrix} 1 & 0 \\ 0 & 1 \end{Vmatrix} = \frac{1}{\sqrt{2}} \begin{Vmatrix} 1 & 0 & 1 & 0 \\ 0 & 1 & 0 & 1 \\ 1 & 0 & \omega^4 & 0 \\ 0 & 1 & 0 & \omega^4 \end{Vmatrix}. \tag{6.60}$$

Note that $\omega^4 = -1$ and $\omega^2 = i$. The controlled-S' gate[5] is represented by the matrix (see Eq. (6.28)):

$$S' = \begin{Vmatrix} 1 & 0 & 0 & 0 \\ 0 & 1 & 0 & 0 \\ 0 & 0 & 1 & 0 \\ 0 & 0 & 0 & \omega \end{Vmatrix}. \tag{6.61}$$

Denoting by V, the product of the matrices $S'(H \otimes I)$, we find

$$V = \frac{1}{\sqrt{2}} \begin{Vmatrix} 1 & 0 & 1 & 0 \\ 0 & 1 & 0 & 1 \\ 1 & 0 & \omega^4 & 0 \\ 0 & \omega^2 & 0 & \omega^6 \end{Vmatrix}. \tag{6.62}$$

Thus V maps the states of the computational basis into the states (omitting the $\otimes$ symbol):

---

[5]For a single-qubit phase-shift gate, $R(\phi) = \begin{Vmatrix} 1 & 0 \\ 0 & e^{i\phi} \end{Vmatrix}$, which includes S, T, and Z, we find $R(\phi)|0\rangle = |0\rangle$, $R(\phi)|1\rangle = e^{i\phi}|1\rangle$. For the corresponding two-qubit controlled gate $R'(\phi)$, it follows that $R'(\phi)|00\rangle = |00\rangle$, $R'(\phi)|01\rangle = |01\rangle$, $R'(\phi)|10\rangle = |10\rangle$, $R'(\phi)|11\rangle = e^{i\phi}|11\rangle$, whichever is the control qubit.

Fig. 6.37 Partial circuit 2

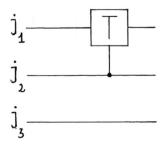

$$
\begin{aligned}
V\big(|0\rangle|0\rangle\big) &= \tfrac{1}{\sqrt{2}}\big(|0\rangle|0\rangle + |1\rangle|0\rangle\big) = \tfrac{1}{\sqrt{2}}\big(|0\rangle + |1\rangle\big)|0\rangle, \\
V\big(|0\rangle|1\rangle\big) &= \tfrac{1}{\sqrt{2}}\big(|0\rangle|1\rangle + \omega^2|1\rangle|1\rangle\big) = \tfrac{1}{\sqrt{2}}\big(|0\rangle + \omega^2|1\rangle\big)|1\rangle, \\
V\big(|1\rangle|0\rangle\big) &= \tfrac{1}{\sqrt{2}}\big(|0\rangle|0\rangle + \omega^4|1\rangle|0\rangle\big) = \tfrac{1}{\sqrt{2}}\big(|0\rangle + \omega^4|1\rangle\big)|0\rangle, \\
V\big(|1\rangle|1\rangle\big) &= \tfrac{1}{\sqrt{2}}\big(|0\rangle|1\rangle + \omega^6|1\rangle|1\rangle\big) = \tfrac{1}{\sqrt{2}}\big(|0\rangle + \omega^6|1\rangle\big)|1\rangle,
\end{aligned}
\tag{6.63}
$$

as in the four cases $4j_1 + 2j_2 = 0, 2, 4, 6$, respectively; thus the lemma is proved.

Let us consider now the inverted controlled-T gate of Fig. 6.38. If the partial circuit had been the circuit 2 of Fig. 6.37, then the corresponding matrix would be

$$
\begin{Vmatrix}
1 & 0 & 0 & 0 \\
0 & 1 & 0 & 0 \\
0 & 0 & 1 & 0 \\
0 & 0 & 0 & \omega
\end{Vmatrix}
\otimes
\begin{Vmatrix}
1 & 0 \\
0 & 1
\end{Vmatrix}
=
\begin{Vmatrix}
1 & 0 & 0 & 0 & 0 & 0 & 0 & 0 \\
0 & 1 & 0 & 0 & 0 & 0 & 0 & 0 \\
0 & 0 & 1 & 0 & 0 & 0 & 0 & 0 \\
0 & 0 & 0 & 1 & 0 & 0 & 0 & 0 \\
0 & 0 & 0 & 0 & 1 & 0 & 0 & 0 \\
0 & 0 & 0 & 0 & 0 & 1 & 0 & 0 \\
0 & 0 & 0 & 0 & 0 & 0 & \omega & 0 \\
0 & 0 & 0 & 0 & 0 & 0 & 0 & \omega
\end{Vmatrix}.
\tag{6.64}
$$

To get the correct result, just swap the j_2, j_3 qubits:

$$
\begin{Vmatrix}
1 & 0 & 0 & 0 & 0 & 0 & 0 & 0 \\
0 & 1 & 0 & 0 & 0 & 0 & 0 & 0 \\
0 & 0 & 1 & 0 & 0 & 0 & 0 & 0 \\
0 & 0 & 0 & 1 & 0 & 0 & 0 & 0 \\
0 & 0 & 0 & 0 & 1 & 0 & 0 & 0 \\
0 & 0 & 0 & 0 & 0 & \omega & 0 & 0 \\
0 & 0 & 0 & 0 & 0 & 0 & 1 & 0 \\
0 & 0 & 0 & 0 & 0 & 0 & 0 & \omega
\end{Vmatrix}.
\tag{6.65}
$$

The matrix (6.65) maps the state $|j_1\rangle|j_2\rangle|j_3\rangle$ into the state $\big(|0\rangle + \omega^{j_3}|1\rangle\big)|j_2\rangle|j_3\rangle$. Therefore, the partial circuit of Fig. 6.38 maps $|j_1\rangle$ into the qubit

Fig. 6.38 Partial circuit 3

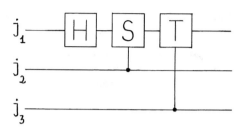

$$\frac{1}{\sqrt{2}}\left(|0\rangle + \omega^{4j_1+2j_2+j_3}|1\rangle\right). \tag{6.66}$$

Consider now the remaining part of the circuit, partial circuit 4 of Fig. 6.39. Making use of the Lemma 6.1, we see that $|j_2\rangle$ is mapped into

$$\frac{1}{\sqrt{2}}\left(|0\rangle + \omega^{4j_2+2j_3}|1\rangle\right). \tag{6.67}$$

Finally, applying H to the third qubit, we get the following output:

$$\frac{1}{\sqrt{2}}\left(|0\rangle + \omega^{4j_3}|1\rangle\right). \tag{6.68}$$

The tensor product of Eqs. (6.66)–(6.68) is, using Eq. (6.55),

$$\frac{1}{\sqrt{8}}\left(|0\rangle + \omega^{4j_1+2j_2+j_3}|1\rangle\right) \otimes \left(|0\rangle + \omega^{4j_2+2j_3}|1\rangle\right) \otimes \left(|0\rangle + \omega^{4j_3}|1\rangle\right)$$
$$= \frac{1}{\sqrt{8}}\left(|0\rangle + \omega^{j}|1\rangle\right) \otimes \left(|0\rangle + \omega^{2j}|1\rangle\right) \otimes \left(|0\rangle + \omega^{4j}|1\rangle\right). \tag{6.69}$$

Using (6.57) in (6.69), we see that the order of the first and third qubits is reversed. After applying the SWAP gate, we obtain the final result:

$$U|j\rangle = \frac{1}{\sqrt{8}}\left(|0\rangle + \omega^{4j}|1\rangle\right) \otimes \left(|0\rangle + \omega^{2j}|1\rangle\right) \otimes \left(|0\rangle + \omega^{j}|1\rangle\right). \tag{6.70}$$

Thus, the circuit of Fig. 6.35 implements the quantum Fourier transform in the case $n = 3$, $N = 8$. The circuit can be generalized to arbitrary n, using $\omega_N = e^{2\pi i/2^n}$.

Fig. 6.39 Partial circuit 4

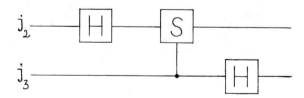

6.9.1 Phase Estimation

An important application of the quantum Fourier transform is *phase estimation*. Let U be a $2^m \times 2^m$ unitary matrix and $|u\rangle \in \mathbb{C}^{2^m}$, an eigenvector. Suppose that both U and $|u\rangle$ are known. We can then write

$$U |u\rangle = e^{i\varphi} |u\rangle, \tag{6.71}$$

recalling that the eigenvalues of a unitary matrix are complex numbers of unit magnitude. The objective now is to estimate the phase φ to an accuracy of n bits. The method is to subject the first n qubits of the eigenvector $|u\rangle$ to a series of controlled operators, involving powers of U, followed by an inverse quantum Fourier transform. As shown in Fig. 6.40, the input of the first register consists of n qubits, all prepared in the state $|0\rangle$. The input of the second register contains the row of m qubits, which represent the state $|u\rangle$. For simplicity, the circuit implementing the phase estimation shown in Fig. 6.40 pertains to the particular case $m = 3$, $n = 3$. The H gates map the qubits $|0\rangle$ into $\frac{1}{\sqrt{2}}(|0\rangle + |1\rangle)$; then the controlled U^{2^j} (for $j = 0, 1, 2$) maps the qubits $|0\rangle + |1\rangle$ into the qubits $|0\rangle + e^{i2^j\varphi}|1\rangle$. In this way the output of the first register is the product state:

$$|\text{OUT}\rangle = \frac{1}{\sqrt{8}}\left(|0\rangle + e^{4i\varphi}|1\rangle\right) \otimes \left(|0\rangle + e^{2i\varphi}|1\rangle\right) \otimes \left(|0\rangle + e^{i\varphi}|1\rangle\right). \tag{6.72}$$

Denoting by y the binary number $[y_2 y_1 y_0]_2 = 4y_2 + 2y_1 + y_0$, and by $|y\rangle$, the corresponding state in the computational basis, the state $|\text{OUT}\rangle$ can be written as

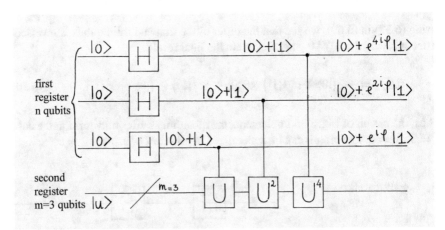

Fig. 6.40 A circuit implementing phase estimation

$$|OUT\rangle = \frac{1}{\sqrt{8}} \sum_{y_2=0}^{1} \sum_{y_1=0}^{1} \sum_{y_0=0}^{1} e^{i(4y_2+2y_1+y_0)\varphi} |y_2\rangle \otimes |y_1\rangle \otimes |y_0\rangle = \frac{1}{\sqrt{8}} \sum_{y=0}^{7} e^{i\varphi y} |y\rangle.$$

(6.73)

In the general case of n qubits, we would have obtained

$$|OUT\rangle = \frac{1}{\sqrt{2^n}} \sum_{y=0}^{2^n-1} e^{i\varphi y} |y\rangle.$$ (6.74)

Given the integer n, the real number $\frac{\varphi}{2\pi}$ $(0 \leq \varphi < 2\pi)$ can be written as a multiple of $\frac{1}{2^n}$ plus a remainder:

$$\frac{\varphi}{2\pi} = \frac{p}{2^n} + \delta,$$ (6.75)

where p is an integer $(0 \leq p < 2^n)$ and

$$|\delta| \leq \frac{1}{2^{n+1}}.$$ (6.76)

In essence, p gives an estimation of the phase accurate to 1 part in 2^n. Note that δ can be positive or negative (or zero). Consider the following two cases:

(1) $\delta = 0$. Then the state $|OUT\rangle$ is given by

$$|OUT\rangle = \frac{1}{\sqrt{2^n}} \sum_{y=0}^{2^n-1} e^{2\pi i p y/2^n} |y\rangle = \frac{1}{\sqrt{2^n}} \sum_{y=0}^{2^n-1} \omega_{2^n}^{py} |y\rangle.$$ (6.77)

The angle φ can be found by applying the *inverse* of the quantum Fourier transform U_F, with the matrix

$$(U_F^{-1})_{jk} = (U_F^\dagger)_{jk} = \frac{1}{\sqrt{2^n}} \omega_{2^n}^{-jk}.$$ (6.78)

Using Dirac notation and the identities $\sum_{j=0}^{2^n-1} |j\rangle\langle j| = \sum_{k=0}^{2^n-1} |k\rangle\langle k| = I$, we can write

$$U_F^{-1}|OUT\rangle = \sum_{j=0}^{2^n-1} \sum_{k=0}^{2^n-1} |j\rangle \langle j|U_F^{-1}|k\rangle \langle k|OUT\rangle.$$ (6.79)

Since $\langle k|OUT\rangle = e^{i\varphi k} = \omega_{2^n}^{pk}$, we have

$$U_F^{-1}|OUT\rangle = \sum_{j=0}^{2^n-1} \sum_{k=0}^{2^n-1} |j\rangle (U_F^{-1})_{jk} \frac{1}{\sqrt{2^n}} \omega_{2^n}^{pk} = \frac{1}{2^n} \sum_{j=0}^{2^n-1} \sum_{k=0}^{2^n-1} \omega_{2^n}^{(p-j)k} |j\rangle. \quad (6.80)$$

If $j = p$, $\frac{1}{2^n} \sum_{k=0}^{2^n-1} 1 = 1$; if $j \neq p$ the sum $\sum_{k=0}^{2^n-1} \omega_{2^n}^{(p-j)k}$ vanishes because the unit vectors $\omega_{2^n}^{(p-j)k}$ form a star polygon in the complex plane, as shown in Fig. 6.34. Thus

$$U_F^{-1}|\text{OUT}\rangle = |p\rangle. \tag{6.81}$$

In the circuit of Fig. 6.40, the number p is represented by the bits $p_2 p_1 p_0$ labeling the qubits in the computational basis. Thus, the output of the "first register" gives the bits $p_2 p_1 p_0$. This simplified example shows how the circuit approximates the phase as $\varphi \simeq 2\pi p/2^n$ when $\delta = 0$.

(2) $\delta \neq 0$. From Eqs. (6.74) and (6.75) we get

$$|\text{OUT}\rangle = \frac{1}{\sqrt{2^n}} \sum_{y=0}^{2^n-1} e^{2\pi i(p/2^n+\delta)y}|y\rangle = \frac{1}{\sqrt{2^n}} \sum_{y=0}^{2^n-1} \omega_{2^n}^{ky} e^{2\pi i\delta y} |y\rangle. \tag{6.82}$$

Applying the inverse quantum Fourier transform,

$$U_F^{-1}|\text{OUT}\rangle = \frac{1}{2^n} \sum_{j=0}^{2^n-1} \sum_{k=0}^{2^n-1} \omega_{2^n}^{(p-j)k} e^{2\pi i\delta k} |j\rangle. \tag{6.83}$$

Projecting this state onto the vector $|p\rangle$, only the term with $j = p$ survives: thus the probability amplitude $|c_p| = |\langle p|U_F^{-1}|\text{OUT}\rangle|$ is given by

$$|c_p| = \frac{1}{2^n} \left| \sum_{k=0}^{2^n-1} e^{2\pi i\delta k} \right| = \frac{1}{2^n} \left| \sum_{k=0}^{2^n-1} \alpha^k \right|, \tag{6.84}$$

where $\alpha = e^{2\pi i\delta}$.

The sum of the geometric series $\sum_{k=0}^{2^n-1} \alpha^k$ equals $\frac{\alpha^{2^n}-1}{\alpha-1}$. Therefore,

$$|c_p| = \frac{1}{2^n} \left| \frac{\alpha^{2^n} - 1}{\alpha - 1} \right| = \frac{1}{2^n} \left| \frac{e^{2\pi i\delta 2^n} - 1}{e^{2\pi i\delta} - 1} \right|. \tag{6.85}$$

For any value of θ (see Fig. 6.41),

$$|e^{i\theta} - 1| = 2 \left| \sin\frac{\theta}{2} \right|, \tag{6.86}$$

so that Eq. (6.85) becomes

$$|c_p| = \frac{1}{2^n} \left| \frac{\sin(\pi\delta 2^n)}{\sin(\pi\delta)} \right|. \tag{6.87}$$

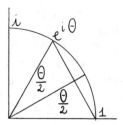

Fig. 6.41 Geometrical proof of Eq. (6.86); $|e^{i\theta} - 1|$ is the distance between the points $(\cos\theta, \sin\theta)$ and $(1, 0)$ of $\mathbb{R}^2$

A rigorous evaluation of the function (6.87) for small θ and large values of $N = 2^n$ would be tedious. However, since the denominator $\sin\pi\delta$ can be well approximated by $\pi\delta$, the resulting function $\frac{\sin(\pi\delta N)}{\pi\delta N}$ can be expressed as a series expansion:

$$|c_p| \simeq 1 - \frac{\pi^2\delta^2 N^2}{3!} + \frac{\pi^4\delta^4 N^4}{5!} - \cdots, \tag{6.88}$$

which is rapidly convergent since $|\delta|N \leq \frac{1}{2}$, by Eq. (6.76). From Eq. (6.88), we see that the width of the maximum at $\delta = 0$ is very small, of the order of $\frac{1}{N}$. Of course, since the numerator of (6.87) is periodic, a graph would show other extraneous maxima, corresponding to $|\delta| = \frac{1}{2N}, \frac{3}{2N}, \frac{5}{2N}, \cdots$.

6.10 Some Results from Number Theory

One of the conjectured applications of quantum computers is *quantum cryptography*, which would enable absolutely secure communication using unbreakable codes. All of the methods of quantum cryptography are based on results from number theory, possibly the "purest" branch of pure mathematics. To the dismay of many mathematicians, notably G.H. Hardy, number theory has surprisingly turned out to have a significant number of practical applications.

6.10.1 The Euclidean Algorithm

Given two positive integers a, n, we denote by $\gcd(a, n)$ the greatest common divisor of a, n. For example, $\gcd(8, 12) = 4$, $\gcd(24, 60) = 12$, $\gcd(8, 27) = 1$.

Definition 6.2 Two numbers a, n are said to be *coprime* if they do not have any common divisors. This is equivalent to saying that $\gcd(a, n) = 1$.

For example, the pairs (8, 27), (9, 28), (13, 19) consist of coprime numbers. Two *prime* numbers are always coprime. A method for finding the gcd of two integers has been known since antiquity, for 2300 years and possibly longer. This is the *Euclidean algorithm*, based on successive divisions with remainder. Let n and a be two integers, with $n > a > 0$. Let q_1 equal the number of times n contains a; thus

$$n = q_1 a + r_1 \text{ with } 0 \le r_1 < a. \qquad (6.89)$$

If the *remainder* r_1 vanishes, n is a *multiple* of a and $\gcd(a, n) = a$. But if r_1 does not vanish, we perform a second division, dividing a by r_1. Every factor that divides n and a, also divides r_1, since $r_1 = n - q_1 a$. But then $\gcd(r_1, a) = \gcd(a, n)$; since $r_1 < a < n$, finding $\gcd(r_1, a)$ is easier than finding $\gcd(a, n)$. Once performed, the division of a by r_1 gives the new remainder r_2:

$$a = q_2 r_1 + r_2 \text{ with } 0 \le r_2 < r_1 \qquad (6.90)$$

and again if $r_2 = 0$, $\gcd(r_1, a) = \gcd(a, n) = r_1$. But even if $r_2 \ne 0$, the task is easier, and

$$\gcd(r_2, r_1) = \gcd(r_1, a) = \gcd(a, n). \qquad (6.91)$$

Iterating the procedure, we have $r_1 = q_3 r_2 + r_3$ with $0 \le r_3 < r_2 < r_1$, etc. Finally, the value of $\gcd(a, n)$ is given by the last nonzero remainder. For example, you want to find the $\gcd(9207, 4203)$. From Table 6.12, we see that $r_4 = 0$. Thus $\gcd(9207, 4203) = r_3 = 9$.

6.10.2   Bezout's Lemma

This is a byproduct of the Euclidean algorithm for the gcd of two integers. Suppose a, b are two positive integers; then there exist two integers X, Y (one positive, one negative) such that

$$\gcd(a, b) = Xa + Yb. \qquad (6.92)$$

Proof Denote by Σ the set of integers $Xa + Yb$, with X, Y integers. Now carry out the procedure given in Table 6.12, for the case $r_4 = 0$. In the general case, the sequence $r_1 > r_2 > r_3 \ldots$ cannot be infinite since $r_i \ge 0$ ($i = 1, 2, 3, \ldots$). Let us prove that $r_3 = \gcd(a, b) \in \Sigma$. Note that Σ is a linear set: any linear combination of elements of Σ with integer coefficients also belongs to Σ. From Table 6.12 (1), it follows that $r_1 = b - q_1 a \in \Sigma$. From (2), $r_2 = a - q_2 r_1 \in \Sigma$. From (3), $r_3 = r_1 - q_3 r_2 \in \Sigma$. Since, in our case, $r_4 = 0$, $r_3 = \gcd(a, b)$. This line of reasoning can

Table 6.12 Euclidean algorithm	(1)	$n = q_1 a + r_1$	$9207 = 4203 \times 2 + 801$
	(2)	$a = q_2 r_1 + r_2$	$4203 = 801 \times 5 + 198$
	(3)	$r_1 = q_3 r_2 + r_3$	$801 = 198 \times 4 + 9$
	(4)	$r_2 = q_4 r_3 + 0$	$198 = 9 \times 22 + 0$

be generalized to any number of remainders $r_1, r_2, r_3, \ldots$, which proves *Bezout's lemma*.[6]

6.10.3 Modular Arithmetic

Think about a clock display: the motion of the hour hand is periodic, because if s denotes the "distance" traveled, and r the position (hour) of the hand, we can write (taking 1 h as unit) as

$$s = r + 12k, \qquad k \geq 0, \tag{6.93}$$

where k is an integer. This is illustrated in Fig. 6.42. For example, the values $s = 14$, $s = 26$, and $s = 38$ correspond to the same position $r = 2$. Eq. (6.93) can also be written as

$$s = r \,(\text{mod } 12) \qquad \text{or} \qquad s \cong r \ (\text{mod } 12). \tag{6.94}$$

We say "s is equal to r modulo 12" or "s is *equivalent* (or *congruent*) to r modulo 12," and we will use the symbol $\cong$ in place of $=$ to denote this equivalence. We led to consider *clock arithmetic* when carrying out the usual operations of sum and product of the paths s, and then reading the result on the clock display. For example, we can write

$$3 + 9 \cong 0(\text{mod } 12), \ \ 3 + 8 \cong 11(\text{mod } 12), \ \ 6 + 15 \cong 9(\text{mod } 12),$$
$$4 \times 4 \cong 4(\text{mod } 12), \ \ 4 \times 5 \cong 8(\text{mod } 12), \ \text{etc.} \tag{6.95}$$

These relations are special cases of *modular arithmetic*, which is a generalization of clock arithmetic in which 12 is replaced by an arbitrary integer $n > 0$. Given two integers a, b, we will write

$$a \cong b \ (\text{mod } n). \tag{6.96}$$

if the difference $a - b$ is an integer multiple of n. For example $5 + 5 \cong 3(\text{mod } 7)$, $15 \cong 39(\text{mod } 6)$, etc.

---

[6]It is also possible to prove that gcd(a, b) is the *smallest* positive integer belonging to Σ. Furthermore, given an element of Σ, the coefficients X, Y are not unique. For example, if $a = 2, b = 4$, the number 6 can be written as $6 = a + b$, $6 = 2b - a$, $6 = 3b - 3a$, etc.

Fig. 6.42 s and r are equivalent in modular arithmetic

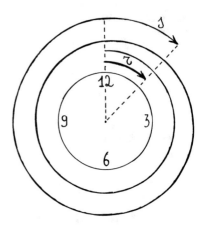

Many of the usual properties of addition, subtraction, and multiplication still hold in modular arithmetic: for example, if

$$a_1 \cong b_1 \;(\text{mod } n) \quad \text{and} \quad a_2 \cong b_2 \;(\text{mod } n), \tag{6.97}$$

then

$$a_1 + a_2 \cong b_1 + b_2 \;(\text{mod } n) \quad \text{and} \quad a_1 a_2 \cong b_1 b_2 \;(\text{mod } n). \tag{6.98}$$

Given an integer r, the set $\{r, r \pm n, r \pm 2n, r \pm 3n, \ldots\}$ is called the *congruence class* or *residue class* of r modulo n, and is denoted by $\bar{r}$ or $\bar{r}_n$. In the above case (clock arithmetic), the elements of r_{12} are all values of s such that $s - r$ is a multiple of 12. The set of these classes is denoted by $\mathbb{Z}_n$ (other notations are $\mathbb{Z}/n\mathbb{Z}$ or $\mathbb{Z}/n$). For example, $\bar{3}_{12} = \{3, 15, -9, 27, -21, \ldots\}$, $\bar{5}_4 = \{5, 9, 1, 13, -3, 17, \ldots\}$. Since a class is determined by one of its elements, we have

$$\mathbb{Z}_n = \{\bar{1}, \bar{2}, \ldots, \bar{n}\}. \tag{6.99}$$

Sums and products of two elements of $\mathbb{Z}_n$ must, of course, be performed modulo n. This is shown in the multiplication tables of $\mathbb{Z}_{12}$ and $\mathbb{Z}_7$.

From Table 6.13 we see that

$$\bar{1} \times \bar{1} \cong \bar{5} \times \bar{5} \cong \bar{7} \times \bar{7} \cong \overline{11} \times \overline{11} \cong \bar{1}(\text{mod } 12). \tag{6.100}$$

Therefore, the (multiplicative) inverses of $\bar{1}, \bar{5}, \bar{7}, \overline{11}$ modulo 12 exist and are given by

$$\bar{1}^{-1} \cong \bar{1}(\text{mod } 12), \; \bar{5}^{-1} \cong \bar{5}(\text{mod } 12), \; \bar{7}^{-1} \cong \bar{7}(\text{mod } 12), \; \overline{11}^{-1} \cong \overline{11}(\text{mod } 12). \tag{6.101}$$

However, the remaining elements $\bar{2}, \bar{3}, \bar{4}$, etc. do *not* admit inverses. The only numbers equal to 1 in the multiplication table of $\mathbb{Z}_{12}$ lie on the diagonal and correspond to $1 \times 1, 5 \times 5, 7 \times 7,$ and 11×11.

Table 6.13 Multiplication table for $\mathbb{Z}_{12}$

1	2	3	4	5	6	7	8	9	10	11	12
2	4	6	8	10	12	2	4	6	8	10	12
3	6	9	12	3	6	9	12	3	6	9	12
4	8	12	4	8	12	4	8	12	4	8	12
5	10	3	8	1	6	11	4	9	2	7	12
6	12	6	12	6	12	6	12	6	12	6	12
7	2	9	4	11	6	1	8	3	10	5	12
8	4	12	8	4	12	8	4	12	8	4	12
9	6	3	12	9	6	3	12	9	6	3	12
10	8	6	4	2	12	10	8	6	4	2	12
11	10	9	8	7	6	5	4	3	2	1	12
12	12	12	12	12	12	12	12	12	12	12	12

Consider now the multiplication Table 6.14 of $\mathbb{Z}_7$; here the situation is different. We have

$$\bar{1} \times \bar{1} \cong \bar{2} \times \bar{4} \cong \bar{3} \times \bar{5} \cong \bar{4} \times \bar{2} \cong \bar{5} \times \bar{3} \cong \bar{6} \times \bar{6} \cong \bar{1}(\text{mod } 7). \qquad (6.102)$$

All of the elements of $\mathbb{Z}_7$ admit inverses (except 7 since $7 \cong 0$):

$$\bar{1}^{-1} \cong \bar{1}(\text{mod } 7), \quad \bar{2}^{-1} \cong \bar{4}(\text{mod } 7), \quad \bar{3}^{-1} \cong \bar{5}(\text{mod } 7),$$
$$\bar{4}^{-1} \cong \bar{2} \ (\text{mod } 7), \quad \bar{5}^{-1} \cong \bar{3} \ (\text{mod } 7), \quad \bar{6}^{-1} \cong \bar{6} \ (\text{mod } 7). \qquad (6.103)$$

We are led to the following theorem:

Theorem 6.3 *Let n be an integer greater than 1. A necessary and sufficient condition for an element $\bar{a} \in \mathbb{Z}_n$ to admit an inverse modulo n is that a and n are coprime.*

Proof Let us first show that the condition is necessary. Assume, to the contrary, that a and n have a common factor n_1:

Table 6.14 Multiplication table for $\mathbb{Z}_7$

1	2	3	4	5	6	7
2	4	6	1	3	5	7
3	6	2	5	1	4	7
4	1	5	2	6	3	7
5	3	1	6	4	2	7
6	5	4	3	2	1	7
7	7	7	7	7	7	7

$$n = n_1 n_2, \qquad a = r n_1. \tag{6.104}$$

Since $n \cong 0 (\mathrm{mod}\ n)$ it follows that

$$\bar{n}_1 \overline{a n_2} \cong 0 (\mathrm{mod}\ n), \qquad \overline{a n_2} \cong \overline{r n_1} \bar{n}_2 \cong 0 (\mathrm{mod}\ n). \tag{6.105}$$

Multiplying the last equality by $\bar{a}^{-1}$, we would get $\bar{n}_2 \cong 0$. Thus a and n cannot have a common factor n_1. Let us now prove sufficiency: we assume that $\gcd(a, n) = 1$; from the discussion of the Euclidean algorithm we learned that $\gcd(a, n)$ can be written as a linear combination with integer coefficients X, Y:

$$Xa + Yn = 1, \qquad Xa = 1 - Yn. \tag{6.106}$$

It follows that $\overline{Xa} \cong \bar{1} (\mathrm{mod}\ n)$; thus the inverse $\overline{X} \cong \bar{a}^{-1} (\mathrm{mod}\ n)$ exists and the proof is complete.

From Theorem 6.3, it follows that if $n = p$ is a prime number, *all* the elements of $\mathbb{Z}_p$ admit an inverse: for example, we have seen from the multiplication table of $\mathbb{Z}_7$ that all classes $\bar{1}, \bar{2}, \bar{3}, \bar{4}, \bar{5}, \bar{6}$ admit an inverse modulo 7.

6.10.4 Fermat's Little Theorem

One of the fundamental results of modular arithmetic (derived by Pierre de Fermat, 1640) is the following[7]:

Theorem 6.4 *Let p be a prime number; for any integer a,*

$$a^p \cong a (\mathrm{mod}\ p). \tag{6.107}$$

If a is not a multiple of p (so that $\bar{a} \not\cong 0$), then

$$a^{p-1} \cong 1 (\mathrm{mod}\ p). \tag{6.108}$$

Proof (for simplicity, we will omit the overlining.) Identifying an element a and its class $\bar{a}$, we can regard a as an element of $\mathbb{Z}_p$. Consider the multiples of a:

$$a, 2a, 3a, \ldots, (p-1)a, \tag{6.109}$$

and let $r \leq p - 1$, $s \leq p - 1$, with two integers $r \neq s$; it is not possible that $ra \cong sa\ (\mathrm{mod}\ p)$, since this would imply $r \cong s\ (\mathrm{mod}\ p)$, so that $r = s$. Thus the classes

---

[7]This is not as famous as "Fermat's last theorem" (concerning $x^n + y^n = z^n$), but his "little" theorem has been far more useful in applications of number theory. The converse of Fermat's little theorem is not true. It holds for many values of p which are *not* prime; these are known as *Fermat pseudoprimes*.

Table 6.15 Multiplication table for $\mathbb{Z}_{12}^*$

1	5	7	11
5	1	11	7
7	11	1	5
11	7	5	1

a, $2a$, $3a$, $\ldots$, $(p-1)a$ are all different (we can imagine that the corresponding hours of a clock with p hours are all different): $ra \neq sa \pmod{p}$. It follows that the above classes coincide, in some order, with the classes $1, 2, 3, \ldots, (p-1)$, so that a, $2a$, $3a$, $\ldots$, $(p-1)a$ occupy *all* the hour marks of the clock display. Then there exists a permutation $k_1, k_2, \ldots, k_{p-1}$ of the numbers $1, 2, 3, \ldots, (p-1)$ such that

$$k_1 a \cong 1 \pmod{p}, \quad k_2 a \cong 2 \pmod{p}, \quad k_{p-1} a \cong (p-1) \pmod{p} \qquad (6.110)$$

Multiplying all these equalities, we get

$$k_1 k_2 k_{p-1} \cong 1, 2, 3, \ldots, (p-1) \pmod{p}, \qquad (6.111)$$

so that

$$(p-1)! \, a^{p-1} \cong (p-1)! \pmod{p}. \qquad (6.112)$$

Thus Eq. (6.108) is proved. Multiplying both sides of Eq. (6.108) by a, we get $a^p \cong a \pmod{p}$. For example, let $a = 3$, $p = 7$; from the multiplication Table 6.15 of $\mathbb{Z}_7$ we find

$$a \cong 3, \quad a^2 \cong 2, \quad a^3 \cong 6, \quad a^4 \cong 4, \quad a^5 \cong 5, \quad a^6 \cong 1. \qquad (6.113)$$

Fermat's little theorem clarifies the structure of $\mathbb{Z}_p$, when p is prime. What can we say about the structure of $\mathbb{Z}_n$, when n is an arbitrary integer? We know that in order for an element $a \in \mathbb{Z}_n$ to admit an inverse modulo n, it is necessary that a and n be *coprime*. Obviously then, we remove from $\mathbb{Z}_n$ all elements that have some common factor with n, and limit ourselves to the set $\mathbb{Z}_n^*$, of the numbers a coprime with n. More precisely, $\mathbb{Z}_n^*$ is the set of congruence classes $\bar{a}$ modulo n such that n and a are coprime. The set $\mathbb{Z}_n^*$ thus obtained is a "clock" more interesting than $\mathbb{Z}_n$ from the mathematical point of view. As we shall see in more detail below, the set $\mathbb{Z}_n^*$ is a *multiplicative group*. For example, let $n = 12$ and consider only the numbers 1, 5, 7, 11 that are coprime with 12. These numbers have the multiplication Table 6.15 (modulo 12):

This table can be obtained from that of $\mathbb{Z}_{12}$ simply by deleting rows and columns corresponding to multiples of 2 or 3. In this particular case, each element is its own inverse, although this is not true in general.

The number of elements of $\mathbb{Z}_n^*$ is called the *totient* or *Euler function*, denoted by $\phi(n)$. Note that the element 1 is, by definition, always included. For example, if $n = 12$, $\phi(n) = 4$. If p is *prime*, all the elements of $\mathbb{Z}_p$ are coprime with p and therefore $\phi(p) = p - 1$. For example, $\phi(7) = 6$. The multiplication table of $\mathbb{Z}_7^*$ can be obtained from the table of $\mathbb{Z}_7$ by deleting the last row and the last column. If

p is prime, the positive integers less than p^α and coprime to p^α are all the $p^\alpha - 1$ elements of $\mathbb{Z}_{p^\alpha}$ except the multiples $p, 2p, 3p, \ldots, (p^\alpha - 1)p$. For example, if $p = 3$, $\alpha = 2$ the elements not coprime with 9 are 3 and 6. If $p = 3$, $\alpha = 3$ the elements not coprime with 27 are $3, 6, 9, 12, 15, 18, 21, 24$. Thus $\phi(3^2) = 8 - 2 = 6$, $\phi(3^3) = 26 - 8 = 18$. In general the number of elements of $\mathbb{Z}_{p^\alpha}$ is $p^{\alpha-1}$; therefore,

$$\phi(p^\alpha) = (p^\alpha - 1) - (p^{\alpha-1} - 1) = p^\alpha - p^{\alpha-1}. \tag{6.114}$$

An important property of the totient function ϕ is that, if a and b are coprime, then (see next section)

$$\phi(ab) = \phi(a)\phi(b). \tag{6.115}$$

6.10.5   Chinese Remainder Theorem

This classic result is needed to prove Eq. (6.115):

Theorem 6.5 *Let $m_1, m_2, \ldots, m_n$ be positive integers that are mutually coprime, and let M be the product $M = m_1 m_2 \ldots m_n$. The system of equations*

$$\begin{aligned} x &= a_1 (\text{mod } m_1), \\ x &= a_2 (\text{mod } m_2), \\ &\vdots \\ x &= a_n (\text{mod } m_n), \end{aligned} \tag{6.116}$$

admits solutions. If x, y are two solutions, $x \cong y$ (mod M).

Proof (for simplicity, we consider the case $n = 3$.) Set $M_i = M/m_i, (i = 1, 2, \ldots, n)$; in this case, $M_1 = m_2 m_3$, $M_2 = m_1 m_3$ and $M_3 = m_1 m_2$. Note that m_i and M_i are relatively prime. Therefore, we can define the inverses M_i^{-1}, such that $M_i M_i^{-1} = 1$ (mod m_i): We can now verify that a solution to Eq. (6.116) can be expressed in the following form:

$$x = a_1 M_1^{-1}(\text{mod } M_1) + a_2 M_2^{-1}(\text{mod } M_2) + a_3 M_3 M_3^{-1}(\text{mod } M_3). \tag{6.117}$$

Re-expressing this as a congruence (mod m_1) we have

$$x \cong a_1 M_1 M_1^{-1}(\text{mod } m_1) + a_2 M_2 M_2^{-1}(\text{mod } m_1) + a_3 M_3^{-1}(\text{mod } m_1). \tag{6.118}$$

But $M_1 M_1^{-1}(\text{mod } m_1) = 1$, while the two last terms vanish (mod m_1), since each contains a factor m_1. Therefore,

$$x = a_1 (\text{mod } m_1). \tag{6.119}$$

Similarly, it can be verified that $x = a_2 (\mathrm{mod}\ m_2)$, $x = a_3 (\mathrm{mod}\ m_3)$. Now let x and y be two solutions of the system of Eq. (6.116). The difference $x - y$ satisfies the equations:

$$
\begin{aligned}
x - y &= 0(\mathrm{mod}\ m_1), \\
x - y &= 0(\mathrm{mod}\ m_2), \\
&\ \vdots \\
x - y &= 0(\mathrm{mod}\ m_n).
\end{aligned}
\tag{6.120}
$$

Then $x - y$ is a multiple of $m_1, m_2, \ldots, m_n$ and therefore a multiple of $M = m_1 m_2 \ldots m_n$. The theorem is proved.

As an example, suppose we want to solve the system of two equations:

$$
x = 2(\mathrm{mod}\ 12), \qquad x = 5(\mathrm{mod}\ 7). \tag{6.121}
$$

Then $m_1 = 12$, $m_2 = 7$, $M = 12 \times 7 = 84$, $M_1 = M/m_1 = 7$, $M_2 = M/m_2 = 12$. From the multiplication Table 6.13 for $\mathbb{Z}_{12}$ we see that $M_1^{-1} = 7^{-1}(\mathrm{mod}\ 12) = 7$. Since $12 \cong 5(\mathrm{mod}\ 7)$, from Table 6.14 for $\mathbb{Z}_7$, we find that $M_2^{-1} \cong 12^{-1} \cong 5^{-1} \cong 3(\mathrm{mod}\ 7)$. Then

$$
x = 2 \times 7 \times 7 + 5 \times 3 \times 12 = 278\ (\mathrm{mod}\ 84). \tag{6.122}
$$

We can remove multiples of 84 from the solution. Then we find

$$
x = 278 - 3 \times 84 = 26\ (\mathrm{mod}\ 84). \tag{6.123}
$$

Indeed, $26 = 2 + 12 \times 2 = 5 + 7 \times 3$.

Let us now prove Eq. (6.115) for the totient ϕ. For simplicity, we first consider the same case as the last example: in the "clock" $\mathbb{Z}_{12}^*$ there are four elements (1, 5, 7, 11), and in the "clock" $\mathbb{Z}_7^*$ there are six elements (1, 2, 3, 4, 5, 6), so that $\phi(12) = 4$ and $\phi(7) = 6$. There is a one-to-one correspondence between the pairs $a_1 \in \mathbb{Z}_{12}^*$, $a_2 \in \mathbb{Z}_7^*$ and the elements of the "big clock" $\mathbb{Z}_{12 \times 7}^* = \mathbb{Z}_{84}^*$. In fact, since 12 and 7 are coprime, if an element $\bar{y} \in \mathbb{Z}_{12 \times 7}^*$, this means that $\gcd(y, 12) = 1$, $\gcd(y, 7) = 1$, so $y \in \mathbb{Z}_{12}^*$ and $y \in \mathbb{Z}_7^*$, and vice versa, if $y \in \mathbb{Z}_{12}^*$, $y \in \mathbb{Z}_7^*$, $\gcd(y, 12) = \gcd(y, 7) = 1$, then $\bar{y} \in \mathbb{Z}_{12 \times 7}^*$. Some doubt might remain that given $a_1 \in \mathbb{Z}_{12}^*$, $a_2 \in \mathbb{Z}_7^*$, there are many elements $\bar{y} \in \mathbb{Z}_{12 \times 7}^*$ satisfying the following equations:

$$
y = a_1\ (\mathrm{mod}\ 17), \qquad y = a_2\ (\mathrm{mod}\ 7). \tag{6.124}
$$

But, by the Chinese remainder theorem, this is not possible. Since the number of possible pairs a_1, a_2 is $\phi(12)\phi(7)$, it follows that $\mathbb{Z}_{12 \times 7}^*$ has $\phi(12)\phi(7)$ elements. Clearly, the proof can be generalized to *any* pair a, b of coprime numbers, showing that Eq. (6.115) has general validity. If $n = p_1^{\alpha_1} p_2^{\alpha_2} \ldots p_k^{\alpha_k}$ is the prime factorization of an arbitrary integer n, Eqs. (6.114) and (6.115), imply

$$\phi(n) = \prod_{j=1}^{k} \phi(p_j^{\alpha_j}) = \prod_{j=1}^{k} \left(p_j^{\alpha_j} - p_j^{\alpha_j - 1} \right). \tag{6.125}$$

We will next discuss an important property of the totient function ϕ, which leads to a generalization of Fermat's little theorem. This involves some results from group theory.

6.10.6 More Group Theory

Recall that for a finite group G, the *order* of the group is the number of elements of G. A *subgroup* H is a proper subset of G which is also a group. A beautiful result due to Lagrange is the following:

Theorem 6.6 *If H is a subgroup of a finite group G, the order of H divides the order of G.*

Proof Let G be the group

$$G = \{e, g_2, g_3, \ldots, g_n\} \tag{6.126}$$

and H the subgroup

$$H = \{e, h_2, h_3, \ldots, h_m\}, \quad m < n, \tag{6.127}$$

where e denotes the identity element of H, which coincides with the identity element of G. Now let g be an element of G *not* belonging to H. The following set

$$Hg = \{g, h_2 g, h_3 g, \ldots, h_m g\} \tag{6.128}$$

is called a *right coset Hg*. The set

$$gH = \{g, gh_2, gh_3, \ldots, gh_m\} \tag{6.129}$$

is correspondingly called a *left coset gH*. Of course, in the case of a commutative group, the right cosets and the left cosets coincide.[8] Since we are dealing with commutative groups, $gH = Hg$, we can omit the modifiers "right" and "left." Two elements gh_j and gh_k ($j \neq k$) of a coset cannot coincide since $gh_j = gh_k$ would imply that $h_j = h_k$. Therefore, all cosets have m *distinct* elements. A coset is never a group since it does not contain the identity e (if $gh_j = e_j$, then $g = h_j^{-1} \in H$). The coset gH has void intersection with H, so that $gh_k \notin H$.

---

[8] More generally, this is true if H is a normal subgroup.

Fig. 6.43 Diagram showing a subgroup H of the group G and the cosets gH and $g'H$. The *dashed arrow* shows the mapping $f : gh_k \mapsto g'h_k$ between cosets

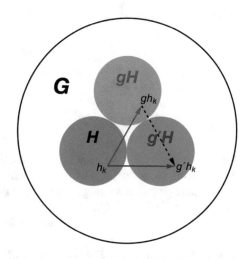

Now consider a second element $g' \in G$ *not* belonging to H; define the following mapping f from the coset gH to the coset $g'H$:

$$f : gH \to g'H. \tag{6.130}$$

Let us take an element $gh_k \in gH$ (see Fig. 6.43). We have

$$f(gh_k) = g'g^{-1}gh_k = g'h_k. \tag{6.131}$$

The inverse mapping f^{-1} exists since

$$f^{-1}(g'h_k) = gg'^{-1}g'h_k = gh_k. \tag{6.132}$$

Both gH and $g'H$ have m elements; therefore f is bijective (one-to-one). Furthermore, if $g'H$ and gH have a common element, they must coincide. Suppose $g'h_k = gh_j$, so $g' = gh_jh_k^{-1}$, and $g'h_r = g(h_jh_k^{-1}h_r)$. The right-hand side of the last equality $\in gH$ (since the expression in parenthesis $\in H$), therefore the left-hand side does too. Thus we have shown that $g'H = gH$. Evidently then, the group G can be partitioned into the subgroup H and all the cosets; both H and its cosets contain m elements. Thus, the order of G must be a multiple of m, which completes the proof of Lagrange's theorem. For example, consider the multiplication Table 6.16 of $\mathbb{Z}_9^*$, which consists of the elements $\{1, 2, 4, 5, 7, 8\}$ which are coprime to 9. The elements $\{1, 4, 7\}$ form a subgroup of 3 elements; and indeed 3 is a submultiple of 9.

The totient ϕ satisfies *Euler's totient theorem*, stated as follows:

Theorem 6.7 *If a, n are coprime positive integers, then*

$$a^{\phi(n)} \cong 1(\text{mod } n). \tag{6.133}$$

Table 6.16 Multiplication table for $\mathbb{Z}_9^*$

1	2	4	5	7	8
2	4	8	1	5	7
4	8	7	2	1	5
5	1	2	7	8	4
7	5	1	8	4	2
8	7	5	4	2	1

Proof The proof follows easily from Lagrange's theorem. Choose any $a \in \mathbb{Z}_n^*$, then the elements $\{a, a^2, a^3, \ldots, a^m\}$(mod n) form a *subgroup* of m elements of $\mathbb{Z}_n$. Therefore there exists an integer k such that $\phi(n) = km$, and we can write

$$a^{\phi(n)} = a^{km} \cong 1^k = 1 \tag{6.134}$$

and the theorem is proved. For example, for $\mathbb{Z}_9^*$, we choose $a = 2$, so that $2^{\phi(9)} = 2^6 = 64 \cong 1$(mod 9). When n is prime, say equal to p, Euler's totient theorem reduces to Fermat's little theorem, since $\phi(p) = p - 1$, $a^{p-1} \cong 1$(mod p).

6.10.7 Factorization of Large Numbers

While it is reasonably easy to decompose a number N, a few digits long, into its prime factors, the computational complexity increases superexponentially as N increases in size.[9] The following derivations are quite intricate and the reader has the option of proceeding directly to the conclusion and summary at the end of this subsection.

In principle, factorization is equivalent to "order finding."

Definition 6.3 Given $n > 0$, and $a \in \mathbb{Z}_n^*$, the least positive integer r such that $a^r \cong 1$(mod n) is called the "order of a modulo n."

Examples: if $n = 12$, the order of 5 is 2, since $5^2 = 25 \cong 1$(mod 12); if $n = 9$, the order of 7 is 3 since the powers of 7 are 7, $7^2 \cong 4$, $7^3 \cong 1$(mod 9). Clearly, the order of an element a divides the order of the group $\mathbb{Z}_n^*$, while the powers of a form a subgroup of $\mathbb{Z}_n^*$. The equation

$$x^2 \cong 1 (\text{mod } n) \tag{6.135}$$

has the "trivial solutions" $x \cong 1$(mod n) and $x \cong -1$(mod n). The relation between *factoring numbers* and *order finding* can be seen from the following lemma:

Lemma 6.2 *A nontrivial solution of Eq. (6.135) has the property that at least one of gcd(x-1,n) and gcd(x+1,n) divides n.*

---

[9]It is easy to multiply prime numbers but much more difficult to factorize a product. As a simple exercise, find the prime factors of 323, then check the result by multiplying the factors.

Proof Equation (6.135) is equivalent to

$$(x+1)(x-1) \cong 0 (\text{mod } n). \qquad (6.136)$$

Let us limit ourselves to the first "clock display." Having excluded the solutions $x = 1$ and $x = -1$, we have $0 < x - 1 < x + 1 < n$. Therefore, $\gcd(x - 1, n) < n$ and $\gcd(x + 1, n) < n$. Since Eq. (6.135) is equivalent to $(x + 1)(x - 1) = kn$, (with $k \geq 1$), $\gcd(x - 1, n)$ and $\gcd(x + 1, n)$ cannot both be equal to 1 because this would imply $n = 1$. Therefore at least one of the numbers, $\gcd(x - 1, n)$ or $\gcd(x + 1, n)$, is a nontrivial factor of n, possibly both. Example: Let $x = 11$, $n = 15$; $x + 1 = 12$ and $x - 1 = 10$. In fact, $\gcd(12, 15) = 3$ and $\gcd(10, 15) = 5$ are both nontrivial factors of 15. Of course, if n is a prime number, $\gcd(x - 1, n)$ and $\gcd(x + 1, n)$ are equal to 1 for all $1 < x < n - 1$. Therefore, if n is large, the search for a factor using *only* Eq. (6.136) can be very time consuming. We should therefore seek a more efficient probabilistic method.

Definition 6.4 A group G of order n is called *cyclic* if there is an element $g \in G$ such that $G = \{g, g^2, g^3, \ldots, g^n\}$, with $g^n = e$.

The element g is called a *generator* of the group. Clearly, a cyclic group is Abelian since $g^m g^n = g^n g^m$. The group $\mathbb{Z}_p^*$, where p is prime, is cyclic, with $p - 1$ elements $\{g, g^2, g^3, \ldots, g^{p-1}\}$, since $g^p = g$. Next, we need the following theorem.

Theorem 6.8 *Let p be an odd prime greater than 2, and α a positive integer. Then the group $\mathbb{Z}_{p^\alpha}^*$ is cyclic.*

An elegant proof of this beautiful Theorem is given in the appendix: for $p = 3$, $\alpha = 2$, the group $\mathbb{Z}_9^*$ consists of six elements 2, 2^2, 2^3, $2^4 \cong 7$, $2^5 \cong 5$, $2^6 \cong 1$. For an odd prime, the group $\mathbb{Z}_{p^\alpha}^*$, with the generator g, has $\phi(p^\alpha)$ elements: $\{g, g^2, g^3, \ldots, g^{\phi(p^\alpha)} \cong 1\}$.

In what follows, we are given two integers a, b, if a divides b we write $a|b$, if this is not true, we write $a \nmid b$. Consider an element $g^k \in \mathbb{Z}_{p^\alpha}^*$, and let r be the *order* of g^k modulo p^α. We need the following lemma.

Lemma 6.3 *The set $\mathbb{Z}_{p^\alpha}^*$ can be partitioned into two equal sets: the first consisting of the elements g^k with k odd and such that $2^d|r$; the second consisting of the elements g^k with k even and $2^d \nmid r$.*

Proof Consider the element g^k with k odd. From $(g^k)^r = g^{kr} \cong 1 (\text{mod } p^\alpha)$ and from $g^{\phi(p^\alpha)} \cong 1 \ (\text{mod } p^\alpha)$, it follows that $kr \geq \phi(p^\alpha)$ since $\phi(p^\alpha)$ is the *order* of $g \in \mathbb{Z}_{p^\alpha}^*$. Now if $kr > \phi(p^\alpha)$, we could perform the division of kr by $\phi(p^\alpha)$ and write $kr = q\phi(p^\alpha) + s$ with $0 \leq s < \phi(p^\alpha)$; it would follow that $g^{kr} = g^{q\phi(p^\alpha)} g^s \cong g^s \cong 1 \ (\text{mod } p^\alpha)$. Then $s = 0$ because $\phi(p^\alpha)$ is of the order of g. So $kr = q\phi(p^\alpha)$ and, since k is odd, the factor 2^d must be contained in r: $2^d|r$. Consider now the case when k is even. Then $g^{\frac{k}{2}\phi(p^\alpha)} \cong 1 (\text{mod } p^\alpha)$; since $\frac{\phi(p^\alpha)}{2} \geq r$, we divide $\frac{\phi(p^\alpha)}{2}$ by r and write $\frac{\phi(p^\alpha)}{2} = qr + s$, with $s < r$. It follows that $g^{k(qr+s)} \cong (g^k)^s \cong 1 (\text{mod } p^\alpha)$.

Thus, $s = 0$, since the order of g^k is r and $r | \frac{\phi(p^\alpha)}{2}$. If now $2^d | r$, from $\phi(p^\alpha) = 2qr$ it would follow that $\phi(p^\alpha)$ is a multiple of 2^{d+1}, contrary to the hypothesis. We conclude then that $2^d \nmid r$.

The simplest example is provided by the multiplication Table 6.17 of $\mathbb{Z}_9^*$. Since $9 = 3^2$, we have $p = 3, \alpha = 2$ $\phi(3^2) = 6$, $g = 2$, $2^d = 2$, $d = 1$. The order r of the element g^k (mod 9) is shown in Table 6.17 (of course, $2^4 \cong 7$ and $2^5 \cong 5$ modulo 9, etc.). From the table we see that if k is odd, r is even, and if k is even, r is odd; therefore if k is odd, $2^d = 2$ and $2|r$, if k is even $2^d \nmid r$. In the general case, choosing an element $g^k \in \mathbb{Z}_{p^\alpha}$ at random, the probability is $\frac{1}{2}$ that 2^d divides r, and $\frac{1}{2}$ that 2^d does not divide r.

The following theorem concerns the prime factorization $N = p_1^{\alpha_1} p_2^{\alpha_2} \ldots p_m^{\alpha_m}$ of an odd integer N. For simplicity, we will treat the case $m = 2$, but generalization to arbitrary values of m is straightforward.

Theorem 6.9 Let $N = p_1^{\alpha_1} p_2^{\alpha_2}$ with primes p_1, p_2. Choosing at random a number $x \in \mathbb{Z}_N^*$, the probability that the order r of x is even is $p \geq \frac{3}{4}$.

Proof Consider the following equations:

$$x = a_1 (\text{mod } p_1^{\alpha_1}), \qquad x = a_2 (\text{mod } p_2^{\alpha_2}). \qquad (6.137)$$

Applying the Chinese remainder theorem, a random choice $x \in \mathbb{Z}_N^*$ is equivalent to a random choice $a_1 \in \mathbb{Z}_{p_1^{\alpha_1}}^*$, $a_2 \in \mathbb{Z}_{p_2^{\alpha_2}}^*$. Denoting by r_1, r_2 the orders of a_1, a_2, respectively, let us prove first that $r_1 | r$ and $r_2 | r$. There exist three integers a, b, c such that

$$\begin{aligned} x^r &= 1 + p_1^{\alpha_1} p_2^{\alpha_2} = 1 + bp_1^{\alpha_1}, & b &= ap_2^{\alpha_2}, \\ x^r &= 1 + p_1^{\alpha_1} p_2^{\alpha_2} = 1 + cp_2^{\alpha_2}, & c &= ap_1^{\alpha_1}. \end{aligned} \qquad (6.138)$$

However $a_1^r \cong x^r (\text{mod } p_1^{\alpha_1})$ and $a_2^r \cong x^r (\text{mod } p_2^{\alpha_2})$. Therefore, from Eq. (6.138) it follows that

$$a_1^r \cong 1 (\text{mod } p_1^{\alpha_1}), \qquad a_2^r \cong 1 (\text{mod } p_2^{\alpha_2}). \qquad (6.139)$$

Then r must be a multiple of r_1 and r_2, that is, $r_1 | r$, $r_2 | r$. It follows that if r is *odd*, both r_1 and r_2 are *odd*. Now, what is the probability that r_1 is odd? At first sight, from Lemma 6.3 it would appear that this probability is $\frac{1}{2}$, but this is true only if $\phi(p_1^{\alpha_1})$ is even but not divisible by 2^d with $d > 1$. Note that the same property follows for $p_1 - 1$ which is a factor of $\phi(p_1^{\alpha_1}) = p_1^{(\alpha_1-1)}(p-1)$. If, for example, $2^d = 4$, and $4|(p-1)$, it is possible that r_1 is even but not a multiple of 4. In this way, the number

Table 6.17 Multiplication table for $\mathbb{Z}_9^*$

k	1	2	3	4	5	6
g^k (mod 9)	2	4	8	7	5	1
r	6	3	2	3	6	1
kr	6	6	6	12	30	6

Table 6.18 Multiplication
table for $\mathbb{Z}_5^*$

1	2	3	4
2	4	1	3
3	1	4	2
4	3	2	1

of cases for which r_1 is even would be greater than $\phi(p_1^{\alpha_1})/2$. A simple example showing this possibility is the following: choose $p_1 = 5$, $\alpha_1 = 1$, then $\phi(p_1) = 4$. The multiplication Table 6.18 shows that the orders of 1, 2, 3, 4 are, respectively, $r_1 = 1, r_2 = 4, r_3 = 4, r_4 = 2$. In this elementary example, we see that the order r is even in 3 cases out of 4, but only in 2 cases $4|r$. Because the argument is of general nature (applying also to $p_2^{\alpha_2}$ and r_2), we have

$$\text{Probability}\{r_1 \text{ odd}\} \leq \frac{1}{2}, \quad \text{Probability}\{r_2 \text{ odd}\} \leq \frac{1}{2} \qquad (6.140)$$

and therefore Probability$\{r$ odd$\} \leq \frac{1}{4}$, thus proving the theorem.

In conclusion and summary, to factorize a large number N, the following algorithm has a high probability of success:
(1) If N is even, a factor is 2 and we start from $N/2$.
(2) Randomly choose x in the interval $1 < x < N - 1$.
(3) Compute $y = \gcd(x, N)$; if $y > 1$, then y is a factor of N.
(4) If $y = 1$, x and N are coprime, then compute the order r of x modulo N.
(5) If r is odd, choose another x.
(6) If r is even, compute $\gcd(x^{r/2} - 1, N)$ and $\gcd(x^{r/2} + 1, N)$. According to Lemma 6.2, at least one of these two numbers is a nontrivial factor of N.

Two elementary examples: Example (1): $N = 21$, $p_1 = 3$, $p_2 = 7$. Table 6.19 shows, for each value of x in range $1 < x < 20$, the values of $r, r/2$, and $x^{r/2} \pmod{21}$, when r is even.

Note that for $x = 2, x^{r/2} \cong 8$, $x^{r/2} - 1 = 7$, and $7|N$; analogously for $x^{r/2} \cong 13$, $x^{r/2} + 1 = 14$, and $\gcd(14, 21) = 7$; instead, if $x^{r/2} \cong 20 \cong -1 \pmod{N}$, $x^{r/2} + 1 = N$ is a trivial factor of N. Example (2): $N = 35$, $p_1 = 5$, $p_2 = 7$. Table 6.20 shows, for $1 < x < 34$, the values of $r, r/2$ and $x^{r/2} \pmod{35}$, when r is even.

Table 6.19 Factorization for $N = 21$

x	2	4	5	8	10	11	13	16	17	19	20
r	6	3	6	2	6	6	2	3	6	6	2
$r/2$	3		3	1	3	3	1		3	3	1
$x^{r/2} \pmod{21}$	8		20	8	13	18	13		20	13	20

Table 6.20 Factorization for $N = 35$

x	2	3	4	6	8	9	11	12	13	16	17	18	19	22	23	24	26	27	29	31	32	33
r	12	12	6	2	4	6	3	12	4	3	12	12	6	4	12	6	6	4	2	6	12	12
$r/2$	6	6	3	1	2	3		6	2		6	6	3	2	6	3	3	2	1	3	6	6
$x^{r/2}$ (mod 35)	29	29	29	6	29	29		29	29		29	29	34	29	29	34	6	29	29	6	29	29

From Table 6.20, we see that, quite frequently, $x^{r/2} \cong 29$; then $x^{r/2} - 1 = 28$, and $\gcd(28, 35) = 7$. In three cases $x^{r/2} \cong 6$: then $x^{r/2} + 1 = 7$, $x^{r/2} - 1 = 5$, both factors of 35. Finally, for $x^{r/2} \cong 34$ we obtain the trivial factor $x^{r/2} + 1 = 35$.

When, as is necessary in effective cryptography, N is very large, $\gcd(x^{r/2} \pm 1, N)$ can be computed using the Euclidean algorithm. Then the algorithm (1) ... (6) can be used only if we have an efficient method of finding the order r.

6.10.8 Quantum Order Finding

We consider next a quantum algorithm (essentially Shor's algorithm[10]) for finding the order; this algorithm is, at least in principle, much more efficient than any algorithm performed by a classical computer.

Let N be a positive integer that can be written in binary notation with L digits. More precisely, we assume $2^{L-1} < N < 2^L$. Consider the M-dimensional Hilbert space $\mathbb{C}^M$, with $M = 2^L$. In this space we take a canonical basis whose elements are labeled by an integer $y \leq M$. For any x coprime with N and such that $1 < x < N$, we define a unitary operator U_x from $\mathbb{C}^M$ to $\mathbb{C}^M$ as follows:

$$U_x|y\rangle = \begin{cases} |xy(\text{mod } N)\rangle & \text{if } 1 \leq y \leq N - 1, \\ |y\rangle & \text{otherwise.} \end{cases} \quad (6.141)$$

Example: Let $N = 5$, $x = 2$, $L = 3$, $M = 8$. The relevant part of the matrix U_2 is the first 4×4 block:

$$U_2 = \begin{Vmatrix} 0 & 0 & 1 & 0 & 0 & 0 & 0 & 0 \\ 1 & 0 & 0 & 0 & 0 & 0 & 0 & 0 \\ 0 & 0 & 0 & 1 & 0 & 0 & 0 & 0 \\ 0 & 1 & 0 & 0 & 0 & 0 & 0 & 0 \\ 0 & 0 & 0 & 0 & 1 & 0 & 0 & 0 \\ 0 & 0 & 0 & 0 & 0 & 1 & 0 & 0 \\ 0 & 0 & 0 & 0 & 0 & 0 & 1 & 0 \\ 0 & 0 & 0 & 0 & 0 & 0 & 0 & 1 \end{Vmatrix}. \quad (6.142)$$

Since our "clock" has now $N = 5$ elements, we have $U_2|1\rangle = |2\rangle$, $U_2|2\rangle = |4\rangle$, $U_2|3\rangle = |6 \,(\text{mod } 5)\rangle = |1\rangle$, $U_2|4\rangle = |8 \,(\text{mod } 5)\rangle = |3\rangle$, $U_2|y\rangle = |y\rangle$ for

[10]Shor P W (1997) *Polynomial-time Algorithms for Prime Factorization and Discrete Logarithms on a Quantum Computer* Siam J Comp 26:1484–1509.

$5 \leq y \leq 8$. We see in this example (which we claim is generally valid) that the matrix U_x performs a permutation of the basis vectors. Therefore, when N is large, these matrices represent permutations of the numbers of a huge "clock with N hours."

At first glance it might seem that the calculation of the order r implies the computation of very large powers of enormous matrices. Fortunately, there is a better procedure, which exploits some properties of the eigenstates of U_x. For any $0 \leq s \leq r - 1$, we define the state:

$$|u_s\rangle = \frac{1}{\sqrt{r}} \sum_{k=0}^{r-1} e^{-\frac{2\pi i s k}{r}} |x^k (\text{mod } N)\rangle. \tag{6.143}$$

We require the following lemma.

Lemma 6.4 $|u_s\rangle$ *is an eigenstate of* U_x *corresponding to the eigenvalue* $e^{\frac{2\pi i s}{r}}$.

Proof First note that

$$U_x|u_s\rangle = \frac{1}{\sqrt{r}} \sum_{k=0}^{r-1} e^{-\frac{2\pi i s k}{r}} U_x |x^k (\text{mod } N)\rangle = \frac{1}{\sqrt{r}} \sum_{k=0}^{r-1} e^{-\frac{2\pi i s k}{r}} |x^{k+s} (\text{mod } N)\rangle. \tag{6.144}$$

Setting $h = k + 1$,

$$\begin{aligned}
U_x|u_s\rangle &= \frac{1}{\sqrt{r}} \sum_{h=1}^{r} e^{-\frac{2\pi i s h}{r}} e^{\frac{2\pi i s}{r}} |x^h (\text{mod } N)\rangle = \\
&e^{\frac{2\pi i s}{r}} \frac{1}{\sqrt{r}} \sum_{h=1}^{r} e^{-\frac{2\pi i s h}{r}} |x^h (\text{mod } N)\rangle = \\
&e^{\frac{2\pi i s}{r}} \frac{1}{\sqrt{r}} \sum_{h=0}^{r-1} e^{-\frac{2\pi i s h}{r}} |x^h (\text{mod } N)\rangle = e^{\frac{2\pi i s}{r}} |u_s\rangle,
\end{aligned} \tag{6.145}$$

where, in the last sum, we used the fact that the terms with $h = 0$ and $h = r$ are both equal to $|1\rangle$, which proves the lemma.

The operator U_x is unitary, since it performs a permutation of an orthonormal basis; in fact, the eigenvalues $\lambda_s = e^{\frac{2\pi i s}{r}}$ have moduli $|\lambda_s| = 1$ and the eigenvectors $|u_s\rangle$ are orthonormal. A useful identity satisfied by the eigenvectors $|u_s\rangle$ is the following:

$$\frac{1}{\sqrt{r}} \sum_{s=0}^{r-1} |u_s\rangle = |1\rangle. \tag{6.146}$$

To prove Eq. (6.146), note that

$$\frac{1}{\sqrt{r}} \sum_{s=0}^{r-1} |u_s\rangle = \frac{1}{\sqrt{r}} \sum_{s=0}^{r-1} \frac{1}{\sqrt{r}} \sum_{k=0}^{r-1} e^{-\frac{2\pi i s k}{r}} |x^k \text{ mod } N\rangle. \tag{6.147}$$

We first do the sum $S_k = \sum_{s=0}^{r-1} e^{-\frac{2\pi i s k}{r}}$ and denote by $z_k = e^{-\frac{2\pi i k}{r}}$, $k = 0, 1, 2, \ldots,$ $r - 1$, the r solutions of the equation $z^r = 1$. If $k \neq 0$, the sum S_k vanishes:

$$S_k = 1 + z_k + z_k^2 + z_k^3 + \cdots + z_k^{r-1} = \frac{z_k^r - 1}{z_k - 1} = \frac{0}{z_k - 1} = 0. \quad (6.148)$$

If $k = 0$, $S_0 = \sum_{s=0}^{r-1} 1 = r$, and we can write

$$\frac{1}{\sqrt{r}} \sum_{s=0}^{r-1} |u_s\rangle = \frac{1}{\sqrt{r}} \frac{1}{\sqrt{r}} r |x^0 (\text{mod } N)\rangle = |1\rangle. \quad (6.149)$$

Thus Eq. (6.146) is proved. However, the state $|1\rangle$ is not an eigenvector of U_x but rather a sum of eigenvectors. Let us see what happens if we apply the phase estimation algorithm and implement the circuit shown in Fig. 6.40 with $U = U_x$, $|u\rangle = |1\rangle$, and $\phi = \frac{2\pi s}{r}$; then $e^{i\phi y} = e^{i\frac{2\pi s}{r} y}$. Thereby Eq. (6.74) gives the following output:

$$|OUT_2\rangle = \frac{1}{\sqrt{r}} \sum_{s=0}^{r-1} \frac{1}{\sqrt{2^n}} \sum_{y=0}^{2^n - 1} e^{i\frac{2\pi s}{r} y} |y\rangle \otimes |u_s\rangle. \quad (6.150)$$

Let us see how we can now get information about the order r. Setting,

$$|v_{(\frac{s}{r})}\rangle = \frac{1}{\sqrt{2^n}} \sum_{y=0}^{2^n - 1} e^{i\frac{2\pi s}{r} y} |y\rangle, \quad (6.151)$$

Equation (6.150) becomes

$$|OUT_2\rangle = \frac{1}{\sqrt{r}} \sum_{s=0}^{r-1} |v_{(\frac{s}{r})}\rangle \otimes |u_s\rangle. \quad (6.152)$$

Now let us apply the inverse Fourier transform U_F^{-1} to $|v_{(\frac{s}{r})}\rangle$. Using Eq. (6.79) we find

$$U_F^{-1}|v_{(\frac{s}{r})}\rangle = \sum_{y=0}^{2^n - 1} \sum_{k=0}^{2^n - 1} |y\rangle \langle y|U_F^{-1}|k\rangle \langle k|v_{(\frac{s}{r})}\rangle =$$
$$\sum_{y=0}^{2^n - 1} \sum_{k=0}^{2^n - 1} \frac{1}{\sqrt{2^n}} \omega_{2^n}^{-yk} \frac{1}{\sqrt{2^n}} e^{i\frac{2\pi s}{r} k} |y\rangle = \sum_{y=0}^{2^n - 1} \frac{1}{2^n} \sum_{k=0}^{2^n - 1} e^{2\pi i k(\frac{s}{r} - \frac{y}{2^n})} |y\rangle, \quad (6.153)$$

since $\omega_{2^n} = e^{\frac{2\pi i}{2^n}}$. Expression (6.153) is the output of the first register of the quantum circuit shown in Fig. 6.44.

Suppose now that we measure the state (6.153) by taking its projection onto the state $|b\rangle$, where $1 \le b \le 2^n$ is a binary number. The probability amplitude $|c_b|$ for obtaining the result b is given by

$$|c_b| = |<b|U_F - 1|v_{(\frac{s}{r})}\rangle| = \frac{1}{2^n} \left| \sum_{k=0}^{2^n - 1} e^{2\pi i k(\frac{s}{r} - \frac{b}{2^n})} \right|. \quad (6.154)$$

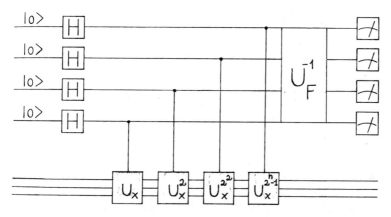

Fig. 6.44 The second register (the lower lines) corresponds to the eigenstate $|1\rangle = \frac{1}{\sqrt{r}}\sum_{s=0}^{r-1}|u_s\rangle \in C^M$

Setting $\delta = \frac{s}{r} - \frac{b}{2^n}$, $\alpha = e^{2\pi i\delta}$, the last expression becomes identical with Eq. (6.84), and can be approximated in the same way. If $\frac{b}{2^n} = \frac{s}{r}$, $|c_b|$ is equal to 1. Otherwise, when $|\frac{b}{2^n} = \frac{s}{r}| = \delta' < \frac{1}{2^n}$, the behavior of $|c_b|$ is given by Eq. (6.88):

$$|c_b| \simeq 1 - \frac{1}{6}\pi^2\delta'^2. \tag{6.155}$$

This method would then give a good approximation for $\frac{s}{r}$. The width of the maximum of $|c_b|$ for $\delta' = 0$ is of the order of $\frac{1}{2^n}$.

Before proceeding with the description of an algorithm (continued fractions) to find the order r, let us briefly review the various quantities involved:
(1) The number of the qubits measured by the quantum circuit is n.
(2) The measured result b satisfies $b \le 2^n$.
(3) Our "clock" has N "hours" since the operator U_x sends the vector $|y\rangle$ to $|xy(\text{mod } N)\rangle$.
(4) N can be written with L binary digits, so that $2^{L-1} \le N \le 2^L$.
(5) The order r obeys the inequality $r \le n$. Therefore, $r \le N \le 2^L$. Our choice of n depends on the computing power of the apparatus, the larger n, the more accurate the result. From (5), we have $2r^2 \le 2^{2L+1}$. If we choose $n = 2L + 1$, we find the following inequality, which will be useful below:

$$2r^2 \le 2^n. \tag{6.156}$$

6.10.9   Continued Fractions

These are expressions of the following form:

$$x = a_0 + \cfrac{1}{a_1 + \cfrac{1}{a_2 + \cfrac{1}{a_3 + \cdots}}} \tag{6.157}$$

where $a_0, a_1, a_2, \ldots$ are positive integers. Continued fractions have the following properties (stated without proof):
(1) A continued fraction is finite if and only if x is a rational number.
(2) The continued fraction of an irrational number is unique.
(3) The continued fraction of a rational number is unique except for alternative ways of writing the last term.

The continued fraction representing a number x can be constructed by iteration with the following steps:
(a) Take the integer part a of the number x.
(b) Write the fractional part d of x as a fraction $1/\frac{1}{d}$.
For example, the irrational number $\pi = 3.1415926\ldots$, which we approximate by 3.141, is to be expressed as a finite continued fraction. We carry out the successive steps:

$$\pi \simeq 3.141 = 3 + \frac{141}{1000} = 3 + \frac{1}{\frac{1000}{141}} = 3 + \frac{1}{7 + \frac{13}{141}} =$$

$$3 + \cfrac{1}{7 + \cfrac{1}{\frac{141}{13}}} = 3 + \cfrac{1}{7 + \cfrac{1}{10 + \frac{11}{13}}} = 3 + \cfrac{1}{7 + \cfrac{1}{10 + \cfrac{1}{1 + \cfrac{1}{5 + \frac{1}{2}}}}} \tag{6.158}$$

Neglecting the last fraction $\frac{1}{2}$, we obtain the approximation $\pi = \frac{1448}{461} \simeq 3.14099$. Note that, since the continued fractions expansion is essentially unique, it can be applied to reduce a fraction to lowest terms.

We state, again without proof, a theorem on continued fractions:

Theorem 6.10 *Suppose s/r is a rational number such that*

$$\left| \frac{s}{r} - \phi \right| \le \frac{1}{2r^2}, \tag{6.159}$$

then the continued fraction expansion of ϕ is equal to $\frac{s}{r}$.

Let us apply the theorem, setting $\phi = \frac{b}{2^n}$ where b is the number measured by the quantum circuit of Fig. 6.44. Changing b by one unit, changes ϕ to $\frac{1}{2^n}$; then from

Eq. (6.156), it follows that $\frac{1}{2^n} \leq \frac{1}{2r^2}$, so the condition (6.159) is satisfied, which allows us to determine the fraction $\frac{s}{r}$ and then the *order r*.

6.10.10   Prime Number Theorem

Consider now the following question: Does there exist an expression for the asymptotic distribution of prime numbers? Or, roughly speaking, for large x, can we evaluate the probability $p(x)$ that a number chosen in a neighborhood of x is prime? This question is addressed by the *prime number theorem*, which we state without proof:

Theorem 6.11 *Let x a positive real number, and let $\pi(x)$ equal the number of prime numbers less than x. We have, in the limit as $x \to \infty$:*

$$\lim_{x \to \infty} \frac{\pi(x)}{\left(\frac{x}{\log_e x}\right)} = 1. \tag{6.160}$$

Therefore, for large x we can write

$$\pi(x) \simeq \frac{x}{\log_e x}, \tag{6.161}$$

an approximation conjectured independently by Legendre and Gauss.

We recognize, of course, that $\frac{x}{\log_e x}$ is a continuous differentiable function, while $\pi(x)$ is stepwise discontinuous. For example, for $x = 10^9$, it is known that $\pi(x) = 50847534$. Since $\log_e x \simeq 20.72326584$, $\pi(x)/\frac{x}{\log_e x} \simeq 1.0537$, which does not differ very much from 1. Let x be an integer randomly chosen in a neighborhood of 10^9 (we do not choose 10^9 itself, which is not a prime). We want to evaluate approximately the probability $p(x)$ that x is prime. Consider the function $y(x) = \frac{x}{\log_e x}$; we know from the prime number theorem that $y(x)$ is a good approximation of $\pi(x)$ for large x. We have

$$\frac{dy}{dx} = \frac{1}{\log_e x} + x \frac{d}{dx}\left(\frac{1}{\log_e x}\right) = \frac{1}{\log_e x} - \frac{1}{(\log_e x)^2}, \simeq \frac{1}{\log_e x}, \tag{6.162}$$

since for large x, $(\log_e x)^{-2}$ is negligible compared to $(\log_e x)^{-1}$ (for $x = 10^9$, the two numbers are, respectively, 0.0023 and 0.0482). We can imagine the prime numbers as points lying on the positive real axis. A first approximation of the density of these points is given by Eq. (6.162). Choosing an interval $\{x, x + \Delta x\}$ such that $1 << \Delta x << x$, we can say that an approximate value of the number $\Delta \pi$ of primes contained in the range Δx is given by

$$\Delta \pi = \frac{\Delta \pi}{\Delta x} \Delta x \simeq \frac{dy}{dx} \Delta x \simeq \frac{1}{\log_e x} \Delta x. \tag{6.163}$$

Therefore the probability $p(x)$ that x is prime is

$$p(x) \simeq \frac{\Delta\pi}{\Delta x} \simeq \frac{1}{\log_e x}. \tag{6.164}$$

For example, for $x \simeq 10^9$, $p(x) \simeq 0.048$, and for $x \simeq 2^L = 2^{100}$, $p(x) \simeq 0.0144$.

There exist *primality tests* to determine whether a number is prime or not. One of the best known is the Miller–Rabin test,[11] which takes about $O(L^3)$ operations for a number with L bits. Assuming that $L = 100$, $O(L^3) = O(10^6)$, we see that the order of magnitude of the number of trials necessary to find a prime with 100 bits is $\frac{10^6}{p(2^{100})} \simeq \frac{10^6}{0.0144} \simeq 6.9 \times 10^7$, so such a calculation is within current capability. It is much more difficult to find a prime with 1000 bits, which involves about 10^{11} operations.

There exist methods that produce large primes, such as the progression (Perichon, 2010)[12]:

$$p = f(n) = 43142746595714191 + 5283234035979900\, n \ \text{ for } n = 0, 1, 2, \dots, 25 \tag{6.165}$$

but these are dangerous to use as prime factors to encrypt a message, because they are too well known.

6.11 Quantum Cryptography

Since time immemorial, people have been exchanging encrypted messages, which are devised so as to be understood only by the sender and recipient; even if a message were intercepted, it would hopefully remain incomprehensible to a third person. The science of cryptography deals with the design of efficient methods to create such ciphers. During the Second World War, the use of encrypted messages became essential for tactical military and naval communications. Thanks to the brilliant work of the Cambridge mathematician Alan Turing, messages encrypted by the Germans (using the Enigma machine) were deciphered by the Allies, which certainly shortened the war.

A message can be conveniently schematized as an integer M smaller than a maximum N_{MAX} (recall that the letters a, b, c, ... can be coded as integers, such as 1, 2, 3, ...). The encrypted message $E(M)$ is an integer, obtained by applying to M a complicated operator E, usually exceedingly nonlinear. The inverse operation D is defined by

$$D[E(M)] = M, \qquad \forall M < N_{\text{MAX}}. \tag{6.166}$$

---

[11]Miller GL (1976) Journal of Computer and System Sciences 13:300; Rabin MO (1980) Journal of Number Theory 12:128.

[12]Euler noted in 1772 that the quadratic polynomial $p_n = n^2 + n + 41$ is prime for all $n < 40$.

The inverse operator D must exist, but must be very difficult to find, making it almost impossible to decipher the message. In the past several years, methods have been proposed in which both the encrypted message $E(M)$ and the operator E can be made publicly available, while the plaintext M remains almost impossible to retrieve. The best known of these methods was invented in 1978 by Rivest, Shamir, and Adleman (RSA) (Rivest et al. 1978). The RSA algorithm is so powerful because a classical computer might be able to determine the operation D only after running an astronomically long period of time. However, a quantum computer could, in principle, decrypt the message much more rapidly.

6.11.1 RSA Cryptography

A message can be compactly modeled as an integer M written in binary digits. We also agreed to denote by $E(M)$ the encrypted message. A trivial method to encrypt M consists in choosing an integer C and put $E(M) = M + C$; clearly in this case the decryption is obtained simply by $M = E(M) - C$.

The RSA method is based on the existence of two ciphers, or distinct *keys*, one *public* P and one *secret* S. Everyone can have access to the key P. Suppose that among them is Bob, who wants to send a secret love letter to Alice. He wants only her to be able to decipher the message; to do this, Alice knows the secret key S (and only she knows it). As a less romantic example, Bob might be an account holder in a bank of which Alice is the director.

Let us see how Alice can make it almost impossible for a "spy" who has gained possession of the ciphertext $E(M)$ to read the original message M. Alice can do the following:

(1) Choose two extremely large primes p, q.
(2) Compute their product $n = pq$, and the totient $\phi(n) = (p - 1)(q - 1)$.
(3) Choose at random a small odd integer e, such that $\gcd(e, \phi(n)) = 1$ (that is, $e, \phi(n)$ must be coprime).
(4) Compute d, the multiplicative inverse of $e \in \mathbb{Z}_n^*$, such that $de = ed = 1 \pmod{\phi(n)}$. The *public* key P is the pair (e, n); the *secret* key is simply the number d. The message M is encrypted by Bob in the following way:

$$E(M) \cong M^e \pmod{n}. \tag{6.167}$$

From the above definition, it follows that if the integer e were very large, the calculation of $E(M)$ would be almost impossible by classical computation.

Let us prove that Alice can decrypt $E(M)$, simply raising it to the power d modulo n. Consider the simplest case, in which M is coprime with n. We have

$$E(M)^d \cong M^{ed} \pmod{n}. \tag{6.168}$$

But, by virtue of step (4) above, $ed = 1 + k\phi(n)$, where k is any integer. Therefore

$$E(M)^d \cong M^{1+k\phi(n)}(\mathrm{mod}\ n) = M M^{k\phi(n)}\ (\mathrm{mod}\ n). \qquad (6.169)$$

Now apply Eq. (6.133) for the totient, with $a = M$. We obtain $M^{\phi(n)} \cong 1(\mathrm{mod}\ n)$; then Eq. (6.169) becomes

$$E(M)^d \cong M 1^k(\mathrm{mod}\ n) = M, \qquad (6.170)$$

since $M < n$. This can also be written as

$$M \cong E(M)^d(\mathrm{mod}\ n). \qquad (6.171)$$

In the case in which M and n are *not* coprime, for example $p|M$ but $q \nmid M$, then $M \cong 0(\mathrm{mod}\ p)$ and therefore,

$$M^{ed} \cong 0 \cong M(\mathrm{mod}\ P). \qquad (6.172)$$

From $q \nmid M$ follows that $M^{q-1} \cong 1$ by Fermat's little theorem. Then,

$$M^{ed} = M^{1+k\phi(n)} = M M^{k(p-1)(q-1)} \cong M(\mathrm{mod}\ q). \qquad (6.173)$$

From Eqs. (6.172) and (6.173), and applying the Chinese remainder theorem, we deduce that

$$M^{ed} \cong M(\mathrm{mod}\ pq) \cong M(\mathrm{mod}\ n). \qquad (6.174)$$

We cannot have both $p|M$, $q|M$, because that would imply $n|M$, while we know that $n > M$. Therefore, the simple operation of raising to the d power can decipher the message $E(M)$. Such an operation would be easy for a modern computer. Suppose, for example, that the number of binary digits of n is 100. So $n \leq 2^L = 2^{100} \simeq 1.2676 \times 10^{30}$. How many operations will we have to perform? Since $d < n$, d has no more than 100 binary digits. We can write

$$d = d_0 + 2d_1 + 4d_2 + \cdots + 2^{L-1}d_{L-1} = (d_0, d_1, d_2, \ldots, d_{L-1}). \qquad (6.175)$$

Therefore, setting $x = E(M)$, we can readily compute the powers x^2, $x^4 = (x^2)^2$, $x^8 = (x^4)^2, \ldots$ modulo n. In this way we can compute, for any $x < n$:

$$x^d(\mathrm{mod}\ n) = x^{d_0}(\mathrm{mod}\ n)x^{2d_1}(\mathrm{mod}\ n)x^{4d_2}(\mathrm{mod}\ n)\ldots x^{2^{L-1}d_{L-1}}(\mathrm{mod}\ n), \quad (6.176)$$

performing $L-1$ multiplications modulo n. Now, each multiplication of two numbers of L bits involves approximately L^2 operations. The order of magnitude of the number of operations to compute the expression (6.176) is then $O(L^3) = 10^6$, which is easy for a modern computer. Note that even if $E(M)$ is 10^3 bits long, the decryption can still be done without difficulty. As pointed out by Nielsen and Chuang:

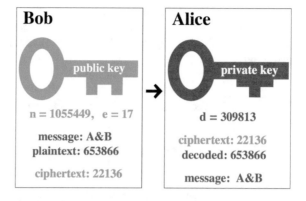

Fig. 6.45 Transmission of message from Bob to Alice using RSA encryption. Publicly available information is shown in *green*, private information, in *red*

"the main bottleneck is the generation of the numbers p and q," that must be very, very large.

A simple example of the RSA procedure is illustrated in Fig. 6.45. Let the secret prime factors be $p = p_{200} = 1223$ and $q = p_{150} = 863$. Then, $n = 1223 \times 863 = 1055449$ and $\phi(n) = (p-1)(q-1) = 1053364$. Also choose $e = 17$. Thus the public key is (1055449, 17). The private key d is the smallest integer satisfying $de \cong 1 \pmod{\phi(n)}$, which gives $d = 309813$. Suppose Bob wishes to transmit the plaintext message "A&B"; converted to ASCII code this might be represented by $M = 653866$. The encrypted message (ciphertext) is then given by $E(M) \cong M^e \pmod{n} = 22136$, which can be openly transmitted to Alice. She decodes the message using $M \cong E(M)^d \pmod{n} = 653866$, which is the ASCII code for A&B.

6.11.2   Code Breaking

It is, of course, relevant to explore the possibilities of breaking RSA encryption: how can we decrypt a message $M^e \pmod{n}$ knowing only e and n? We will describe two possible methods, the first based on order finding, the second on factoring the number n; both methods are very difficult to implement on a classical computer (recalling that n is very very large).

(1) Method based on order finding. Suppose we know the order r, the smallest integer such that

$$(M^e)^r \cong 1 \pmod{n}. \tag{6.177}$$

Assume M^e and n are coprime. Otherwise, n can be factored and we can go to step (2). By Euler's theorem for the totient, we have

$$(M^e)^{\phi(n)} \cong 1 \pmod{n}. \tag{6.178}$$

From Eqs. (6.177) and (6.178) it follows $r|\phi(n)$. We also know by Alice's choice (3) that $\gcd(e, \phi(n)) = 1$. Since r is a factor of $\phi(n)$, it will also be true that $\gcd(e, r) = 1$. Noting Eqs. (6.171) and (6.174), e admits an inverse $d'(\mathrm{mod}\ r)$, given by

$$ed' = 1 + kr. \tag{6.179}$$

Then to decrypt the message M^e, just raise it to the power d'; in fact we can write

$$(M^e)^{d'}(\mathrm{mod}\ n) = M^{1+kr}(\mathrm{mod}\ n) \cong MM^{kr}(\mathrm{mod}\ n) \cong M(M^r)^k(\mathrm{mod}\ n) \cong M1^k = M. \tag{6.180}$$

(2) Method based on factoring. Suppose we succeed in factoring n, then we know p and q. By simple multiplication we find $\phi(n) = (p-1)(q-1)$ and then we get d by solving the equation $de \cong 1(\mathrm{mod}\ \phi(n))$. Knowing d, we can decrypt the original message by raising M^e to the d power (see Eq. 6.174)[13]:

$$(M^e)^d \cong M(\mathrm{mod}\ n). \tag{6.181}$$

Perhaps the reader will wonder whether it is possible to find r simply raising M^e to a large number of powers $m(\mathrm{mod}\ n)$, which involves computing:

$$(M^e)^m(\mathrm{mod}\ n) \qquad \text{for} \qquad m = 2, 3, \ldots, n, \tag{6.182}$$

until the result $1(\mathrm{mod}\ n)$ is obtained. In principle, this is possible, but in practice, when n is very large, the calculation becomes prohibitively long. If, however, we know the exponent d in advance and we want to compute $(M^e)^d$, just use the method of repeated squaring (see Eq. 6.176), so that the number of operations will be reduced from order n to order $\log_2 n$.

Based on what we discussed in the preceding pages on factoring numbers and quantum order finding, if we had a quantum computer able to calculate the powers $U_x, U_x{}^2, U_x{}^4, \ldots, U_x{}^{2L}$ with $L \simeq 1000$, we could solve the problem of deciphering encrypted messages in use today.

Appendix

We now present a proof of Theorem 6.8 on p.245. As a preliminary:

Remark 6.1 The condition for g to be a generator is equivalent to the equality of the orders of g and G. Moreover, if g is an element of order d then the group H generated by g is isomorphic to $\mathbb{Z}/d$.

---

[13] It is possible to prove that $d = d'(\mathrm{mod}\ r)$.

Proof We denote by $|G|$ the order of G, and by $o(g)$ the order of g. Let g be any element of G, and let $m = o(g)$. Choosing two integers a, b, such that $1 \le a < b \le m$, the powers g^a, g^b are distinct, since from $g^a = g^b$ it will follow $g^b g^{-a} = g^{b-a} = e$. But this is absurd since $b - a < m$. In other words, the powers $g, g^2, \ldots, g^m = e$ are all distinct elements of the group H generated by g, which therefore has at least m elements. On the other hand, if we continue to multiply by g, the higher powers $g^{m+1} = g$, $g^{m+2} = g^2, \ldots$ repeat cyclically. This means that the elements listed above cover the whole H, so that $|H| = m$. Saying that g generates G is equivalent to the identity $H = G$, and because G is finite this is equivalent to $|G| = |H| = m$.

Moreover, if $a = b$ (mod m), then one can write $a = b + hm$ for some integer h, so $g^a = (g^m)^h g^b$, and because $g^m = e$ one has $g^a = g^b$. This is to say that, if we think of a as an element of $\mathbb{Z}/m$, the element g^a is well posed. This gives a bijection between H and $\mathbb{Z}/m$. Also, because $g^{a+b} = g^a g^b$, the bijection respects the operations, so it is an isomorphism.

Next, we address the theorem itself:

Theorem 6.8 *Suppose that $p > 2$ is a prime, and that α is a positive integer. Then the multiplicative group $\left(\mathbb{Z}/p^\alpha\right)^*$ is cyclic.*

In order to prove the theorem we will need an additional property of $\mathbb{Z}/p$. This object belongs to a particular class of rings, called fields, which enjoy many useful properties, one of which is given by the next lemma. (Roughly speaking, a field is a set on which are defined addition, subtraction and multiplication, for example, the real numbers: $\mathbb{R}$ or and the complex numbers $\mathbb{C}$. See also footnote on p. xxx.)

Lemma 6.5 *If F is a field and r is a positive integer, then there exist at most r distinct values of $x \in F$ satisfying $x^r = 1$.*

In $\mathbb{R}$ the equation $x^r = 1$ has exactly one solution if r is odd, and two solutions if it is even; in $\mathbb{C}$ it is well known that the equation has exactly r solutions, namely the complex roots of unity. When $F = \mathbb{Z}/p$, the lemma states that there exist at most r distinct (modulo p) solutions of the equation

$$x^r \equiv 1 \quad (\text{mod } p) \tag{6.183}$$

In the course of the proof of Theorem 6.8 we will also make use of the following two lemmas:

Lemma 6.6 *Let G be an abelian group, g and h elements of G, of order respectively a and b. If a and b are coprime, the order of gh is ab.*

Proof Since $(gh)^{ab} = g^{ab}h^{ab} = (g^a)^b(h^b)^a = e \cdot e = e$, ab is a multiple of the order m of gh. We need to prove that $m = ab$. From $(gh)^m = e$ it follows $g^m = h^{-m}$. We denote by X the order $o(g^m) = o(h^{-m})$. From Lagrange's theorem (Theorem 6.6) and $1 = (g^a)^m = (g^m)^a$ it follows that $X|a$. Analogously $X|b$, and if a and b are coprime, $X = e$. Then:

$$e = (g^m)^X = g^m, \quad e = (h^{-m})^X = h^{-m}, \tag{6.184}$$

that is $g^m = h^m = e$, thus $a|m$ and $b|m$. Therefore, m must contain all prime factors of a and all prime factors of b, and because all these factors are distinct, $ab|m$, thus $m \geq ab$. On the other hand, since $m|ab$, $m \leq ab$. It follows that $m = ab$ and the lemma is proved.

Lemma 6.7 *Let G be an abelian group, $g \in G$ and $o(g) = m$, and d a divisor of m, then $o[g^{m/d}] = d$.*

Proof $[g^{m/d}]^d = g^m = e$. We need to prove that it is not possible for $[g^{m/d}]^l = e$ with $l < d$. Indeed if $1 \leq l < d$, from $g^{ml/d} = e$, it would follow $g^{m'} = e$ with $m' = ml/d < m$, i.e. $o(g) < m$, in contradiction to the assumption. The lemma is proved.

Finally, we can turn to the proof of Theorem 6.8:

Proof Our goal is to show that there exists an integer z whose order modulo p^α equals the order of $(\mathbb{Z}/p^\alpha)^*$, which is $\phi(p^\alpha) = (p-1)p^{\alpha-1}$. We shall first argue that the conclusion holds if the two following conditions are fulfilled:

- There exists an element $x \in (\mathbb{Z}/p^\alpha)^*$ of order $p^{\alpha-1}$;
- There exists an element $y \in (\mathbb{Z}/p^\alpha)^*$ of order $p-1$.

In this case, the orders of x and y are coprime, because the only prime factor of $p^{\alpha-1}$ is p, which does not divide $p-1$. Lemma 6.6 implies then that xy has order $(p-1)p^{\alpha-1}$ and is therefore a generator, as desired.

Now we need to prove the two conditions above. For the first, we argue by induction that, for every non-negative integer h, the highest power of p that divides $(p+1)^{p^h} - 1$ is p^{h+1}. If this is true, for $h = \alpha - 1$ it follows that p^α divides $(p+1)^{p^{\alpha-1}} - 1$, which is to say that $(p+1)^{p^{\alpha-1}} = 1 \pmod{p^\alpha}$. On the other hand, for $h < \alpha - 1$ this implies that p^α does *not* divide $(p+1)^{p^h} - 1$. Therefore, the order of $p+1$ modulo p^α is $p^{\alpha-1}$, and one can choose $x = p+1$.

Note that the statement above is trivial for $h = 0$, as the highest power of p dividing $(1-p)^{p^0} - 1$ is clearly p itself. If the statement holds for h, then there exists some integer a, prime with p, such that

$$(p+1)^{p^h} = 1 + ap^{h+1}. \tag{6.185}$$

One can then use Newton's binomial theorem for the expansion:

$$(p+1)^{p^{h+1}} = (1 + ap^{h+1})^p = \sum_{j=0}^{p} \binom{p}{j} a^j p^{j(h+1)}. \tag{6.186}$$

The summand corresponding to $j = 0$ equals 1, and by separating the first few terms of the sum one can write:

$$(p+1)^{p^{h+1}} - 1 = pap^{h+1} + \frac{p(p-1)}{2}a^2 p^{j(h+1)} + \sum_{j=3}^{p} \binom{p}{j} a^j p^{j(h+1)} =$$

$$= ap^{h+2} + \frac{a(p-1)}{2} p^{j(h+1)+1} + \sum_{j=3}^{p} \binom{p}{j} a^j p^{j(h+1)}.$$

(6.187)

It is easily seen that p^{h+3} divides all the summands in this expression except for the first, which is only divisible by p^{h+2}. This proves that the whole sum is divisible by p^{h+2}, but not by p^{h+3}, which completes the induction, and the existence of x.

The last point to check is that there exists y of order $(p-1)$. We shall first prove this for $\alpha = 1$, and then deduce the general result from this case. Our aim is now to show that there exists an element of $(\mathbb{Z}/p)^*$ of order $p-1$: to this end we shall count the elements of lower order and show that they are fewer than $|(\mathbb{Z}/p)^*|$. First of all, note that if there are elements in $(\mathbb{Z}/p)^*$ of order d, then by Lagrange's theorem, $d|(p-1)$. Given such a d, we claim that the number h_d of elements of order d is at most $\phi(d)$. The claim is clearly verified if there are no such elements; otherwise let g be an element of order d. By our remark above, the subgroup $\langle g \rangle$ generated by g has exactly d elements, and since they all belong to a group of order d they all satisfy the equation $x^d = e$. Lemma 6.5, on the other hand, implies that this equation has at most d solutions in $\mathbb{Z}/p$, and therefore they are exactly the elements of $\langle g \rangle$. This shows that all the elements of order d in $(\mathbb{Z}/p)^*$ are all contained in $\langle g \rangle$, and therefore they can be characterised as the generators of this subgroup. Again according to the remark, $\langle g \rangle$ can be seen as a copy of $\mathbb{Z}/d$, which has exactly $\phi(d)$ generators. This proves that either $h_d = 0$ or $h_d = \phi(d)$; in either case $h_d \leq \phi(d)$. Therefore, the total number of elements of $(\mathbb{Z}/p)^*$ which are not generators is:

$$\sum_{\substack{d|(p-1) \\ d<p-1}} h_d \leq \sum_{\substack{d|(p-1) \\ d<p-1}} \phi(d)$$

(6.188)

For example, if $p = 11$, $p - 1 = 10$, the divisors d are $1, 2, 5, 10$, and the orders h_d are:

$h(1) = 1$ (only the element 1)
$h(2) = 1$ (only the element 10)
$h(5) = 4$ (elements $3, 4, 5, 9$)
$h(10) = 4$ (elements $2, 6, 7, 8$)

The generators of G are the elements $2, 6, 7, 8$ as their order is 10.

To conclude the case $\alpha = 1$ it is enough to establish that the sum on the right hand side is strictly smaller than $p - 1$. This can be done by comparison with the cyclic group $\mathbb{Z}/(p-1)$. Indeed, fix $d < p - 1$, which divides $p - 1$, and let $a = (p-1)/d$. The subgroup $\langle a \rangle$ of $\mathbb{Z}/(p-1)$ generated by a is a cyclic group of order d, and has therefore $\phi(d)$ generators. On the other hand, if k is any other element of $\mathbb{Z}/(p-1)$ of order d one has that $dk = 0 \pmod{p-1}$, which means $k = 0 \pmod{a}$, so that $a|k$. Therefore $\langle a \rangle$ contains all the elements of order d in

$\mathbb{Z}/(p-1)$, so there are exactly $\phi(d)$. This means that the sum on the right-hand side of (6.188) gives an exact count of the number of elements of $\mathbb{Z}/(p-1)$ of order lower than $p-1$, which is to say the elements of this group which do not generate it. But since this group is cyclic, and hence it does have generators, the value of this sum has to be smaller than the order of $\mathbb{Z}/(p-1)$. This proves that the number of elements of $(\mathbb{Z}/p)^*$ which are not generators is smaller than the order of the group, thus a generator has to exist. To conclude the proof for $\alpha > 1$, let $\tilde{y}$ be an integer which generates $(\mathbb{Z}/p)^*$, which has been established to exist. Note that the condition $\tilde{y}^n = 1 \pmod{p^\alpha}$ means that $p^\alpha | (\tilde{y}^n - 1)$, so in particular $p | (\tilde{y}^n - 1)$ and therefore $\tilde{y}^n = 1 \pmod{p}$. This proves that the order of $\tilde{y}$ modulo p^α is a multiple of that modulo p, which is $p-1$. Furthermore, $o(\tilde{y})$ divides the order of $(\mathbb{Z}/p^\alpha)^*$, which is $(p-1)p^{\alpha-1}$. It follows that $o(\tilde{y})$ is of the form $(p-1)p^l$ for some $0 \le l < \alpha$. If we put $y = \tilde{y}^{p^l}$, it is clear that $y^{p-1} = \tilde{y}^{(p-1)p^l} = 1$ modulo p^α, and if $n < (p-1)$ then $y^n \ne 1$ modulo p^α because $np^l < (p-1)p^l$, which is the order of $\tilde{y}$. This proves that $o(y) = p - 1$ and establishes the second point. Thus, Theorem 6.8 is proved.

Bibliography

Accardi L (1995) Can mathematics help solving the interpretational problems of quantum theory? Il Nuovo Cimento 110B:685–721

Aspect A, Grangier P, Roger G (1981) Experimental tests of realistic local theories via Bell's theorem. Phys Rev Lett 47:460–463; Aspect A, Dalibard J, Roger G (1982) Experimental test of Bell's inequalities using time-varying analyzers. Phys Rev Lett 49:1804–1807

Avogadro A (1834) Nouvelles recherches sur la chaleur specifique des corps solides et liquides. Crochard, Paris

Bell JS (1964) On the Einstein–Podolsky–Rosen Paradox. Physics 1:195–2001

Bell JS (1987) Speakable and unspeakable in quantum mechanics. Cambridge University Press, Cambridge

Bohm D (1951) Quantum Theory. Prentice-Hall, Englewood Cliffs NJ, Sections 5:19

Bohr N (1913) On the constitution of atoms and molecules. Philos Mag 26:1–25, 476–502, 857–875

Boltzman L (2005) Fisica e probabilit (a cura di Massimiliano Badino). Edizioni Melquades, Milano

Born M (1926) Zur Quantenmechanik der Stossvorgänge. Zeits Phys 37:863–867

Blinder SM (2004) Introduction of quantum mechanics in chemistry, materials science, and biology. Elsevier, Amsterdam

Bragg WH, Bragg WL (1913) The reflection of X-rays by crystals. Proc Roy Soc A 88:428–438

Cartan E (1966) English translation: theory of spinors. MIT Press, Cambridge

Clauser JF, Horne MA, Shimony A, Holt RA (1969) Proposed experiment to test local hidden-variable theories. Phys Rev Lett 23:880–884

Compton AH (1923) A quantum theory of the scattering of x-rays by light elements. Phys Rev 21:483–502

Daneri A, Loinger A, Prosperi GM (1962) Quantum theory of measurement and ergodicity conditions. Nucl Phys 33:297–319

Davisson CJ, Germer LH (1927) Diffraction of electrons by a crystal of nickel. Phys Rev 30:705–740

De Broglie L (1924) Recherches sur la théorie des quanta. Ph.D. thesis, University of Paris

Deutch D (1997) The fabric of reality. Viking, New York

Dirac PAM (1958) The Principles of quantum mechanics. Oxford University Press, Oxford Chap. II, Sec. 10

Doplicher S (2012) The Measurement process in local quantum theory and the EPR paradox. arXiv:0908.0480v1

Einstein A (1906) Zur Theorie der Lichterzeugung und Lichtabsorption. Annalen der Physik 20:199–206

Einstein A, Podolski B, Rosen N (1935) Can quantum-mechanical description of physical reality be considered complete? Phys Rev 47:777–780

© Springer International Publishing AG 2017
G. Fano and S.M. Blinder, *Twenty-First Century Quantum Mechanics:*
Hilbert Space to Quantum Computers, UNITEXT for Physics,
DOI 10.1007/978-3-319-58732-5

Fano G (1971) Mathematical methods of quantum mechanics. McGraw-Hill, New York

Feynman RP (1948) Space-time approach to non-relativistic quantum mechanics. Rev Mod Phys 20:367–387

Feynman RP (1951) the concept of probability in quantum mechanics. In: Proceeding II Berkeley symposium in mathematical statistics and probability, pp 533–541

Feynman RP, Leighton RB, Sands M (1965-1966) Lectures on physics, vol III. Addison-Wesley, Reading, Chaps 2-6

Feynman RP (1982) Simulating physics with computers. Int J Theor Phys 21:467–488

Feynman RP (1985) Quantum mechanical computers. Opt News 11:11–20, reprinted in Foundations of Physics (1986) 16:507-531

Fraunhofer J (1817) Bestimmung des Brechungs- und des Farbenzerstreuungs- Vermögens verschiedener Glasarten, in Bezug auf die Vervollkommnung achromatischer Fernröhre. Denkschriften der Munich Akademie der Wissenschaften, München 5:193–226

Ghirardi GC, Rimini A, Weber T (1986) Unified dynamics for microscopic and macroscopic systems. Phys Rev D 34:470–491

Griffiths RB (2002) Consistent quantum theory. Cambridge University Press, Cambridge

Heisenberg W (1925) Über quantentheoretische Umdeutung kinematischer und mechanischer Beziehungen. Zeits Phys 33:879–893

Kellert SH (1993) In the wake of chaos: unpredictable order in dynamical systems. University of Chicago Press, Chicago

Kirchhoff RB (1894) Vorlesungen über matematische Physik. Teubner, Leipzig

Klein F (1872) Vergleichende Betrachtungen über neuere geometrische Forschungen. Erlangen 1872; English translation: A comparative review of recent researches in geometry. arXiv:0807.3161

Kuhn TS (1996) The structure of scientific revolutions, 3rd edn. University of Chicago Press, Chicago

Landau LD, Lifshitz EM (1959) Statistical physics. Pergamon Press, London

London F, Bauer E, (1939) La theorie de l'observation en mécanique quantique. English translation in Wheeler JA, Zurek WH, (eds) (1983) Quantum Theory and Measurement. Princeton University Press, Princeton, pp 217–259

Merli PG, Missiroli GF, Pozzi G (1976) On the statistical aspect of electron interference phenomena. Am J Phys 44:306–307

Mermin ND (1985) Is the moon there when nobody looks? Reality and the quantum theory. Phys Today 38:38–47

Millikan RA (1913) On the elementary electric charge and the avogadro constant. Phys Rev 2:109–143

Nielson MA, Chuang IL (2000) Quantum computation and quantum information. Cambridge University Press, Cambridge

Nozick R (2001) Invariances: the structure of the objective world. Harvard University Press, Cambridge

Omnes R (1994) The interpretation of quantum mechanics., Princeton series in physicsPrinceton University press, Princeton

Pais A (1982) Subtle is the lord: the science and the life of Albert Einstein. Oxford University Press, New York, p 456

Paz JP, Zurek WH (2002) Enviroment-induced decoherence and the transition from quantum to classical. Lect Notes Phys 587:77–148

Penrose R (1996) On gravity's role in quantum state reduction. General Relat Gravit 28:581–600

Planck M (1900) Zur Theorie des Gestzes der Energieverteilung im Normalspectrum. Verhandlungen der Deutchen Physikalisken Gesellschaft 2:237–45

Rayleigh JWS (1905) The dynamical theory of gases and radiation. Nature 72:54–55, reprinted in Rayleigh's Scientific Paper, vol 5, Cambridge, UK

Rivest R, Shamir A, Adleman L (1978) A method for obtaining digital signatures and public-key cryptosystems. Commun ACM 21:120–126

Rutherford E (1911) The scattering of α and β particles by matter and the structure of the atom. Philos Mag Ser 6 21(125):669–688

Rydberg JR (1890) On the structure of the line-spectra of the chemical elements. Philos Mag 29:331–337

Schrödinger E (1926) An undulatory theory of the mechanics of atoms and molecules. Phys Rev 28:1049–70

Schiff LI (1955) Quantum mechanics, 3rd edn. McGraw-Hill, New York

Stapp HP (1975) Bell's theorem and world process. Nuovo Cim B 29:270–276

Stapp HP (2001) Quantum theory and the role of mind in nature. Found Phys 31:1465–1499

Thomson JJ (1893) Notes on recent researches in electricity and magnetism: intended as a sequel to Professor Clerk-Maxwell's 'Treatise on Electricity and Magnetism'. Oxford University Press, Oxford, UK, pp.xvi and 578

Young T (1807) A course of lectures on natural philosophy and the mechanical arts. Thoemmes Press, London, Republished 2002

Von Neumann J, (1932) Mathematische Grundlagen der Quantenmechanik. Springer, Berlin; English translation, (1955) Mathematical foundations of quantum mechanics. Princeton University Press, Princeton NJ

Von Laue M (1913) Röntgenstrahlinterferenzen. Physikalische Zeitschrift 14:1075–1079

Zurek WH (1991) Decoherence and the transition from quantum to classical. Phys Today 44:36–44

Zurek WH (2003) Decoherence, eigenselection, and the quantum origins of the classical. Rev Mod Phys 75:715

Index

A

Adder
  Full-adder, 202–207
  Half-adder, 202, 205, 207
Angular momentum, 96
  operators, 98
  quantum number, 101
  total, 99
Annihilation operator, 93, 96
Antibonding orbital, 39
Anticommutator, 94
Antipodal points, 150
Argument of a complex number, 74
Aspect's experiment, 182, 183
Atomic orbital, 109
Average
  of random variable, 21
Average of observable, 59
Avogadro's number, 21, 192
Azimuthal angle θ, 106

B

Bell
  Bell's inequality, 118, 164, 172, 183
  Bell states, 155, 218
  Bell's theorem, 153, 168, 184
  generalized Bell's inequality, 169
  Wigner's version, 165
Bezout's lemma, 234
Binary
  function, 196
  numbers, 195
Blackbody, 5
Bloch sphere, 208

Bohr
  atomic model, 11, 12
  radius a_0, 108
Boltzmann
  constant k_B, 22
  distribution, 30
Bonding orbital, 38
Boolean algebra, 195, 196
Born interpretation, 33
Bose–Einstein condensate, 192
Bosons, 103
Bound states, 81
Bragg formula, 10
Brianchon's theorem, 124

C

Carry, 201
Central force, 105
Chinese remainder theorem, 240, 242, 246, 256
Class of objects, 122
Classical computers, 201
Clauser–Shimony optimal choice, 172
Clock arithmetic, 235
Closed shell, 111
Code breaking, 257
Commutative property, 122
Commutator, 57, 144
Complement of a set, 198
Complex
  conjugate, 73
  modulus, 73
  plane, 72
  real, imaginary parts, 72

© Springer International Publishing AG 2017
G. Fano and S.M. Blinder, *Twenty-First Century Quantum Mechanics:
Hilbert Space to Quantum Computers*, UNITEXT for Physics,
DOI 10.1007/978-3-319-58732-5

Printed in the United States
By Bookmasters